26th Center for Chemical Process Safety International Conference 2011

(CCPS)

Topical Conference at the 2011 AIChE Spring Meeting and 7th Global Congress on Process Safety

Chicago, Illinois, USA
13 – 17 March 2011

ISBN: 978-1-61782-888-1

Printed from e-media with permission by:

Curran Associates, Inc.
57 Morehouse Lane
Red Hook, NY 12571

CURRAN ASSOCIATES INC. proceedings.com

Some format issues inherent in the e-media version may also appear in this print version.

Printed by Curran Associates, Inc. (2011)

For permission requests, please contact AIChE
at the address below.

AIChE
3 Park Avenue
New York, NY 10016-5991

Phone: (203) 702-7660
Fax: (203) 775-5177

www.aiche.org

Additional copies of this publication are available from:

Curran Associates, Inc.
57 Morehouse Lane
Red Hook, NY 12571 USA
Phone: 845-758-0400
Fax: 845-758-2634
Email: curran@proceedings.com
Web: www.proceedings.com

TABLE OF CONTENTS

Process Safety Performance Tool

Clarissa Sossmeier da Silva
Braskem
clarissa.silva@braskem.com.br

Pablo Ricardo Barrera
Braskem
pablo.barrera@braskem.com.br

Keywords: Process Safety, Indicators, Management.

Abstract

This work aims to present a Process Safety Performance Tool developed to allow managing risks inherent to chemical and petrochemical process, such as fire, explosions or releases of flammable, combustible, or toxic chemicals, through performance monitoring of the Process Safety Management (PSM) implemented in the facility. This tool consists of indicators divided in two categories. The first one includes the lagging indicators, also called, reactive indicators, which are related to the frequency and severity of incidents. The second category includes the leading indicators that measure the evolution in practices related to the prevention or mitigation of Process Safety incidents. For the first group of indicators the ones recommended by the Center of Chemical Process Safety (CCPS) were considered in order to allow benchmark with industries worldwide. The leading indicators were selected in order to consider the most important 29CFR1910.119 OSHA (Occupational Safety and Health Administration) requirements. All indicators have targets or specific references and are weighted according to their relevance, in order to compose a single value for process safety performance. The group of lagging indicators account for 30% of total value and leading ones by 70%. This approach was inspired by the Reason's model of organizational accidents. On the whole, this tool provides a basis for the Process Safety Management, guaranteeing a real evolution of this subject, therefore contributing for the maintenance of plant and people integrity as well as the operational reliability and the organization image.

1. Introduction

One of the main factors that had contributed to the world economy growth in recent decades, according to Robert M. Solow, that received the Nobel Prize in Economics in 1987, was the technological advance of the industrial processes (Crowl and PRAISE, 1989). These technological advances are present in all industrial areas, making processes more efficient, productive and automated. In chemical and petrochemical industries, however, beyond these characteristics, there's another one: the process complexity.

More complex processes involve operation under higher conditions of temperature and pressure and the use of more reactive chemicals. It also involves increasing the production scale and consequently the inventory of hazardous materials and the number of equipments at a unit. Due to this complexity, the severity of an incident, such as fire, explosion or release of products has increased. These situations could not be restricted only to a small area within the company and to small assets losses, but also affects company's reputation and business continuity due to a high number of deaths or injuries, environmental damages and major assets losses.

As industrial processes become more complex, safety standards and technology should become more restrictive in order to avoid major accidents. These hazards however could not be managed by the traditional occupational safety programs. The occupational safety hazards give rise to incidents, such as slips, falls, and vehicle accidents that primarily affect one individual worker for each occurrence. The major accidents mentioned above are related to the Process Safety and need specific attention of the chemical industries in order to avoid greater losses and seriously consequences.

The Process Safety aims to manage technological risks in industrial facilities such as fire, explosion and leakage of chemicals, which may compromise the integrity of facilities and people, operational reliability and business image and continuity. Technological risks are those ones inherent to a specific industrial activity involving the use of toxic, flammable or reactive products.

An important point about the Process Safety Management in industries, on the opposite of the occupational safety, is that there are no broadcast indicators that specifically represent the organization's evolution in Process Safety practices and culture. In 2005, the British Petroleum (BP) refinery accident in Texas City (USA) drew attention to this fact. The final investigation report (CSB, 2007) showed that the company believed in the good performance of their occupational safety indicators as representative for the performance in process safety. The end of this story had shown that this consideration is not true, once these good results had not avoided the tragic accident that resulted in 15 deaths and more than 170 injuries. As a result of these accident, the CSB recommended chemical industries to create and follow indicators that could clearly show the performance in Process Safety.

From this recommendation on, agencies and associations related to the sector around the world have increased their attention to this issue and started to propose in a structured way some standard indicators for monitoring performance in Process Safety. However, we are still far from a standard set of indicators that can be easily applied in all industry segments and government does not impose yet this kind of report. This practice still depends on industries initiatives that need to assess their specific process and determine which indicators may in fact represent the evolution of Process Safety.

In this way, the objective of the present work, developed in the Braskem Basic Petrochemical Unit of South Brazil (BPU-RS), was to develop a performance tool that allows monitoring the performance of the Process Safety Management at an industrial site by means of a single index that group indicators related to the elements and requirements of the safety management systems adopted by Braskem. This tool allows business leaders to critically analyze the results in Process Safety, and thereafter determine actions to continually improve these results, ensuring the integrity of people and assets as well as increasing the process reliability.

2. Process Safety Management Systems

A strong management system is crucial to ensure the effectiveness of any process. The act of managing may be defined as taking actions, based on critical analysis of a scenario, whose primary goal is to maintain a baseline or achieve a certain goal. In case of Process Safety Management the goal is to guarantee industrial assets integrity as the primary condition for Loss Prevention.

An efficient management system that aims to prevent losses involves a clear structure with a well defined matrix of responsibilities and qualified people to carry out each of these assignments. Also some requirements are necessary: existence and use of appropriate standards and procedures to operate the process plant; training and qualification of workers in these procedures; providing process information for everyone involved in operation and maintenance in order to disseminate knowledge and allow risks identification and treatment; good emergency response planning if all previous steps have failed; compliance audits of all these steps, and we could not forget to mention indicators to measure performance of these practices.

Initiatives throughout history allow us to see the evolution of the Process Safety Management until today. In 1964 Dow Chemical Company had published its Process Safety Manual. Three years later, in 1967, the American Institute of Chemical Engineers (AIChE) began a series of symposia on Loss Prevention, followed by the European Federation of Chemical Engineers. On the other hand, in subsequent decades some of the most emblematic and tragic disasters related to process safety had occurred. It's impossible to not mention the Flixborough accident in England (1974), the Seveso accident in Italy (1976) and the Bhopal accident in India (1984). These events highlighted the need for improvements in Loss Prevention (LEES, 1980).

In 1992, strongly motivated by the accidents of the 70's and 80's and also by events in the early 90's, as the 1989 Phillips Petroleum Company accident in Texas, which resulted in 23 deaths and 132 injuries and the accident at BASF in Cincinnati (Ohio) in 1990, which resulted in two deaths, OSHA (Occupational Safety and Health Administration) published the standard 29 CFR 1910.119 - Process Safety Management of Highly Hazardous Chemicals, containing requirements for preventing or minimizing the consequences of catastrophic releases of toxic, reactive, flammable, or explosive chemicals. These releases may result in toxic, fire or explosion hazards. This standard has been compiled in a document entitled OSHA 3132, reprinted in 2000.

The Process Safety Management system defined in 29 CFR 1910.119 is structured based on 14 elements of management as shown in Figure 1.

Braskem adopt the OSHA Process Safety Management (PSM) through its Safety, Health and Environment (SHE) Excellence Program, called SEMPRE (that means "Always" in English). Below, this program is explained.

Figure 1: *OSHA 3132 PSM – Process Safety Management.*

2.1 SEMPRE – SHE Excellence Program

The Excellence Program in Safety, Health and Environment (SHE) of Braskem – SEMPRE – similar to the OSHA PSM consists of 16 elements and was originally developed in partnership with DuPont and DNV. The 16 strategic elements (SE) are the following:

- SE1: Commitment and Leadership
- SE2: Organizational Synergy
- SE3: Risk and Impact Identification and Analysis in safety, health and Environment
- SE4: Standards and Procedures
- SE5: Product and Process Information
- SE6: Behavioral Development
- SE7: Competency and Skill Management
- SE8: Suppliers and Element Services Management
- SE9: Effective Communication
- SE10: Facilities and Process Installation and Commissioning
- SE11: Facilities and Equipment Integrity
- SE12: Product Management
- SE13: Management of Change
- SE14: Emergency Preparation and Assistance
- SE15: Deviation and Losses Treatment
- SE16: Verification in Health, Safety and Environment

It's possible to see that the elements of OSHA and SEMPRE are correlated and this is shown in Figure 2, that represents those elements as pillars that support the Process safety and the Operational reliability.

Both systems presented here, the PSM proposed by OSHA 3132 and SEMPRE, are management systems and therefore need indicators for their evaluation. Indicators evaluate the process performance, in comparison with the short and long-term organizational goals, feed historical database helping organizations to define which is the next step they expect the process to be in the future and lead managers to the development of corrective and preventative actions that should be taken in order to guarantee an improvement in the Management and Culture of Process Safety.

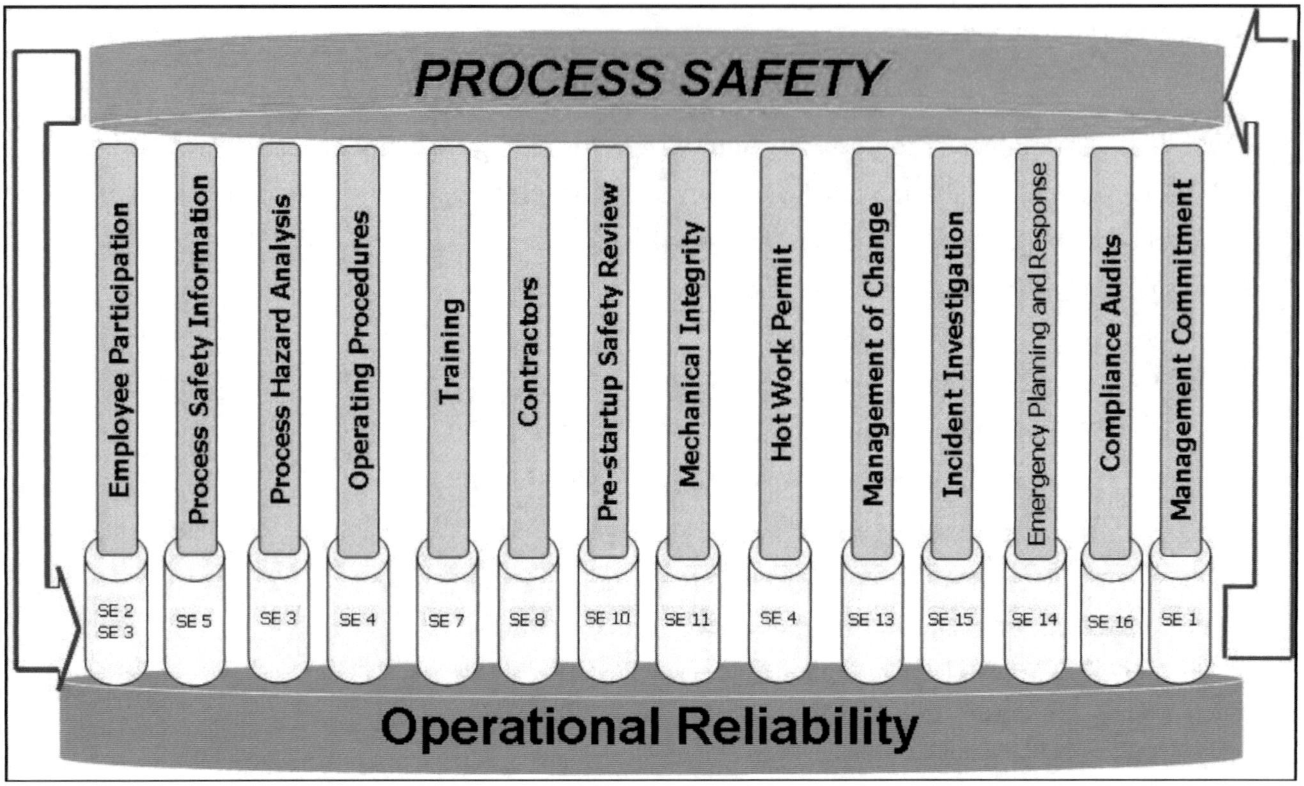

Figure 2: *Correlation between OSHA PSM – Process Safety Management – and Braskem Excellence Program in Safety, Health and Environment (SEMPRE).*

3. Process Safety Key Performance Indicators

On March 23, 2005, the BP Texas City refinery experienced one of the most serious U.S. workplace disasters of the past two decades, resulting in 15 deaths, more than 170 injuries, and significant economic losses. The U.S. Chemical Safety and Hazard Investigation Board (CSB), an independent federal agency charged with investigating industrial chemical accidents, promptly began the accident investigation.

On August 17, 2005, the CSB issued an urgent safety recommendation to the BP Global Executive Board of Directors that it commission an independent panel to assess and report on the effectiveness of BP North America's corporate oversight of safety management systems at its refineries and its corporate safety culture. In making its urgent recommendation, after the first few months of a preliminary investigation, CSB was especially concerned about:

• The effectiveness of the safety management system at the BP Texas City refinery,

• The effectiveness of BP North America's corporate safety oversight of its refining facilities, and

• A corporate safety culture that may have tolerated serious and longstanding deviations from good safety practice.

BP embraced the urgent recommendation of the CSB and on October 24, 2005, and announced the formation of the BP U.S. Refineries Independent Safety Review Panel.

On January 17, 2007, BP issued "The report of the BP U.S. Refineries Independent Safety Review Panel", also known as "Baker Panel". The findings of this panel made chemical industries around the world aware of the importance in developing and implementing a Process Safety Management System together with tools that allow Performance Evaluation and the definition of Corrective Actions, improving Corporate Oversight and strengthening the Corporate Safety Culture.

Braskem Basic Petrochemical Unit of South Brazil (BPU-RS) got aware of all this findings, however it was given priority to a specific recommendation that wasn't well developed at the facility and had already been identified in the Process Safety Meetings with the company leaders: The Performance Evaluation. Maintaining and improving a process safety management system requires the periodic evaluation of performance and the correction of identified deficiencies. In this way, the Baker Panel recommended that BP should develop, implement, maintain, and periodically update an integrated set of leading and lagging performance indicators for more effectively monitoring the process safety performance.

This important and essential recommendation, however, was not so easy to implement, once there was no worldwide set of leading and lagging indicators for process safety performance. To develop a consensus set of indicators for use in the refining and chemical processing industries, working together with the CSB and with industry, labor organizations and other governmental agencies, was a part of this recommendation.

Taking account this recommendation BPU-RS began in March, 2007, the development of Key Performance Indicators, selecting consolidated indicators inside the company that could truly represent the performance of the Process Safety. This indicators were grouped in a panel were their individual results could be seen at any moment with a monthly updating. At the beginning the panel contained indicators related to the Process Safety Management System and the Process Safety Accident rate, and the information was available separated by industrial areas and as a global result for the whole facility.

The panel was presented to the Managers Board of BPU-RS on June 6, 2007, and the idea was approved by all the company leaders. Suggestions were made and considered in the panel and it was released for all company. Monthly, a Committee assessed the panel in order to evaluate the results and also the applicability of the panel and its indicators in the Process Safety daily management.

On December 20, 2007, the CCPS (Center of Chemical Process Safety), issued the initial release of the publication "Process Safety Leading and Lagging Metrics", describing the recommendations assembled by the CCPS Process Safety Metric committee for a common set of company and industry leading and lagging metrics.

This publication defined that indicators fall in the three categories presented below:

• "Lagging" Metrics – a retrospective set of metrics that are based on incidents that meet the threshold of severity that should be reported as part of the industry-wide process safety metric.

• "Leading" Metrics – a forward looking set of metrics which indicate the performance of the key work processes, operating discipline, or layers of protection that prevent incidents

• "Near Miss" and other internal Lagging Metrics – the description of less severe incidents (i.e., below the threshold for inclusion in the industry lagging metric), or unsafe conditions which activated one or more layers of protection. Although these events are actual events (i.e., a "lagging" metric), they are generally considered to be a good indicator of conditions which could ultimately lead to a severe incident.

These three types of metrics can be considered as measurements at different levels of the "safety pyramid" illustrated in Figure 4.

CCPS strongly recommend that all companies incorporate each of these three types of metrics into their internal process safety management system.

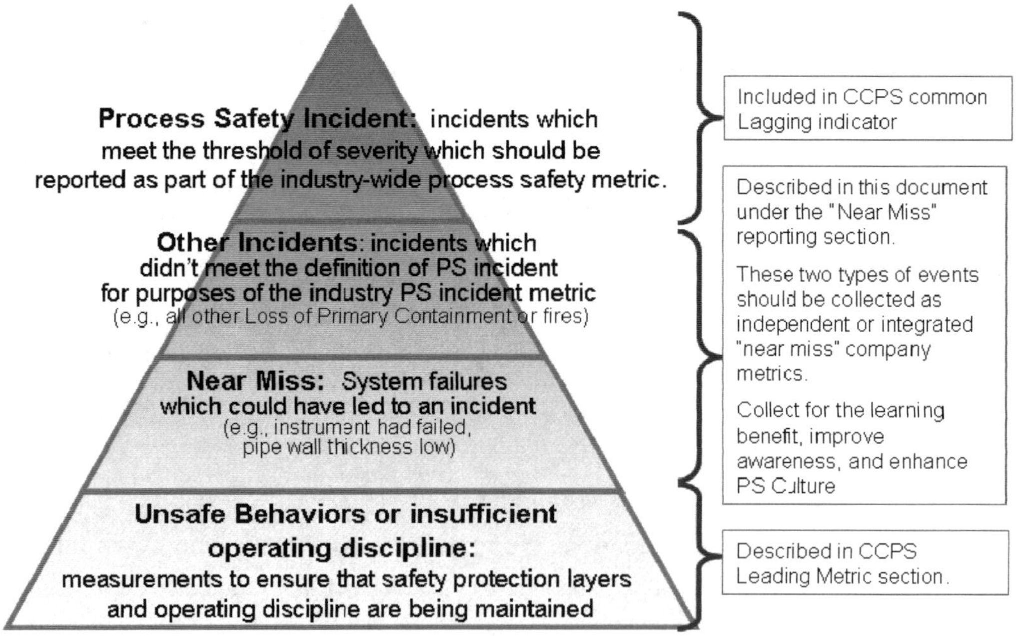

Figure 4: *Safety Pyramid. Source: CCPS, 2007.*

After this publication, BPU-RS analyzed and adapted its KPI's to the concept presented by the CCPS. In the next sections will be described the last review of the selected indicators for Key Performance Indicators Panel already framed on each of the CCPS categories. However, taking account that the "near misses" category was not considered in the first review of the panel due to the fact that monitoring of these events was not well structured in the company and considering also that the objective was to promptly start the systemic use of those indicators, it was decided not to include this category of indicators in the KPI's panel.

3.1 Lagging Indicators

As mentioned before, lagging metrics are based on the historic of process safety incidents or accidents that have happen in the company. Define what is considered a process safety incident or accident is not easy and just make sense when other companies have the same classification criteria to allow performing company-to-company benchmarking.

On this way, CCPS (2007) proposed the standardization of the process safety incident classification criteria, the severity criteria and also the metrics for these indicators. Now, follows those definitions.

A. Process Safety Incident Classification Criteria

To understand the concept of this metrics is necessary to see CCPS definitions for a Process Safety Incident and some other specifically terms. A Process Safety Incident is an incident that meets all four of the following criteria:

1. **Chemical or chemical process involvement:** when a chemical or chemical process is directly involved in the damage caused. The term "process" here is used to include equipments and technology needed for petrochemical production (i.e. reactors, tanks, piping, boilers, cooling towers, refrigeration systems, etc.)
2. **Above minimum reporting threshold:** when a release of material or energy from a chemical process results in fatality or lost time injury for an employee; fires or explosions resulting in greater than or equal to $25,000 of direct cost; or an acute release of flammable, combustible, or toxic chemicals greater than the chemical release threshold quantities defined by CCPS in Table 1.
3. **Location:** when the incident occurs in production, distribution, storage, utilities or pilot plants.
4. **Acute release:** when the release of material reaches or exceeds the reporting threshold in 1 hour or less.

B. Process Safety Incident Severity Criteria

A severity level will be assigned for each consequence category to a process safety incident utilizing the criteria defined by CCPS (2007). For each of these categories there's a weight that will enter on the severity rate calculation.

C. Lagging Process Safety Metrics

Based on these concepts, the proposed metrics for the Lagging indicators are:

- **Total Count of Process Safety Incidents (PSIC):** The count of all Process Safety Incidents (PSI).
- **Process Safety Total Incident Rate (PSTIR):** The cumulative (annual) count of incidents normalized by man-hours within a base of 200.000 hours.
- **Process Safety Incident Severity Rate (PSISR):** The cumulative (annual) severity-weighted rate of process safety incidents normalized by man-hours within a base of 200.000 hours and considering a specifically weight for each level of incident defined on this document.

Table 1: Process Safety Incident Threshold Values

Material Hazard classification as Defined by United Nations Dangerous Goods definitions:	"Process Safety incident" threshold quantity:
TIH Hazard Zone A materials[1]	5 kg (11 lbs.)
TIH Hazard Zone B materials[1]	25 kg (55 lbs.)
TIH Hazard Zone C materials [1]	100 kg (220 lbs.)
TIH Hazard Zone D materials[1]	200 kg (440 lbs.)
Other "Packing Group I" materials[2] & "Flammable Gases/Vapors"	500 kg (1100 lbs.)
Other "Packing Group II" materials[2] & "Flammable Liquids"	1000 kg (2200 lbs.)
Other "Packing Group III" materials[2] & "Combustible Liquids" & Division 2.2 - Nonflammable, Nontoxic Gases	2000 kg (4400 lbs.)

[1] The TIH Hazard Zone A, B, C, and D designations are mentioned specifically in the US Department of Transportation regulations (49 CFR 172.101), and not in the UN Dangerous Goods definitions. However, these definitions do align with toxic vapor categories described in the Harmonized System of Classification and Labeling of Chemicals (GHS). .

[2] Packing Group definitions based upon UN Dangerous Goods definitions. In the USA, these definitions are also described in US Department of Transportation regulations (49 CFR 172.101). These descriptions include several generic descriptions (e.g., "Toxic fluids") or materials which are not germane to the chemical or petroleum industries (e.g., cotton, explosive ammunition). Although the specific chemicals listed in the UN Dangerous Goods listing have been used as a basis for establishing the process safety incident threshold quantities, these generic or non-applicable materials have been excluded.

Due to the reasons presented here, for the lagging indicators the ones recommended by the CCPS were considered in this work.

3.2 Leading Indicators

As mentioned above this section presents a set of metrics which indicate the performance of important safety management system, such as some key work processes, operating discipline, or layers of protection that prevent incidents. This metric allows identifying previously any deterioration in the effectiveness of key safety systems, it means before any accident occurs. In this way, it's also possible to create an action plane with preventative actions to restore de stability of the systems.

For the CCPS (2007), the elements of process safety management for which the preventive indicators have been developed are:

- Maintenance of mechanical integrity;
- Action items follow-up;
- Management of change; and
- Process safety training and competency (and training competency assessment).

The CCPS (2007) recommends that all companies adopt and implement leading process safety metrics, including a measurement of process safety culture. However, given the number of metrics recommended by CCPS it may be impracticable to collect and report data for each of these categories. For this reason, it is recommended that companies identify which of these components are most important for ensuring the safety of their facilities, and select the most meaningful leading metrics, creating other ones as well if applicable.

Considering this, the specific metrics defined by CCPS (2007) to the leading metrics were not adopted exactly as proposed. The definition of the indicators that would be used for this category as well as the determination of the calculation basis of these metrics was an important stage of this work.

The premise adopted during the selection of the leading indicators was to use only existing indicators in the company. There are two reasons why this consideration was done. The first one is the applicability of these indicators, since the creation of new ones would increase demand of information from those areas that generate this data and consequently the calculation of the indicators would have a delay. The second reason concerns about the Process Safety culture evolution. This can be explained by the fact that the existing indicators in the company were used for monitoring the performance of a specific area, such as Maintenance, Production or SHE area, and therefore were not seen as a Process Safety indicator. With this consideration those areas would start understanding the importance of their work to Process Safety.

Considering also the fact that these indicators must reflect the Process Safety Management elements performance, in this case the elements of OSHA 3132 (2000), the selection strategy was to look for available indicators at each of the company's areas that have interface with each element of this standard. These indicators should be related to the specific requirements of this standard.

During this search, however, it could be noted that some elements of the Process Safety Management did not have consolidated indicators that could measure their performance. This was the reason why it was not possible to select indicators for all the requirements established in the 29CFR1910.119 standard.

At the end of the selection step, the leading indicators presented in Figure 5 were chosen. This figure illustrates the relation between each one of the proposed metrics with the elements and requirements of OSHA 3132. These indicators are described below.

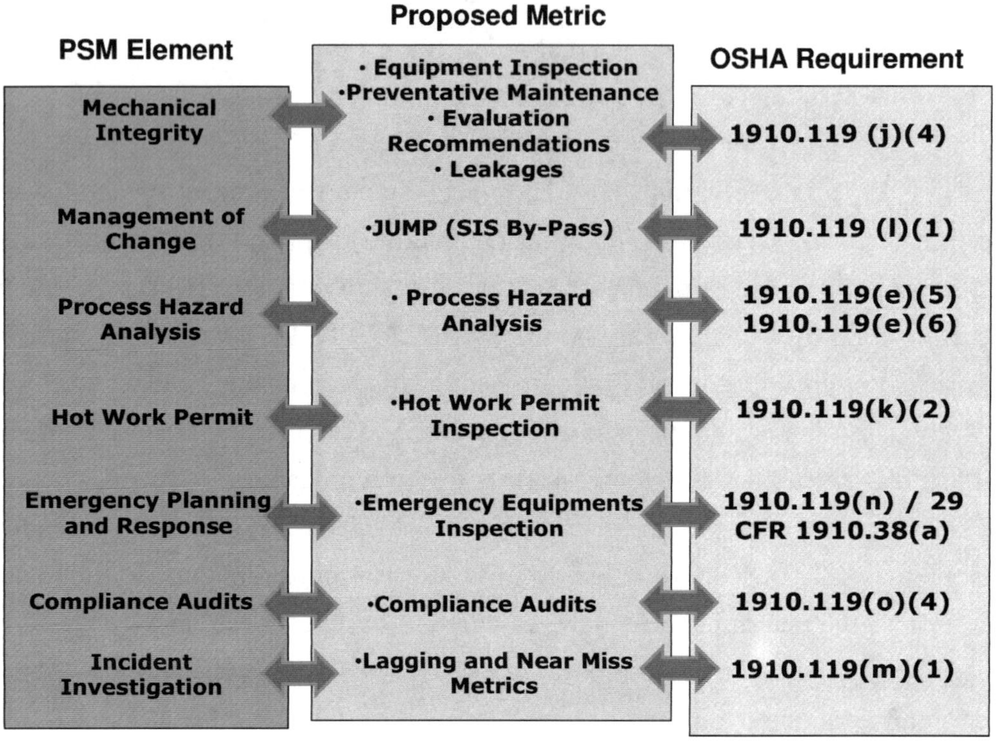

Figure 5: *Correlation between the proposed indicators and the elements and requirements of 29CFR1910.119.*

- **Equipment Inspection**

For Mechanical Integrity, the requirement 1910.119 (j) (4) of OSHA establishes that inspections and tests shall be performed on process equipment and their procedures shall follow recognized and generally accepted good engineering practices as well as the frequency of this inspections and tests (items i, ii, iii). These requirements are applicable for:

- Pressure vessels and storage tanks;
- Piping systems (including piping components such as valves);
- Relief and vent systems and devices;

- Emergency shutdown systems;
- Controls (including monitoring devices and sensors, alarms, and interlocks);
- Pumps;

In order to meet these requirements for pressure vessels, storage tanks, piping systems and relief and vent devices, Braskem is certified to self inspect its own facilities under the Brazilian NR-13 Regulations. This certification allows the Inspection Process to be known as Owner Equipment Inspection System or SPIE (from Portuguese Serviço Próprio de Inspeção de Equipamentos).

This indicator shows the percentage of equipment and pipe with inspection overdue as well as the total number of pressure and temperature safety valves and relief valves that have not been calibrated in the required period. This indicator gives an idea of the critical equipments non-functioning potential and consequently of the possibility of an accident or loss.

The measure of this metrics involves the follow up of an existing annual inspection plan managed by the Maintenance Unit and this number can be monitored monthly. The NR-13 Regulations requires that the proportion of overdue items does not exceed 0.5%.

- **Evaluation Recommendations**

Inspections or tests performed in any equipment, in special those one originated from SPIE inspection, mentioned in before, may generate some recommendations, here called Evaluation Recommendations. These recommendations are classified in 6 categories defined by means of their severity (P1 being the most critical and P6 the less critical), which considers impacts in processes safety and reliability, and their likelihood.

The indicator for this process is the percentage of recommendations overdue separated by their categories. It represents a possible non-integrity of the system or equipment in which the recommendation should have been executed. This deviation may lead to a loss in safety and reliability.

- **Leakages**

Small leaking in systems containing water, steam or hydrocarbons are not rare in petrochemical industries. Braskem classifies leakages into six categories considering the severity and the likelihood of an impact on process safety and reliability. Braskem calls "clamps" all the gadgets utilized to remedy a leaking. This indicator measures the number of critical active leakages as well as the compliance percentage of the inspection plan defined for those devices.

A higher number of critical active leakages in a petrochemical unit represent a greater risk of a major accident. The compliance percentage of the inspection plan is also an important indicator since the Clamp replaces the piping that has its specific inspection plan.

- **Preventative Maintenance Plans**

This indicator measures the compliance percentage of the preventive maintenance plans established for all systems and process equipments in the company. This percentage can be divided into categories, such as instrumentation, electrical and rotating equipments or a global value. The intention of this indicator is to evaluate if compliance of these plans are within an acceptable limit consistent with the criteria of risk acceptability in order to ensure the reliability and integrity of the site.

- **Jump**

The deviation of a protection barrier considered in the project, although temporary, may bring seriously consequence for de process safety. In this way, every bypass or jump done in any Safety Instrumented System (SIS) may be managed by a specific procedure and monitored.
The objective of this indicator is to assess and ensure the integrity of the layers of protection (SIS) provided in the original project or implemented due to recommendations of safety and reliability. This indicator monitors the total number of active jumps and the period they stay active. The higher these two values are, the greater the risk that is being assumed.

- **Process Hazard Analysis**

The process hazard analysis is a thorough, orderly, and systematic approach for identifying, evaluating, and controlling the hazards of processes involving highly hazardous chemicals, OSHA recommends that at least every five years the process hazard analysis must be updated and revalidated. In order to achieve this requirement and considering that Braskem has a systematic flow of process hazard analysis review, this indicator evaluate the progress of the review plan, measuring the number of systems planned to be analyzed versus the number performed.

Another important consideration made by OSHA is that the process hazard analysis is best performed by a team with expertise in engineering and process operations, and specific analysis methods. Du to this recommendation, another monitored indicator is the percentage of this multidisciplinary participation in the analysis.

- **Hot Work Permit Inspection**

A robust process of hot work permit is essential for the safety of works involving ignition sources on or near a covered process. OSHA requirement 1910.119(k)(2) says that the permit shall document that the fire prevention and protection requirements of 29 CFR 1910.252(a) have been implemented prior to beginning the hot work operations. It also shall indicate the date authorized for hot work and identify the object on which hot work is to be performed. Usually the process of Hot Work Permit is well defined in industries and the management of this process can be enhanced by monitoring their compliance to the procedure established. Thus, the purpose of this indicator is to assess and ensure the compliance of the hot work permit procedure, measuring the non-compliance percentage of the inspections.

- **Emergency Equipments Inspection**

Emergency Planning and Response is the last level of defense in case of an accident and for this reason it is essential that emergency equipment are available and in good conditions for immediate use. Therefore, this indicator is divided in three parts:

- Percentage of inspections and tests overdue;
- Number of emergency equipment unavailable;
- Percentage of equipments with corrective maintenance overdue.

- **Compliance Audits**

To be certain process safety management is effective, it is necessary to perform a compliance audit in order to verify that the procedures and practices developed are adequate and are being followed. Thus, the objective of this indicator is to evaluate the degree of compliance in the process safety inspection performed monthly in the company areas in which are observed items as lock-out and tag-out, missing and shorting bolts, gas detectors, etc. The indicator also includes the follow up of recommendations originated from internal inspections and external audits, such as Corporate Audits or Insurance Company Audits.

4. Process Safety Performance Tool

The Process Safety Performance Tool created by this work intends to measure the industrial areas performance in terms of process safety. As it was shown before the success of the loss prevention process depends on a good Management System and, therefore, on a set of preventative and mitigating factors whose are called management system elements. The indicators for each of these factors were shown in the section before.

On the other hand, it is not necessary to have excellent results in each of these indicators in order to avoid an incident. In other words, it's possible to say that a poor performance on a single element of the management system would not represent the risk of an immediately accident.

This can be further elucidated by looking at the Swiss Cheese Model developed by James Reason (Reason, 1990). Reason compares the protective barriers between a hazard and an accident with slices of a Swiss cheese. These slices may have holes that represent the weaknesses of the protective barriers. When there are holes in all barriers and these holes are aligned the hazard energy or chemical is released, resulting in personal, material or environmental losses. The holes in the slices can be understood as failures on the process safety management elements. Figure 6 illustrates this explanation.

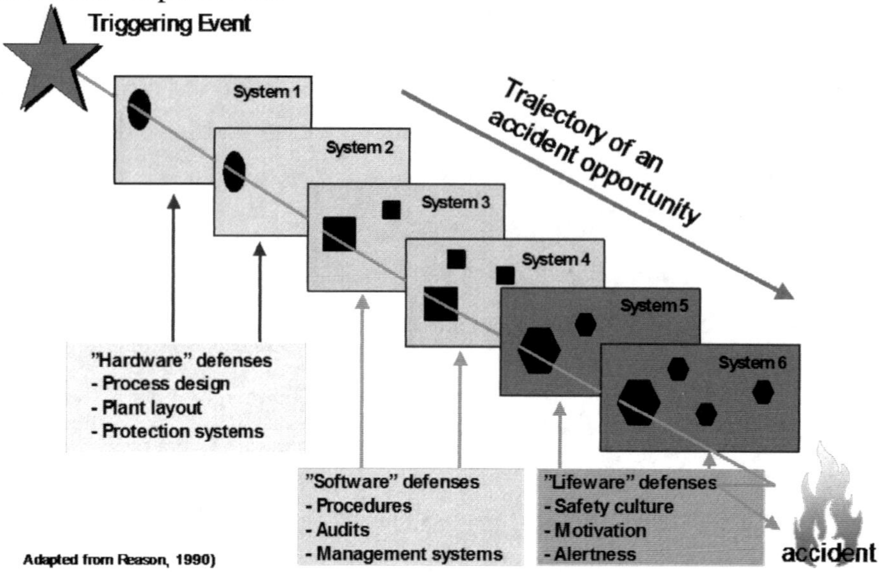

Figure 6: *The Swiss cheese Mode. Adapted from REASON, 1990.*

On this way, to monitor individually the indicators mentioned before is not the best way to represent the overall performance in Process Safety. It is necessary to evaluate all the performance indicators simultaneously in order to have a consistent result. Besides this, for the business leaders, a good performance indicator shall result in a single number that, when compared to a goal, could say if their process are "below standard", "standard" or "above standard".

Considering all that, the process safety performance tool developed in this paper, called I-SEPRO (from Portuguese: Process Safety Index) grouped all indicators related to the process safety elements in a single index. This index is composed as follow:

- 70% related to the leading indicators;
- 30% related to the lagging indicators.

The inclusion of the lagging indicators is important in order to consider whether the improvement in the leading indicators really is reflected in a reduction of frequency and severity rates.

For each of the indicators that compose the index must be assigned a weight in order that the sum of the individual weights reaches the value of 70% for the leading metrics and 30% for lagging metrics. Each indicator should have a goal so that the outcome of the index will be the weighted average of the percentage attendance of each indicator. The indicator weights must be established according to the relevance of the indicator and the element of PSM it represents. Targets should be set based on historical data and, if possible, on values from internal or external references. In Table 2 are presented all the indicators and the respective metrics that are considered in the index.

Table 2: Indicators of the Process Safety Performance Tool.

1.0 JUMP		
Metrics	1.1)	Total number of jump's in a period below 30 days.
	1.2)	Total number of jump's in a period above 30 days.
	1.3)	Number of non-conformities during the monthly process safety inspection related to the JUMP procedure (registration, authorization, corrective actions proposed, etc.)
2.0 Hot Work Permit Inspection		
Metrics	2.1)	% of conformities related to process safety during the Hot Work Permit Inspection.
	2.2)	Number of non-conformities during the monthly process safety inspection related to Hot Work Permit Procedure.
3.0 Equipment Inspection		
Metrics	3.1)	% of compliance to the static equipments inspections plans.
	3.2)	% of compliance to the piping inspections plans.
4.0 Evaluation Recommendations		
Metrics	4.1)	Number of recommendations P1 (Priority 1) overdue.
	4.2)	Number of recommendations P2 (Priority 2) overdue.
5.0 Leakages		
Metrics	5.1)	Active leakages Priority 1 and 2 in systems containing hydrocarbons.
	5.2)	% of compliance to the clamp's inspections plan.
6.0 Preventative Maintenance Plans		
Metrics	6.1)	% of compliance to the preventative instrumentation plan.
	6.2)	% of compliance to the preventative electrical plan.
	6.3)	% of compliance to the preventative rotating equipments plan.
	6.4)	% of compliance to the preventative analytical plan.
7.0 Process Hazard Analysis		
Metrics	7.1)	% multidisciplinary in the Hazard Analysis.
	7.2)	Number of systems planned to be analyzed versus the number performed.
8.0 Emergency Equipments Inspection		
Metrics	8.1)	% of inspections and tests overdue.
	8.2)	Number of emergency equipment unavailable.
	8.3)	% of equipments with corrective maintenance overdue.

9.0 Compliance Audits		
Metrics	9.1)	Number of non-conformities in the monthly Process Safety Inspection.
	9.2)	Number of actions planned actions versus actions concluded.
Total % - Leading		**70%**
10.0 Lagging Indicators		
Metrics	10.1)	Total Count of Process Safety Incidents (PSIC)
	10.2)	Process Safety Total Incident Rate (PSTIR)
	10.3)	Process Safety Total Incident Rate (PSTIR)
Total % - Leading		**30%**

Below, are presented some pictures related to the application of this tool in the BPU-RS site.

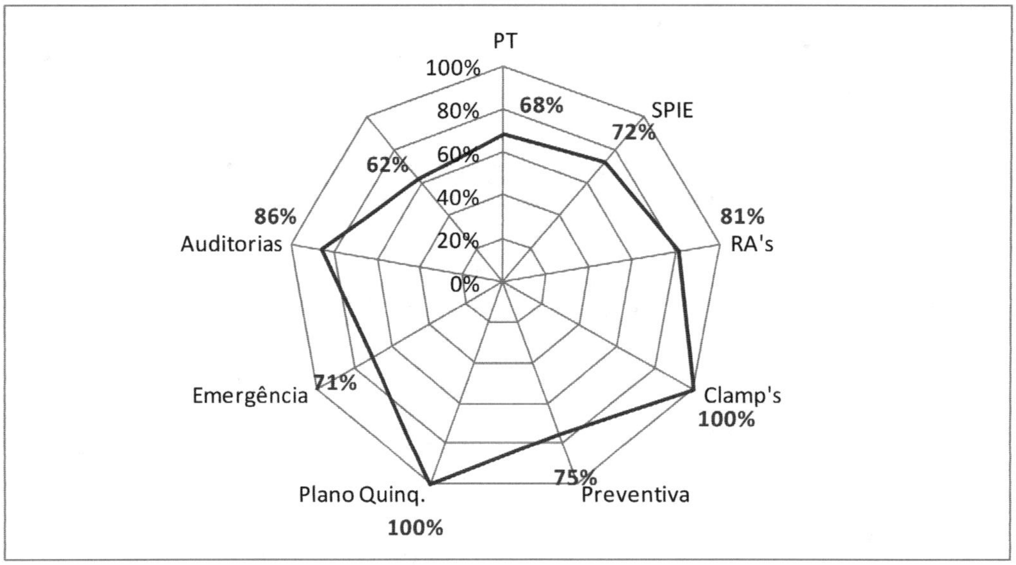

Figure 7: *Spider Net Graphic of the indicators individual results.*

Figure 8: *Monthly Process Safety Inspection.*

5. Conclusion

To have a good Process Safety Management System is the basis to ensure the safety and integrity of chemical processes and having a tool capable of providing accurate information on the actual performance of the management is essential to allow company managers make decisions that effectively reduce risks. This paper presented a tool for easy monitoring, since it combines in a single number or index a series of indicators related to OSHA 3132 elements.

The practical application of this tool was well received both by the leaders of the company where the work was developed as the operational areas that actually applied the tool in their processes. An important factor that contributed to this acceptance of this tool was, as a first step, the criteria of adopting company's existing indicators instead of create new indicators.

This tool for assessing performance in Process Safety can be applied to any industry whose business involves the production, distribution, storage or handling of toxic, flammable or reactive chemicals.

As highlighted in the development of this article, the set of indicators selected to compose this tool did not cover all elements of the Process Safety elements in OSHA 3132. However, this is a limitation on the application of the tool and not a limitation of the tool itself. The methodology developed in this work allows companies during the implementation, be as comprehensive as desired, since the data required are available.

Finally, we strongly believe that this tool will not only reduce the occurrence of Process Safety accidents and incidents, but also helps to strengthen a Loss Prevention culture.

6. References

[1] CCPS. Center of Chemical Process Safety (CCPS®). Process Safety Leading and Lagging Metrics, 2007.

[2] LEES, F. P. Loss Prevention in the Process Industries. Butterworth-Heinemann Ltd, 1980.

[3] CROWL, D. A.; LOUVAR, J. F. Chemical Process Safety: Fundamentals with Applications. Prentice Hall, 1989.

[4] REASON, J. Human Error. New York. Cambridge University Press, 1990.

[5] CSB, U.S. CHEMICAL SAFETY AND HAZARD INVESTIGATION BOARD. Investigation Report No. 2005-04-I-TX: Refinery Explosion and Fire. March 2007.

[6] OSHA 3132. Process Safety Management. U.S. Department of Labor. Occupational Safety and Health Administration, 2000 (Reprinted).

Improving Management of Critical Controls to Avoid Catastrophic Incidents

Angela Wong, P.Eng., Ph.D.
Dyadem International Ltd.
155 Gordon Baker Road, Suite 401
Toronto, Ontario, M2H 3N5
Canada
awong@dyadem.com

Keywords: Risk management, critical controls, catastrophic incidents, key performance indicators (KPI), leading indicators, lagging indicators, risk register, proactive risk management, control effectiveness, control assurance

Abstract

Catastrophic incidents can hinder operations and critical business functions. Indeed, as highlighted time and time again, they can frighten a company's very survival. Catastrophic incidents also often lead to more stringent regulatory requirements, making business environments more challenging and increasing operating costs. In some cases such events can led to industry-wide shutdowns while safety and regulations are debated. BP is not alone. With the average global organization trying to manage thousands of sites around the world, effective risk management becomes a daunting challenge. In this paper, we explore how to identify the critical controls needed to efficiently manage these risks, and how to implement those controls effectively to avoid catastrophic incidents.

1 Introduction

Often, risks associated with catastrophic events are deemed to be low and acceptable, provided effective controls are implemented. However, in many cases, these controls are the last line of defense. If they fail, catastrophic incidents can, and will eventually occur. To ensure controls are effective for managing catastrophic events, it is essential to identify and manage these controls in a systematic and structured manner.

The process of managing critical controls should also be integrated into the risk management process of an organization. In 2009, the International Organization for Standardization (ISO) published *ISO/FDIS 31000 Risk Management – Principles and Guidelines Risk Management Standard* [1]. It presents a generic risk management framework and risk management process for any industry. In this paper, we propose using the ISO 31000 risk management process as a basis for developing a critical control management process. We chose the ISO 31000 process as a basis because of its wide adoption in process industries.

The proposed process includes the following basic steps (illustrated in Figure 1):

1) Define criteria for catastrophic events and critical controls
2) Conduct risk assessment for catastrophic events
3) Manage the critical control life cycle
4) Monitor and review critical control activities
5) Communication and consultation

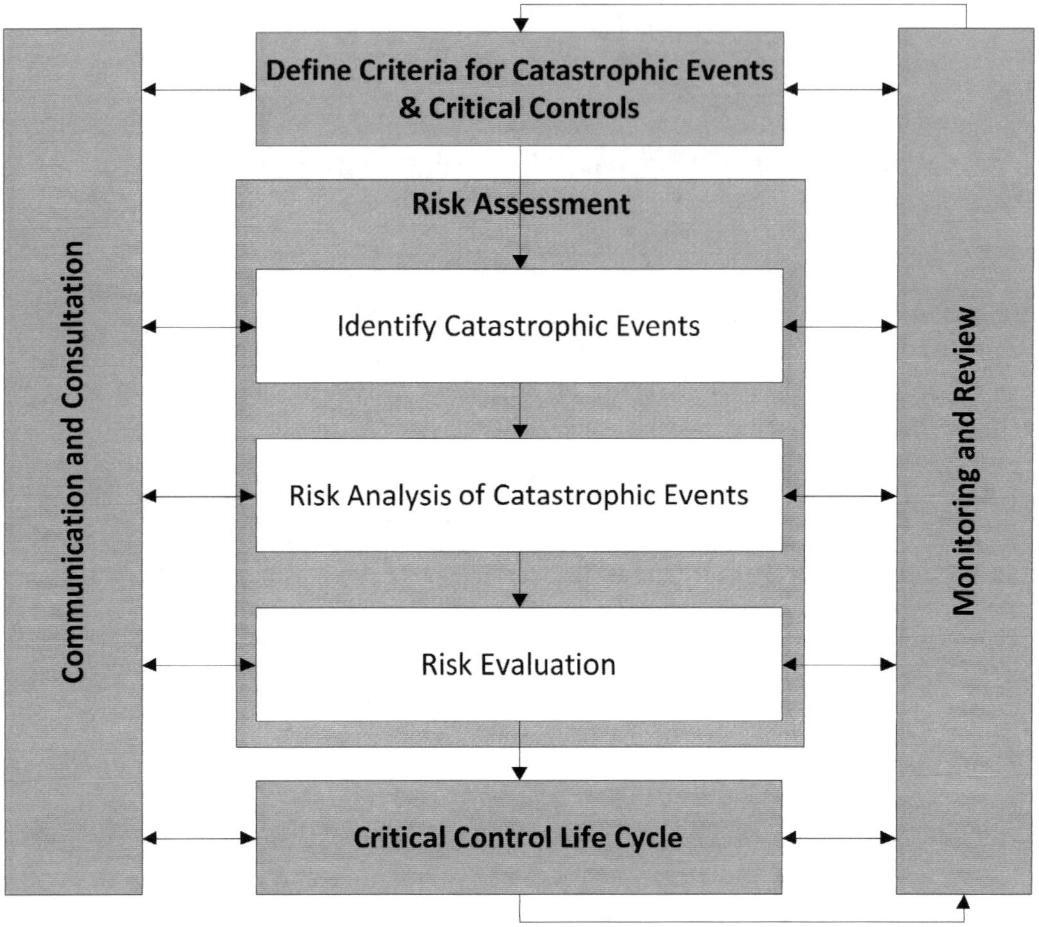

Figure 1: Critical Control Management Process

2 Critical Control Management Process

2.1 Define criteria for catastrophic events and critical controls

In any company, controls exist to manage risk. Risk controls can be categorized per the hierarchy of controls proposed by the UK Health and Safety Executive (HSE). Figure 2 shows the hierarchy and Table 1 gives definitions of each control type.

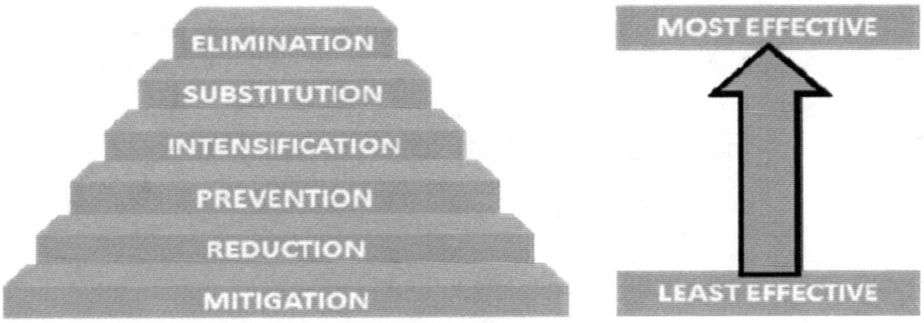

Figure 2: Hierarchy of Controls

Table 1. Control Type Definitions

Hierarchy	Explanation
Elimination	Something that removes a hazard completely. While this is clearly the most effective type of control measure, it is often not practicable to eliminate hazards. For example, if a toxic material is an essential raw material, then removal is most likely not possible.
Substitution	Using a less hazardous material to meet the same need as a highly hazardous material.
Intensification	Reducing the total inventory of a hazardous material.
Prevention	Something that prevents accident scenarios from occurring or significantly reduces the likelihood.
Reduction	Control measures that reduce the magnitude of the consequences, usually by detecting an unwanted condition and acting to stop the scenario. They do not directly combat the consequences of an accident.
Mitigation	Control measures that directly combat the consequences of an accident by reducing their effects on people, plant and the environment. While the least preferred type of control measure, well designed mitigation control measures are essential for safe operation, since they provide the absolute last line of defense.

In the context of critical control management, a critical control can be defined as a control that is part of the risk treatment of a catastrophic event. A catastrophic event is a potential event that can severely harm people, the environment, production of a process facility, reputation and / or market share of an organization—should the full potential of the event be realized. Typically during process hazard analysis (PHA), where undesired process hazardous events are identified and ranked using a risk matrix, such an event would have a very high severity ranking and a very low likelihood ranking. The criteria of a catastrophic event should be developed by the management of an organization with inputs from sites.

Critical controls are usually identified within PHA and assumed to be well managed under the Process Safety Management (PSM) program at the site level. This assumption is true unless issues or concerns are raised by the PHA team, or during other PSM activities such as mechanical integrity management, management of change, incident or near misses investigation and internal / external audits. An OSHA PSM program does not require an integrated approach to managing critical controls, but it is strongly recommended. If any deficiency in the life cycle of a critical control is not readily realized and rectified in a timely manner, this can lead to deterioration of critical controls over time. The catastrophic event that was deemed to be very unlikely to happen in the PHA is now much more likely to happen, consequently exposing an organization to much higher risk than expected. It is therefore beneficial to have an integrated management process for these controls due to their criticality to process facilities and organizations.

2.2 Risk assessment

Per ISO 31000, risk assessment includes the following steps:

1. **Risk Identification** – Identification of potential hazards and associated risks in the area under consideration.

2. **Risk Analysis –** Consideration of the causes and sources of risk, their positive and negative consequences, and the likelihood that those consequences can occur.

3. **Risk Evaluation** – The likelihood and impact of the hazardous scenario taking into consideration existing controls. The product of likelihood and impact is the risk associated with the scenario; see equation below:

   $$\textbf{Risk} = \textbf{Likelihood} \times \textbf{Impact}$$

   The risk is compared with risk criteria established when the context was considered. Based on this comparison, the need for Risk Treatment can be considered.

2.2.1 Identify catastrophic events

PHAs conducted at process facilities to meet OSHA requirements typically includes the three steps listed above. HAZOP and What-If are commonly used PHA methodologies. By applying catastrophic event criteria in PHA, initial screening can be performed to identify catastrophic events for further analysis.

In addition to PHA, identification of potential catastrophic events should be extended to other PSM elements such as:

- Management of Change;
- Incident Management;
- Mechanical Integrity Management; and
- Internal and External Audits.

Other operating modes such as startup / shutdown and maintenance activities should be included in the identification step as the majority of past catastrophic incidents have taken place during abnormal operations.

2.2.2 Risk analysis of catastrophic events

As there can be multiple causes, controls and consequences for a catastrophic event, HAZOP and What-if are not the best tools to illustrate multiple relationships between a catastrophic event, causes, consequences and controls. LOPA, introduced in IEC 61511 as one of the SIL (Safety Integrity Level) determination methodologies, is effective in illustrating relationships between multiple causes and preventive controls, such as interlocks, alarms and relief systems for an unwanted event. However, LOPA does not effectively address mitigating measures and mechanical integrity issues.

Catastrophic events identified in PHA and other PSM activities could be further analyzed using a detailed and structured analysis. Bow-tie analysis, as shown in Figure 3, is an effective method to analyze complex relationships between a catastrophic event, critical controls, causes and consequences.

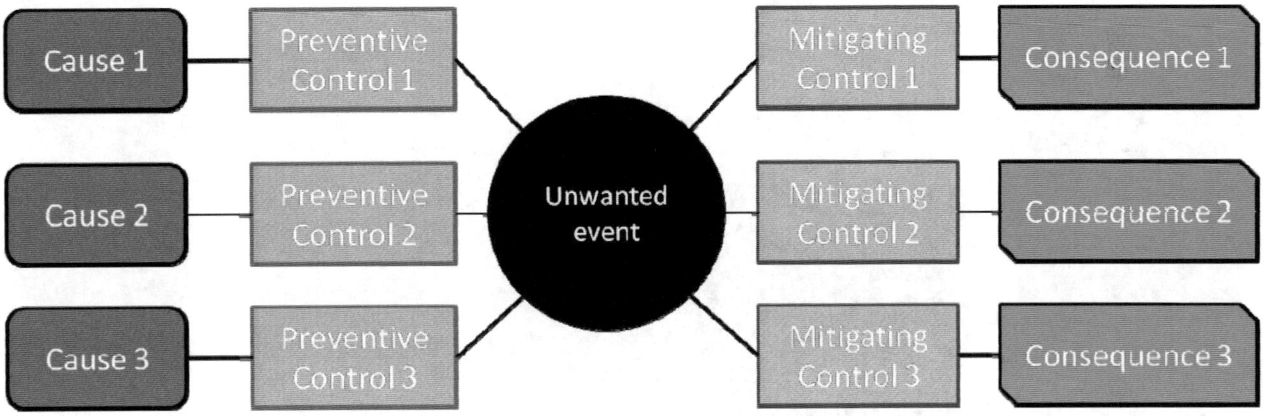

Figure 3: Example of Bow-Tie Analysis

PHA can be used as the basis to create the Bow-Tie diagram. When assigning controls (or safeguards) to a catastrophic event, emphasis should be placed on systematically working through each level of the hierarchy of controls (see Figure 2), from the most effective control to the least effective control.

2.2.3 Risk evaluation

A risk matrix can be used to evaluate the risk for a catastrophic event. If the event considered is relatively complex, Quantitative Risk Assessment (QRA) can be used to quantity the risk associated with the event. QRA involves computational modeling of fire, explosion and toxic gas dispersion modeling for consequence analysis. Frequency analysis can also be employed. Frequency analysis involves use of failure rate data and detailed analysis using techniques such as Fault Tree Analysis (FTA) and Event Tree Analysis (ETA).

2.3 Critical control life cycle management

In order to effectively manage critical controls, it is essential to establish a performance requirement detailing the expected performance of a control. The performance requirement should include the following information from the life cycle of a critical control (see Figure 4):

- **Design** – Design basis of the critical control
- **Operation** – How the critical control should be operated
- **Maintenance** – Maintenance activities required to maintain effectiveness of critical control
- **Change** – Changes that can impact the above activities should go through the management of change process and the critical control performance requirement should be updated accordingly

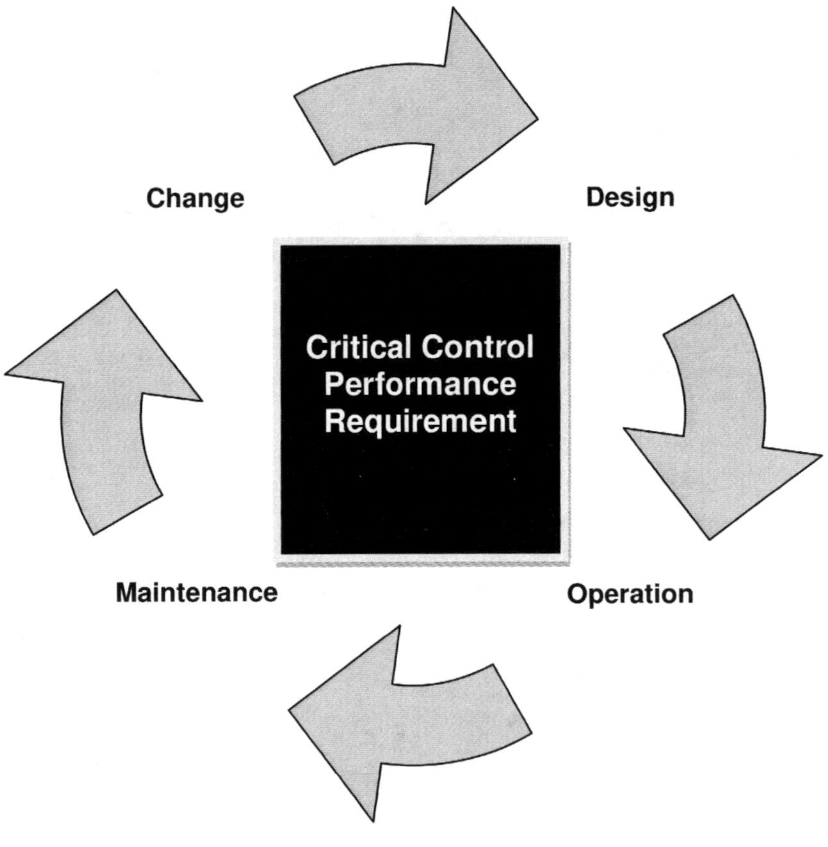

Figure 4: Critical Control Performance Requirement and Management Life Cycle

Table 2 provides examples of information to be included in Critical Control Performance Requirement.

Table 2. Information included in Critical Control Performance Requirement

Control	Design	Operation	Maintenance	Change
Safety interlock	Type of instrument; Logic diagram	Set points for alarm and trip; Shutdown key	Functional testing procedure; Testing frequency	Decrease testing frequency due to change of plant turnaround interval
Relief valve	Sizing case; Equipment the relief valve is sized for	Relief pressure and relief destination	Testing procedure and frequency	Add more load to relief valve through modification of process equipment
Gas detector	Dispersion modeling; Type of detectors	Gas concentration detected by the detector; Location of gas detectors	Testing of gas detector and test frequency	Change of air flow – e.g. doors of a process building are kept open instead of closed
Emergency response	Dispersion modeling of toxic gas release for emergency response planning	Emergency response plan (including contact information of local emergency services)	Verification activities – e.g. check if all emergency response contact information is up to date.	New release scenario has been identified which can impact emergency response plan.

2.4 Monitor and review

The intent of monitor and review is to monitor critical control management performance. There are various ways to monitor and review, such as:

- Risk Register review and reporting
- Audit (internal and external)
- Key Performance Indicators (KPI) (e.g., leading or lagging indicators)

Risk Register

Risk Register is a common tool used in risk management to report and track significant risks to a site or an organization. Typically each risk on the register has a risk owner, risk description, risk controls and rankings. Catastrophic events and associated controls should be reported and tracked in the register throughout the organization. This would increase the transparency of critical control management and accountability.

Audit

The critical control performance requirement document can be used as the basis for internal and external audits. The audit of critical control management should not only focus on whether the management process is being followed per pre-defined procedure, but also on the adequacy and effectiveness of critical control management activities such as:

- Risk Assessment for catastrophic events
  - Is the assessment being done by competent personnel(s)?
  - Are critical concerns being addressed in the assessment?
  - Are all critical controls being identified in the assessment?

- Critical Control Management
  - Is the test procedure adequate to detect all failure modes?
  - Is operating procedure being reviewed by competent personnel?
  - Are all hazards being identified for a control?

An audit is typically done on a longer timeframe, as audits are very resource draining. To ensure performance of critical controls on a day-to-day basis, another approach is required.

Key Performance Indicators

Another way to verify performance is through Key Performance Indicators (KPIs). The critical control performance requirement can be used to create leading and lagging indicators. The HSE (Health and Safety Executive) of the UK published a paper *Developing Process Safety Indicators* [2] to provide guidance on developing leading and lagging indicators for risk controls based on their performance requirements. In this published paper, leading and lagging indicators are defined as such:

- **Leading indicators—**a form of active monitoring focused on a few critical risk control systems to ensure their continued effectiveness. Leading indicators require a routine systematic check that key actions or activities are undertaken as intended. They can be considered as measures of process or inputs essential to deliver the desired safety outcome.

- **Lagging indicators—** a form of reactive monitoring requiring the reporting and investigation of specific incidents and events to discover weaknesses in that system. These incidents or events do not have a to result in major damage or injury or even a loss of containment, providing that they represent a failure of a significant control system that guards against or limits the consequences of a major incident. Lagging incidents show when a desired safety outcome has failed, or has not been achieved.

For example, for safety interlocks, a leading indicator can be created to measure the percentage of inspection being conducted per scheduled inspection. A lagging indicator can indicate the percentage of safety interlocks that fail to perform the intended functions, which are defined in the performance requirement document.

2.5 Communication and consultation

Effective communication is critical for a management system. Due to the importance of critical controls to an organization and process facility, management representatives with defined roles, responsibilities and authorities should be assigned to ensure critical controls are managed effectively. A communication process should be established from site-level to appropriate management levels within an organization. Furthermore, escalation criteria should be established based on KPIs, audit results and risk register information. This is to ensure that issues within the management process are dealt with by the appropriate authority in a timely manner.

3 Enable critical control management

In order to effectively manage critical controls, historical data and life cycle activities of controls need to be tracked and monitored on an ongoing basis to ensure:

- Performances of controls are not deviating from pre-established performance requirements;
- Effectiveness of controls are maintained;
- Risks are effectively managed by critical controls;
- Timely identification of emerging critical controls;
- Easy access to critical controls information by personnel;
- Consolidated view of trends on critical controls, such as failure information, maintenance activities, KPIs

In order to achieve the above, running a paper or Excel-based system can be very challenging and resource draining as a significant amount of time can be spent consolidating critical control life cycle data across multiple systems. For example, maintenance actives and failure rate data of controls are typically managed in an asset management system. Operation documents and drawings are kept in a document control system. Hard and soft copies of risk assessments are typically kept in offices and/or a company intranet. Management of Change at facility level is managed in an independent system, which can be of various forms, such as paper-based, Excel-based or a home-grown system.

To ensure accuracy and efficiency in managing critical controls, one should consider an enterprise management system that can communicate to various systems and pull data into a centralized system for reporting and trending. A critical control management system should also be closely integrated into the risk assessment system in order to provide a dynamic and transparent view of risk associated with catastrophic events. If a control that mitigates a catastrophic event fails to perform per the performance requirements, appropriate stakeholders should be notified and associated corrective actions should be tracked by the system.

Such a system should also be configurable to meet organization needs, allowing users to aggregate data across the organization to create consistency in assessing hazards, modifying existing procedures, and developing best practices.

4 Conclusion

Organizations often find themselves focusing on conformance to industry regulations and standards, rather than how well their risk management process is performing. In this paper, we outline how companies can achieve high performance risk management processes through identifying critical controls, tracking controls via an efficient, robust system, and putting into place a system that ensures risks are dealt with properly when they arise.

5 References

[1] International Standard ISO/FDIS 31000 Risk management – Principles and guidelines, http://www.iso.org/iso/catalogue_detail.htm?csnumber=43170

[2] Developing Process Safety Indicators, Health and Safety Executive, UK http://www.hse.gov.uk/pubns/books/hsg254.htm

A Spaghetti Western: Introduction to Safety Cases for Gulf of Mexico Deepwater Drilling

John F. Kill

North America Risk Practice Group

Environmental Resources Management

15810 Park Ten Place, Suite 300, Houston, TX 77084

john.kill@erm.com

Morris E. Burch, Jeremy Goddard, Marlane J. Kayfes
North America Risk Practice Group
Environmental Resources Management
15810 Park Ten Place, Suite 300, Houston, TX 77084
morris.burch@erm.com
jeremy.goddard@erm.com
marlane.kayfes@erm.com

Kevin G. Kinsella
Environmental Resources Management
Floor 11 5 Exchange Quay
Manchester M5 3EF
United Kingdom
kevin.kinsella@erm.com

Keywords: safety case, HSE case, Gulf of Mexico, deepwater

Abstract

Since the April 2010 rig explosion and oil spill in the Gulf of Mexico, factions of the U.S. oil and gas industry, including regulators, have jockeyed to identify what to do (and not do) in reaction to the event.

While U.S. regulatory officials want to increase safety in deepwater drilling, the industry tends to balk at the thought of more regulation. In May 2010, the Department of Interior recommendations included to initiate a safety case regime such as that in the United Kingdom, Australia, and other countries. Again, January 11, 2011, safety cases were strongly recommended in the final report to the President from National Commission on the BP Deepwater Horizon Oil Spill and Offshore Drilling, "Deep Water: The Gulf Oil Disaster and the future of Offshore Drilling." Some members of oil-and-gas exploration and production industry already are preparing for that eventuality, beginning to create safety cases for U.S. offshore assets. Others already have safety cases prepare, as a matter of their course of doing business.

To address varying degrees of knowledge about safety cases, this paper is a primer of the history and growing pains of safety case regimes worldwide, principles and elements of safety cases, and their implementation. It addresses what makes a safety case effective and ineffective—the good, the bad, and the ugly.

1. Waiting in the Wings

A flurry of activity among and between U.S. regulators and industry leaders to increase the safety of future deepwater exploration and production followed the April 2010 rig explosion and subsequent massive oil spill in the Gulf of Mexico (GoM). In May 2010, the U.S. Department of the Interior proposed recommendations for improving deepwater drilling safety [1]. Those recommendations included adopting "safety case requirements for floating drilling operations on the Outer Continental Shelf." Although many of the recommendations were passed into regulations in September, effective immediately, adopting safety cases was postponed pending further study [2].

The industry knows that safety cases will come to the U.S.; the September rule stated, "BOEMRE is evaluating how a drilling safety case should be most appropriately integrated with an overall Safety and Environmental Management System (SEMS) approach, which BOEMRE may implement through a separate rulemaking process." Two weeks later, the SEMS approach was adopted with incorporation by reference of American Petroleum Institute Recommended Practice (API RP) 75 [3]. And then in January of 2011, the National Commission on the BP Deepwater Horizon Oil Spill and Offshore Drilling strongly recommended initiating safety cases its final report to the President "Deep Water: The Gulf Oil Disaster and the future of Offshore Drilling."[4]

So while the safety case process waits in the wings while regulators play out their research roles, some drilling contractors already are producing safety cases for their facilities. Other companies regularly build safety cases on all their assets—onshore or offshore, required by regulations or not. When safety cases become a reality for the GoM, those companies will be ahead of the game, perhaps with only slight modifications necessary. The remaining affected industry groups should strive to understand what safety cases are, how they are used worldwide, and what makes them excellent tools in a strong safety culture as well as what renders them useless. Then they, too, can become prepared and proactive for the regulatory changes ahead.

2. Safety Case Defined

The philosophy of safety cases is that:

- Those who create risks are responsible for controlling them,
- Safe operations are achieved by setting and achieving goals rather than by following prescriptive rules, and
- All risks must be reduced such that they are below a threshold of acceptability.

In a regulatory regime, a broad definition might be that a safety case is legally defensible documentation proving that all necessary steps were taken to ensure that the individual risk per annum (IRPA) of the facility and its processes is as low as reasonably practicable (ALARP). IRPA is the probability that a person will be killed. The risk has to be weighed against the measures necessary to eliminate it. ALARP "means that the degree of risk has to be balanced against the time, trouble and cost involved in taking the measures necessary to avoid the risk" [5].

Figure 1: ALARP process

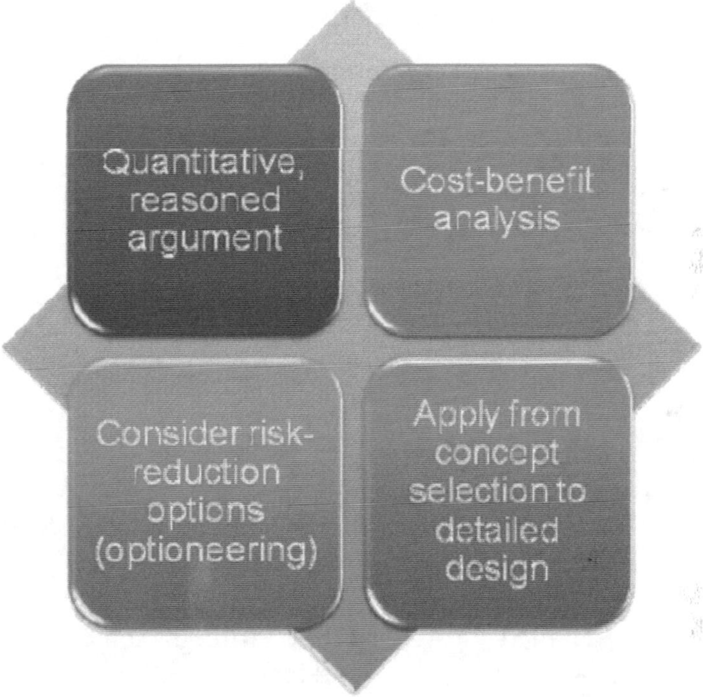

This means that major hazards have been identified as well as how those risks are controlled—the barriers to a major accident event (MAE) occurring and mitigation measures to lessen the effect of an MAE if it does occur. It also means that a strong safety management system is in place to ensure the controls are effectively and consistently applied to instill and support a strong safety culture.

The Australian regulatory body, NOPSA (National Offshore Petroleum Safety Authority), defines a safety case as "a document produced by the operator of a facility which identifies the hazards and risks, describes how the risks are controlled, and describes the safety management system in place to ensure the controls are effectively and consistently applied" [6].

2.1 Safety Case Parts

Although safety cases take slightly different forms across the globe, their contents generally are consistent:

- Introduction, Executive Summary
- Facility Description
- Risk Management (including formal risk assessments [FSAs])
- Safety Management System
- ALARP Statement
- Performance Monitoring
- Emergency Response

The International Association of Drilling Contractors (IADC) is one of many common models used and is known as a Health, Safety and Environment (HSE) Case. The organization has two guidelines for HSE cases—one for mobile offshore drilling units (MODUs) and one for land drilling units. IADC is one organization working with U.S. regulators to define what U.S. safety cases might look like, so it is likely that model will have a least some influence.

2.1 Stages of Safety Case Development

The three common stages of safety case development are during design, operations, and (for offshore) abandonment. They can be done individually, as needed, or contracted for as a group, known as lifecycle safety cases. Most often formal safety assessments are performed during the front-end engineering design stage of a project and then are revised during detailed design. Those formal safety assessments would include hazard identification (HAZID) workshop; fire, explosion, and toxic gas modeling and consequence analysis; ALARP workshop; and quantitative risk assessment.

2.2 Approaches to Safety Case Regulation

When it is said that we do not yet know what form a U.S. safety case might take, the biggest question is whether the regime will be a prescriptive, goal-setting (self regulating), or a combination. Historically, U.S. regulations are prescriptive—explaining in some detail what must be done. Typically, however, safety case regimes follow a goal-setting or combination model.

Consider a simple example—a tract of land that you have to protect.

2.2.1 Prescriptive Approach

With the prescriptive approach, the regulators would lay out exactly what steps you must take. For example, install a fence of a particular construction (wood, chain link, concrete block, etc.) to a specific height and include any number of enhancements (cameras, electronic wiring, barbed wire, etc.).

Operating plans must be submitted for approval and, ultimately, the regulator carries most of the burden.

This type of approach is common in the China, Indonesia, Malaysia, and the United States.

2.2.2 Goal-setting Approach

With the goal-setting approach and that same tract of land, you know that the goal is to protect the land and people on it and you would determine how to best accomplish that. You might choose from determining that no physical protective measures are warranted, to installing a six-foot tall chain link fence, to installing a ten-foot tall chain link fence wired for electric shock upon contact and topped with three rows of barbed wire.

Whatever techniques are employed, the individual (or company) has to prove through the safety case that the level of protection and barriers makes the risks to people, assets, and the environment as low as reasonably practicable. The companies gain the trust of regulators (and other stakeholders such as the public or investors) by conducting a thorough risk analysis and documenting hazard barriers and emergency response procedures.

This type of approach is common in the Australia, Brazil, Canada, Norway, and the United Kingdom.

2.2.3 Other Approaches

The combination approach simply means that it is some combination of the prescriptive and goal-setting concepts.

But a fourth option exists: self-regulation. This type of approach is common in Angola, Ghana, Mexico, and Nigeria. The risks can be much higher unless companies working in this type of environment willfully employ safety measures used in areas that operate by codes and standards or under regulatory regimes.

3. Brief History

A brief history of safety cases in the United Kingdom begins with their implementation in the nuclear industry in 1965. However, before safety cases were required in oil and gas process industries several major accident events occurred, as well as changes to strengthen regulation of hazardous process industries.

3.1 Birth of the U.K. HSE

In 1969, Alfred Robens, Baron Robens of Woldingham, an English trade unionist, Labour politician and industrialist, was appointed to chair a committee on workplace health and safety. The 1972 Robens Report promoted self-regulation by employers and led to the Health and Safety at Work etc. Act 1974 and creation of both the Health and Safety Commission (HSC) and the Health and Safety Executive (HSE). This goal-setting approach to safety was supported by codes of practice and guidance. In 2008, the HSC and HSE merged under the name of Health and Safety Executive (HSE).

3.2 Advent of COMAH

The June 1974 chemical plant explosion near Flixborough, England, killed 28 people and injured 36. Substantial property damage was recorded—the plant was destroyed and structural damage affected neighboring villages up to 8 miles away. The blast itself was heard 30 miles away.

As a result of Flixborough incident, the U.K. government tightened regulations for hazardous industrial processes. It took another major accident event, the Seveso disaster in 1976, for harsh industrial regulations to extend across the European community, and eventually the U.K. introduced COMAH Regulations—Control of Major Accident Hazards Regulations 1999. The COMAH regulations were two-tiered. They required (1) the creation of a Major Accident Prevention Policy and (2) a full safety report demonstrating that all necessary measures have been taken to minimize risks to the environment and the populace.

3.2 Safety Cases Required Offshore U.K.

Several other major accidents occurred throughout the years but the catalyst that brought safety cases to offshore fields was the Piper Alpha disaster in 1988. The rig was completely destroyed and 167 men lost their lives. After a 13-month investigation, the Cullen Report was issued with 106 recommendations to improve the safety of offshore operations.

One of Lord Cullen's recommendations addressed an apparent conflict of interest. The U.K. Department of Energy had overseen offshore production as well as safety, and the Cullen Report recommended moving responsibility for safety to the U.K. HSE. This parallels actions taken in the United States after the Macondo well incident in April 2010; the U.S. Mineral Management Service had had joint responsibility for offshore production and safety, and the safety responsibilities were moved to the newly-formed Bureau of Ocean Energy Management, Regulation and Enforcement (BOEMRE).

The Cullen Report recommendations included formal safety assessment of major hazards and the implementation of Safety Management Systems. This activity also parallels U.S. reactions to the Macondo incident; the BOEMRE passed a rule incorporating by reference the American Petroleum Institute Recommended Practice (API RP) 75 "Recommended Practice for Development of a Safety and Environmental Management Program (SEMP) for Offshore Operations and Facilities." A comparison of SEMP with a typical Safety Management System part of a safety case reveals few differences.

Ultimately the HSE installed a safety case regime for offshore operations—The Offshore Installations Regulations of 1992 (effective May 1, 1993). In the U.S., for now, the BOEMRE continues to investigate and evaluate how to properly integrate a safety case with a safety and environmental management system approach [2].

4. Elements of Safety Cases

As mentioned in Section 2.1, most safety cases have very similar content requirements. In this discussion, the model used is that of the International Association of Drilling Contractors (IADC). The Introduction and Executive summary are followed by six other sections:

- Description of Facilities, Operations, and Surrounding Environment

  This includes ownership and location, operations and facilities, the offshore environment, primary systems and functions, and safety features and emergency systems.

- Description of Safety Management System

  This includes leadership and commitment; policy and strategic objectives; organization, resources, and documentation; competence; risk evaluation and management; operational controls; management of change; emergency planning; and auditing and review.

- Formal Safety Assessment

  This includes summaries of the hazard identification; risk assessments; identification of safety-critical elements (barriers to major accident events); and performance standards.

  A performance standard is a qualitative or quantitative statement of the performance required of a system, piece of equipment, person, or procedure, which is used as the basis for managing the hazard through the life cycle of the installation. Without performance standards, neither operators nor regulators can assess or measure the performance of the (technical or managerial) systems.

- ALARP Demonstration

  This includes adherence to codes and standards; practice of good design; risk-based analysis against criteria; and evaluation of risk-reduction options.

- Performance Monitoring

  The performance standards established during the formal safety assessments state how a safety-critical piece of equipment or system should perform in an emergency. The performance monitoring section of a safety case sets out a verification scheme (written scheme of examination) "to ensure that the identified HSE-critical activities/tasks and equipment/systems are suitable and remain in good repair and condition" [7]. The scheme details the frequency and type of examination (monitoring) established for each safety-critical element as well as compliance with legislation and the health and safety document (the safety case).

- Emergency Response Arrangements

  This includes emergency organization and responsibilities; scenario time-based analysis; escape route evaluation; temporary refuge evaluation; muster points for evacuation; means of evacuation (helicopter, totally enclosed motor propelled survival craft (TEMPSC), etc.); and facilities to monitor and control incidents and for organizing evacuation.

4.1 Bringing It All Together

Perhaps most important is to bring together all parts of a safety case, translate it to the workforce as a practical philosophy and steps to support a company-wide, strong safety culture, and keep it current and alive.

To do this it is important (and usually required) to involved the workforce in the entire safety case process. This step ensures input from perhaps the most valuable people—those familiar with the daily operations and realistic scenarios.

Clearly link the parts of the safety case so the interrelationships are clear. Then ensure competency through a program of building the safety culture in the company and on the facility (including a strong stop-work policy), train the workforce (employees and contractors) and regularly reinforce the training, consistently manage change. Also, maintain a "living" hazard register and safety case through regular review, update when there is a major change, and make it easy for the workforce to use.

Figure 2: Integrating a safety case with operations

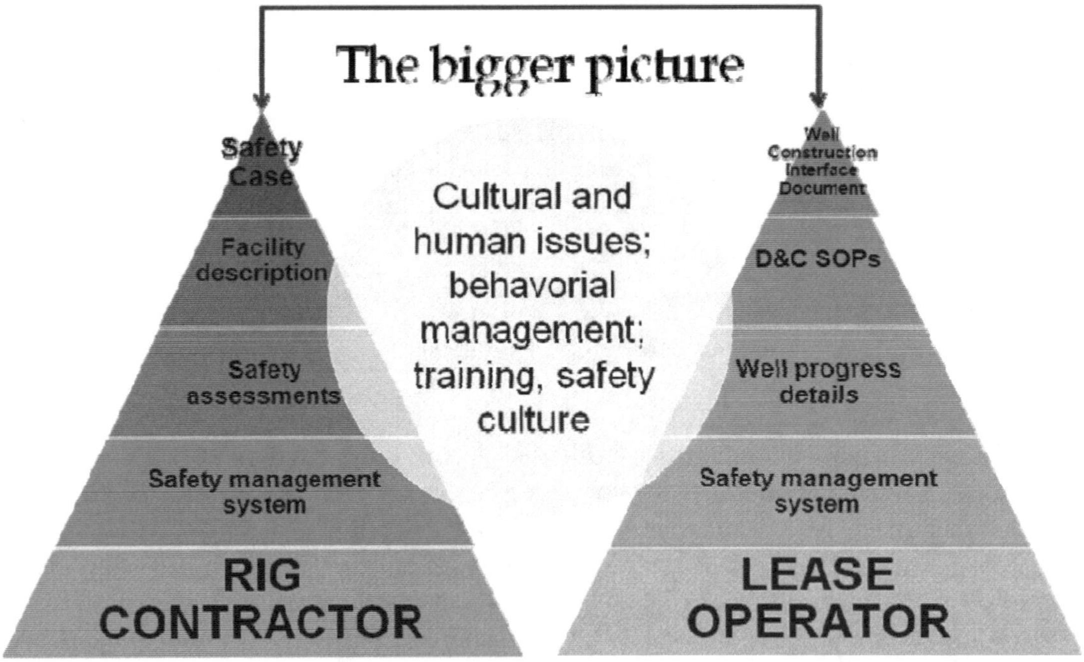

5. The "Spaghetti Western" of Safety Cases

The U.K. and other parts of the world have had 20 years to wrestle and refine the safety case concept—to work through the growing pains. Regulators in safety case regimes have dealt with the challenges of development, implementation (sometimes over-implementation), and follow-up, as well as dealing with potential complacency. Safety cases have their good aspects, the bad, and the downright ugly.

5.1 The Good

- What are the worst five events and do key people know them?
- What are the top five risks and do key people know them?
- Is the company prepared for the worst accidents?
- Is the inherent design sound?
- What equipment and procedures are critical to protect people, the environment, assets, and reputation?
- Are the critical aspects working properly?
- If not, what operational limitations exist?
- Is the overall health, safety, and environment management system working?

5.2 The Bad

The several "bad" aspects of safety cases include:

- Providing too much or too little detail

  Regulators and organizations have experienced any number of difficulties in applying safety cases and getting the level of detail right. Often the size and complexity of a safety case can undermine its usefulness.

- Conflicting needs

  It is important to strike a balance between the needs of the regulators and those of the operator's staff. Increasingly electronic safety cases or very high-quality summary documents help users navigate what often is a labyrinth of information.

- Inappropriate use of probabilistic techniques

  Probabilistic techniques, such as quantitative risk assessment (QRA), are widely used and although those techniques can be very useful and informative, they sometimes are used to stretch beyond what available data will support. For example, using QRA to demonstrate that the risks of a main export pipeline running through a corridor adjacent to the accommodation block does not warrant modifications is inappropriate when good engineering practice suggests a series of measures would be warranted to reduce risks.

Sometimes a safety case is viewed as merely a paper exercise or a checklist exercise. This mindset can create a "do it cheap" attitude, and then the safety case benefits no one, is unusable, and is a dust-collector.

5.3 The Ugly

Poor implementation is the key phrase here. Safety cases can turn ugly when they are poorly implemented and not used. A safety case can be created as a paper exercise and then ignored. Then if a major accident event occurs often it is unmanaged, deadly, and costly to the company's reputation and bottom line—that's the "ugly."

6. Conclusion

It appears that safety case regimes have been established after very severe major accident events when some stakeholder(s) have said "Enough!" and the industry's social license for operating is pulled.

Since April 2010, U.S. regulators have moved steadily forward in considering safety cases for offshore drilling and production. Meanwhile, the questions are many.

Will safety cases be established in a prescriptive style as are many current U.S. regulations or will they take a more cooperative and trusting stance with industry and initiate a goal-setting or combination regime?

Will the U.S. start from scratch in developing a safety case regime or will it learn from and apply lessons of the 20-year history of safety cases elsewhere? Perhaps this will be the case because we know the BOEMRE is consulting with other regulators and moving slowly.

And how will companies in the industry approach safety cases: with an embrace or half-hearted disdain? How long will be the road to acceptance? Industry's approach might well be most prophetic of the success of safety cases in operations offshore the U.S.

The results might be good (demonstrations of safe [or unsafe] facilities and providing answers to key questions that can protect the company overall), bad (a complicated document created cheaply that is unusable), or ugly (a high-dollar investment in a quality document that is poorly implemented).

7. References

[1] Increased Safety Measures for Energy Development on the Outer Continental Shelf, May 27, 2010, U.S. Department of the Interior, http://www.doi.gov/deepwaterhorizon/loader.cfm?csModule=security/getfile&PageID=33598 (accessed Dec. 9, 2010)

[2] 30 CFR Part 250, Oil and Gas and Sulphur Operations in the Outer Continental Shelf—Increased Safety Measures for Energy Development on the Outer Continental Shelf; Final Rule, Part III, October 14, 2010, Department of the

Interior, Federal Register, Vol. 75, No. 198. http://edocket.access.gpo.gov/2010/pdf/2010-25256.pdf (accessed Dec. 9, 2010)

[3] 30 CFR Part 250, Oil and Gas and Sulphur Operations in the Outer Continental Shelf—Safety and Environmental Management Systems; Final Rule, October 15, 2010, Department of the Interior, Federal Register, Vol. 75, No. 199. http://edocket.access.gpo.gov/2010/pdf/2010-25665.pdf (accessed Dec. 10, 2010)

[4] National Commission on the BP Deepwater Horizon Oil Spill and Offshore Drilling. January 11, 2011. Final Report To the President. "Deep Water: The Gulf Oil Disaster and the Future of Offshore Drilling." http://www.oilspillcommission.gov/final-report (accessed January 31, 2011)

[5] Edwards v. National Coal Board. Court of Appeal CA. Tucker, Asquith and Singleton L.JJ. 1949 Mar. 8, 9, 10, 21. Cited by Nick Higginson, 2002, Case Law Database www.healthandsafetytips.co.uk/Downloads/Nick.../Case_Law.doc (accessed Dec 9, 2010)

[6] Safety Case Approach, (n.d.), National Offshore Petroleum Safety Authority http://www.nopsa.gov.au/safety.asp (accessed Dec. 9, 2010).

[7] IADC (3 February 2010) Appendix 1 to Health, Safety and Environmental Case Guidelines for Offshore Drilling Contractors, Issue 3.3.2. http://www.iadc.org/hsecase/MODU%20Offshore%20Appendix%20Issue%203.3.1%20Final.pdf (accessed Dec. 13, 2010).

Operational Discipline
Does Your Organization Do the Job Right Every Time?

Brian D. Rains
Global Process Safety Management Leader
DuPont Sustainable Solutions
4417 Lancaster Pike, BMP 13/1110, Wilmington, DE 19805
brian.d.rains@usa.dupont.com

Keywords: Operational Discipline, Risk Management, Operational Excellence, DuPont

Abstract

This presentation is intended to help safety professionals and leaders consider the robustness of their own organization's Operating Discipline. The session will include details of how Operating Discipline is measured, monitored and improved within DuPont. In addition, tools and recommended actions will be shared so that all attendees will be better equipped to strengthen their own organization's Operating Discipline.

Introduction: Operational Discipline Defined

DuPont defines Operational Discipline (or OD) as "the deeply rooted dedication and commitment by every member of an organization to carry out each task the right way every time." Expressed more succinctly, OD is "Everyone: Do It Right Every Time" (Exhibit 1). As applied to safety, DuPont reinforces the company's desire for a strong Operational Discipline culture by stating that safety is a condition of employment.

Exhibit 1. Operational Discipline leads to operating excellence.

A former employer used the term 10,000% Compliance to describe the expectation that 100% of the people are to comply 100% of the time. And again, failure to meet this expectation was met with negative consequences.

Another company, a DuPont client, describes Operational Discipline in terms of its results, stating that Operational Discipline "delivers stable, low-cost production."

All of these messages have one thing in common: They are statements by management, directed to the entire organization, expressing clearly their expectations for personal conduct and behavior; a standard of "following the rules." In this talk, I will reference the DuPont definition.

When we talk about everyone doing it right every time, we often focus first on defining "everyone" or even "every time," while concurrently attempting to describe what type of culture this might suggest. In my experience of leading manufacturing organizations in many settings and on several continents, I believe that the first step to increasing Operational Discipline is to focus first on the "it," that is, what is it that we expect people to do?

Most organizations, especially those handling hazardous materials or performing hazardous operations, invest considerable effort in preparing written instructions and procedures. Written procedures have many functions, including enabling consistent operations and behaviors, and they serve as a vehicle for codifying technology, best practices and experience. So why do so many companies find that employees are not following these instructions and procedures?

In this presentation, I will share the recipe that I use to lead manufacturing organizations to a state of higher operational discipline. It is a simple – but, I believe, effective – eight-step approach.

Eight Steps Toward Operational Discipline

Step 1: Be Convinced Operational Discipline Is Needed.

The greater the risks in your operations, the greater the need for a strong Operational Discipline culture. Perhaps you believe that relying on the experience of your employees and word-of-mouth training is sufficient for your organization. But if your business would suffer from human errors leading to safety incidents, quality and reliability issues, poor process optimization and higher costs, then it is probably time to implement the changes that create a strong Operational Discipline culture.

A review of some of the operating practices of the most hazardous industries, including airline transportation and nuclear power generation, shows that Operational Discipline is built into operations through check sheets, step-by-step sign-off's and independent verification by a co-worker or supervisor of the most critical steps. In the area of health care, a recent news report interviewed a Harvard University physician who had introduced checklists into the operating room. He reported a dramatic decrease in medical errors and avoidable patient complications and deaths.

In most DuPont manufacturing operations, we do not rely on direct supervisory oversight or direct secondary observation to ensure employees perform tasks correctly, and we often reserve checklists and sign-offs for the most critical tasks. Instead, we depend on a healthy Operational Discipline culture. And upon reflection, even checklists, sign-offs and other observation approaches rely upon workers' doing the job right – without taking shortcuts and following required procedures.

How can you gauge the status of Operational Discipline in your organization? One way is to review your incidents and ask some questions. Do you have repeated procedural violations? Are you analyzing incident trends and categorizing root causes for indications of systemic weaknesses?

When DuPont's Process Safety Incidents for 2000 were analyzed as part of our Operational Discipline initiative, the data revealed that a lack of Operational Discipline was a dominant cause in more than half the cases. This triggered the development of a number of additional tools to assess and strengthen our OD culture. Some of these will be described below.

When management becomes convinced that a strong Operational Discipline culture will be a key determinant in achieving key business metrics and that the prevailing OD culture needs to be strengthened, then the environment is ready to move to Step 2.

Step 2: Assess the Current State.

What follows is an informal process I have conducted with many client groups.

I start by asking the group to consider all the individual tasks that are performed in their facility in a normal week where a documented procedure exists. The number is often very large, on the order of 10,000 tasks. The actual number is unknown, but it is clear that the size of the facility and the workforce plus the complexity of the operation are key determinants. I then ask what percentage of those numerous tasks are completed according to the required procedures and request each individual in the group to write this number down for future sharing. We then go around the room capturing the individuals' responses on a chart pad.

What kind of responses do we get? Simply stated, they are wide-ranging. Individual estimates of the number of tasks completed according to required procedures have ranged from less than 5 percent to 98 percent. Most organizations, however, average less than 50 percent.

I have also observed that managers tend to offer a higher relative percentage than others in a group and those individuals, whose duties are closest to the shop floor, offer lower numbers. For example, in a recent such session, a middle manager suggested that the correct answer for his organization was 10 percent. His 25 colleagues and peers in the room said nothing.

There are many reasons for the range of answers:

1. Many people don't know what procedures exist.
2. There has been no initial training and/or no refresher training provided on the procedure.
3. There are better ways to complete the tasks than what is documented in the procedure.
4. Following the procedures is optional.
5. Some people don't care.

Of course, this question addresses only compliance with the existing procedures. It does not take into account whether procedures actually exist for all of the facility's important operational and maintenance tasks.

Over the years, DuPont has developed a number of diagnostic tools to improve our Operational Discipline. One of these is a set of 11 characteristics that are used in assessing an organization's OD culture. These characteristics are depicted in Exhibit 2.

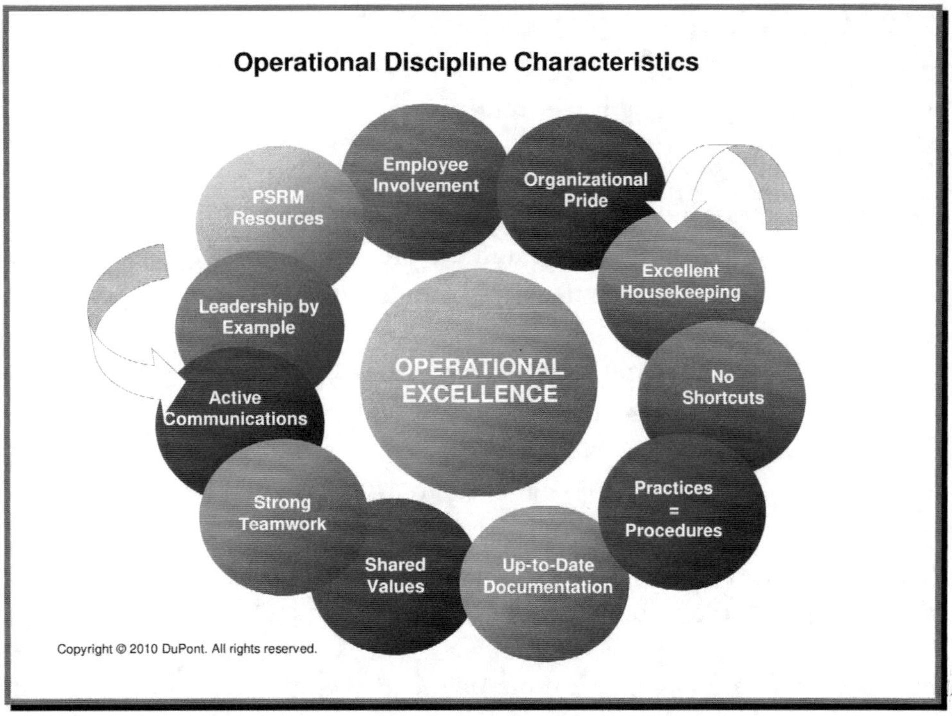

Exhibit 2. These 11 OD characteristics are used to assess OD culture.

In OD assessments, individuals and groups of workers are asked questions such as:

- Are alternative ways to completing a task rather than the documented procedure accepted by the organization? Are shortcuts taken? Do you take shortcuts?
- Given the circumstances that led to a recent process safety incident, were procedures adequate to prevent the incident? Were they followed? What would you have done differently, if anything?
- Are line managers aware of how work is actually done? Do they tolerate shortcuts? What are their standards of performance? How do they communicate their standards?

The assessment team provides site management with feedback for each of the 11 OD characteristics. Strengths and improvement opportunities are highlighted, as are best practices at other sites.

Other assessment tools include a comprehensive self assessment methodology and an on-line survey. Management has the right to apply the tools most appropriate for its situation.

Another example is the First and Second Party Process Safety Management (PSM) Audit Protocol. It includes approximately 300 audit questions asked during a routine one-week audit. Of these questions, 120 are mapped or correlated to one of the 11 OD Characteristics. When the audit scoring is complete, not only is an overall PSM score awarded, but also Operational Discipline scoring that includes a total OD score and an individual score for each OD characteristic.

Assessing OD can be sensitive. It is also highly subjective. A high degree of trust is needed to obtain a realistic picture. With these factors in mind, it is recommended that all organizations use more than one assessment tool and use them frequently.

Step 3: Develop a Comprehensive Roll-Out Plan.

With every change initiative, especially those that affect the organization's fundamental culture, detailed and comprehensive planning is essential.

All good plans describe:

- The plan goal or objective and how it supports and is aligned with overall business strategies
- The gap between the desired future state and the present condition
- The steps that will be taken to close the gap, including the role of management in leading the change
- The assignments that will be given to those who are responsible for completing these tasks
- An agreed-upon timetable for accomplishment of the tasks
- Communications to all those involved
- Metrics and other measurement tools, including periodic reviews and audits that leaders will use to track progress

One company that DuPont worked with developed a 43-step Operational Discipline Enhancement Campaign. It focused on the 11 OD Characteristics that it believed would most effectively "move the culture." – PPE Compliance, Housekeeping, Elimination of Shortcuts and "Absolute Adherence to Procedures." Communication assignments included playing a recorded message by the site leader twice per shift on the site's PA system. Other important steps in the plan were:

- An announcement and information about the OD campaign provided to all personnel at all levels
- The use of skits, displays, mannequins, videos, and other communication vehicles to get the word out
- Training on key procedures and improving the availability and access of these documents to the organization
- On-going communication to the entire organization on the campaign's progress
- Focused housekeeping audits and the implementation of a 5S program

A key component of any such initiative is the celebration of the positive results and the process of communicating and learning from deviations.

Another company took a more direct approach to implementing OD by issuing an "Operational Discipline Standard." Some of their statements were excellent; for example: *Site Management shall establish the foundation for an effective Operational Discipline process through visible and felt leadership and commitment.*

Other statements may have been a bit idealistic: *All employees shall believe that all injuries, occupational illnesses, as well as safety and environmental incidents are preventable and commit to the policy that safety of a person overrides all production targets* and *all employees shall understand and internalize the Health & Safety principles of the company.*

Overall, what is most important in the development of an OD improvement plan is that it is consistent with the company's values and that the management is credible and authentic in its communication and implementation.

Step 4: Focus on the "It."

In our definition of OD – *everyone does it right every time* – "it" refers to the work or day-to-day tasks to be performed. For there to be a single right way to do the job, the "it" is typically described in instructions or procedures. Procedures should be the documented, codified best practices that represent the organization's complete and best knowledge and experience. Procedures should contain the cumulative body of knowledge expressed in writing that instructs how a particular task is to be performed each and every time.

Too often, when an organization attempts to improve its Operational Discipline, it focuses first on the people and "fixing" what is wrong in the culture. In my experience, the right place to start is in the quality of the procedures.

If management instructs its employees to follow procedures when the procedures do not represent the best knowledge and experience of the facility, employees are placed in an untenable position and faced with two options:

1. Obey and follow the bad procedure and get a bad result
2. Ignore the procedure and do the task a "better way," inviting negative consequences for failing to follow management instructions

In this situation, not only is the employee placed in a difficult position, but management will suffer a loss of credibility and loyalty.

A robust process of technical review and employee involvement combined with talented procedure writers and a document management/control system is absolutely essential to building a culture of Operational Discipline. It is a continuous process of improvement.

As shown in Exhibit 3, the state of operational effectiveness is the product of the quality of the management standards and the Operational Discipline. Weakness of either (or both) will result in ineffective operations and safety incidents.

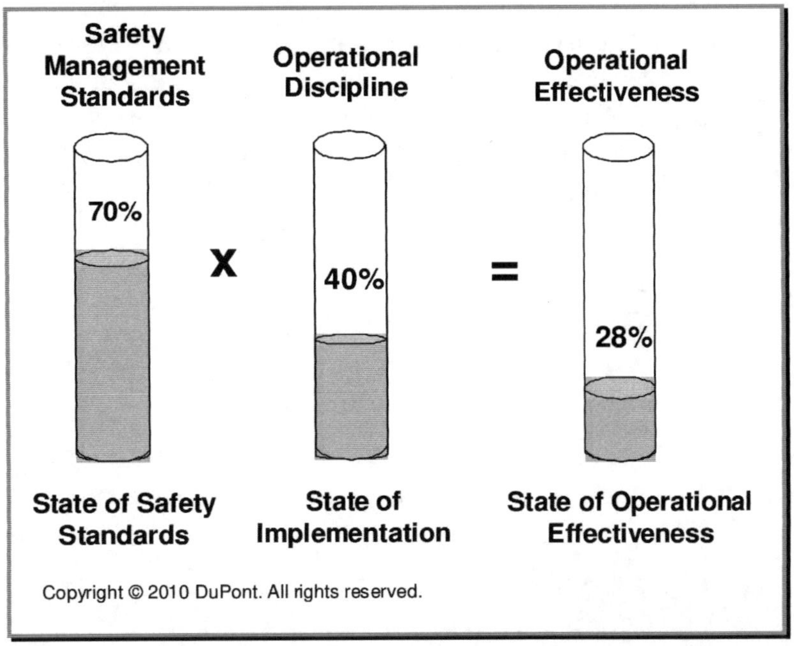

Exhibit 3. Safety performance is a function of the safety management systems and level of execution.

Management should not insist upon compliance to procedures until the procedures are of sufficient high quality to merit compliance.

Step 5: Establish Priorities.

All procedures are *not* created equal. Employees understand that to operate a facility or perform their tasks safely, some procedures and regulations are more important than others. Thus, to strengthen Operational Discipline, a focus on compliance with the most important procedures is the place to start. Taking this approach, DuPont and many other companies compile and publish a list of the most critical safety rules. The list becomes a line in the sand, and there is zero tolerance for willful violation.

Seven examples of these lists appear in the Appendix to this paper. What is important is not so much the rules themselves, as they will vary from company to company and site to site depending on the risks and cultures, but management's complete commitment to seeing them followed without exception. Although the lists in the Appendix deal with safety, other operational and maintenance procedures can be equally critical and equally valuable in an organization's drive for full compliance and greater Operational Discipline. Shutdown and startup procedures, interlock test procedures, relief valve test procedures, and PPE compliance can be given high priority and no-tolerance status.

All workers should know without any doubt the organization's inviolable rules. The rules should be highly visible, with framed copies in conference rooms, control rooms, maintenance shops and laboratories. Contractors should also be fully aware of the rules and should be expected to comply. Every possible effort should be made to avoid the situation where an employee claims, usually during an incident investigation, that he or she was not aware of violating an inviolable standard.

Step 6: Strengthen Discipline Processes.

To encourage cooperation and adherence to Operational Discipline principles, many facilities use a process called Progressive Discipline that establishes negative consequences for failing to comply as expected. Although Operational Discipline can and should be a positive element in an organization's culture, it is important to make a clear statement of what will happen if individuals elect to not follow procedures.

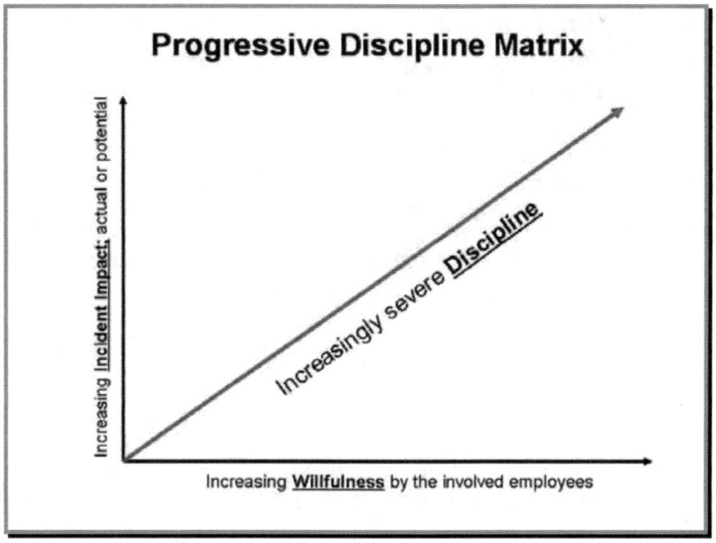

Exhibit 4. As this two-axis matrix shows, the severity of discipline increases as the impact of an incident and the willfulness of intent increases.

Typical steps in a Progressive Discipline system are 1) verbal warning, 2) written warning, 3) probation, 4) suspension with or without pay, and 5) termination.

The discipline system must apply to all employees, from shop floor to shift technicians to supervisors, managers and superintendents. It is most effective when the system is broadly communicated by line management to all employees. Although the human resources function is typically the administrator of the process, line management is the user and should be responsible for its communication and implementation.

If a new or enhanced system is being implemented as part of an overall OD initiative, management must communicate clearly to all employees the date when the new system will go into effect.

Step 7: Implement.

Once the planning is finished, the difficult task of implementation begins. As with all cultural changes, new patterns of behavior are initiated. The implementation plan and expectations for behavior must be communicated clearly to all employees. Managers should solicit feedback from employees to know what progress is being made and where there might be difficulties or misunderstandings. Clear accountability for key tasks should be assigned. Timelines should be established and deadlines met.

Building a strong Operational Discipline culture in an organization has been described by some as Management 101 – in other words, it is good, fundamental management practice. Implementing an Operational Discipline plan should be kept as simple as possible. The most important considerations are management consistency, exhaustive two-way communication and a prompt response to questions or concerns.

Step 8: Audit, Audit, Audit.

Step 2 in the process is to assess the current state. With this data in hand, plans and programs can be put in place to strengthen the organization's OD culture. Similarly, Step 8 revisits the assessment and audit step to trigger the next round of planning and improvement. The same or possibly different assessment/audit tools can be used, but it is important to be able to determine whether true progress is being made by a comparison with prior results.

In a trusting and mature culture, assessing the level of Operational Discipline is important for all members of the organization, regardless of their roles. Defensiveness on the part of individuals or work groups or the failure to report the "whole truth" can usually be avoided if management effectively communicates the critical importance of a strong OD culture and the changes that are being made.

Conclusion

In an organization where a strong Operational Discipline culture exists, there will be a "deeply rooted dedication and commitment by every member of an organization to carry out each task – "the right way, every time." When an organization needs to create or strengthen its OD culture, the eight-step process described above, when led with authenticity and commitment, has been shown to be an effective and successful approach. Continuous effort by management will be required to sustain the gains.

A final comment: It is unfortunate that the term "Operational Discipline" has been confused with "Progressive Discipline" or other forms of punishment. Part of the confusion comes from the role that Progressive Discipline might have in initiatives to strengthen the underlying Operational Discipline culture (see Step 6). The word "discipline" can evoke a strong negative reaction even though when coupled as "Operational Discipline" it is intended to describe a culture of order and predictability. It is everyone's best interest to view Operational Discipline in a positive light; an important enabler in the journey towards Operational Excellence.

Appendix: Examples of Life Saving Safety Rules

Company A:

INVIOLABLE SAFETY STANDARDS

The following standards are inviolable; if followed, they are known to drastically reduce employee exposure, serious injuries and/or death. Thorough job planning will help ensure that these standards are met. These standards apply equally to both employees and contractors.

1. **CONDONING THE VIOLATION OF SITE STANDARDS**
   Supervision will not direct or condone an employee or contractor to violate any safety or environmental standards. It is every person's (but especially supervision's) responsibility to immediately correct any sub-standard safety, health or environmental performance that is observed.

2. **ENERGY CONTROL (formerly LOCK, TAG AND TRY)**
   The Energy Control procedure must be observed and complied. No unauthorized maintenance, servicing, installation, repairs or inspections of machinery or equipment shall be performed.

3. **CONFINED SPACE ENTRY**
The Confined Space Procedure must be observed and compiled with. No unauthorized confined space or vessel entry shall be performed.

4. **LINE BREAK (formerly FIRST BREAK)**
The Line Break Procedure must be observed and complied with. No unauthorized line breaks shall be performed.

5. **HOT WORK**
The Hot Work Procedure must be observed and complied with. No unauthorized hot work shall be performed within an area that contains flammable and/or combustible materials, especially those activities that can result in a naked flame or in sparks being generated.

6. **FALL PROTECTION**
The Fall Protection Procedure must be observed and complied with. No unauthorized elevated work without the prescribed fall protection shall be performed.

7. **ELECTRICAL WORK PERMIT**
The Electrical Work Permit Procedure must be observed and complied with. No unauthorized work on electrically energized equipment shall be performed.

8. **INTERLOCK BYPASS**
The Interlock Bypass Procedure must be observed and complied with. No unauthorized bypass or disabling of a safety or environmental interlock shall be performed.

9. **RELEASE OF HAZARDOUS MATERIALS TO THE ENVIRONMENT**
No person shall perform any unauthorized discharge of hazardous materials to a storm drain, the ground or the atmosphere.

10. **RECKLESS ENDANGERMENT**
No person shall participate in or condone any reckless activity (e.g. horseplay, fighting) that will endanger any persons including himself to potential injury.

WILFUL VIOLATIONS OF OR SERIOUS DISREGARD FOR ANY OF THESE INVIOLABLE SAFETY STANDARDS WILL SUBJECT THE INVOLVED INDIVIDUAL(S) TO IMMEDIATE DISCIPLINARY ACTION WHICH COULD RESULT IN TERMINATION

Company B:

Golden Rules

- You have the time to do the job safely
- You must perform a risk assessment and implement the controls before starting the job
- Procedures must be followed

- Appropriate PPE for the job is a must
- You have to make choices and use judgment
- You must act on all unsafe acts / conditions immediately

Company C:

Cardinal Rules

- It is mandatory to report all "on the job" safety incidents
- Disabling of safety critical and security devices without proper written authorization, is strictly prohibited
- Lockout / Tagout & Try procedures for all isolations are mandatory
- Lighting a flame, without authorization, in Operating Areas and Warehouses is prohibited
- Engaging in or provoking horseplay or fighting within Company premises is prohibited
- Smoking is not allowed in Plant & Office premises except in designated areas
- Willful damage to Company property is prohibited
- Violation of traffic rules with plant boundaries is prohibited
- Any person under the influence of drugs or alcohol will not be permitted to enter Plant & Office Premises
- Weapons, intoxicants, alcoholic beverages and illicit drugs are strictly prohibited in Plant & Office premises
- It is mandatory to follow all Company policies, safety rules and applicable laws as stipulated by the government and the Company

Not following the above rules will be treated as Misconduct.

Company D:

Non-Tolerance Items and Activities

- Use and carrying of narcotics, drugs, alcohols of any kind at site is strictly prohibited
- Bringing in and use of fire-arms and weapons of all kinds at the site is strictly prohibited. Harassment of any kind & horseplay is forbidden at site.
- Usage of defective tools, tackles, slings, lifting gears and full boy harnesses is strictly prohibited.
- Ingress at site of un-checked/un-inspected equipments & plants such as cranes, welding plants and equipments, drill machines, transit mixers is strictly forbidden. Only perfectly fit equipment is allowed at site.
- Cables buried underground without proper protection against heavy traffic is highly dangerous and strictly forbidden.
- Operating electrical equipment/systems without double grounding and ELCB's is not allowed
- Scaffolding built improperly and not red tagged is a highly unsafe offence and strictly forbidden.
- Non-compliance of full body harness protocol. Use full body harness when working at heights.
- Vehicles, equipment driven/operated by un-trained/non-qualified personnel are criminal offences. Anybody doing so shall be immediately turned out of the site.
- Smoking in non-designated areas is strictly forbidden. Violators to be expelled from site
- Non-compliance of Work Permit Procedure is strictly forbidden. Violators to be dealt with strictly and will be expelled from site.
- Violation of Rigging Protocol (rigging of loads less than 10 tons without pre-lift checklist, 10-50 tons without filling rigging card and more than 50 tons without carrying out a detailed rigging study) is strictly forbidden.
- Sleeping at any area within the site is strictly forbidden. Violators will be immediately turned out of site.

Company E

Red Rules

- No smoking inside defined plant areas
- All use of fire fighting equipment must be reported to the Fire Department
- No maintenance in high risk areas without specified PPE

Company F

Life Saving Rules

All safety rules are important and must be followed. A few rules, because of imminent and potentially serious injury or fatality, are categorized as inviolable safety rules. Termination of employment is the first consideration for violation of an inviolable safety rule.

1. **Seat Belts.** Seat belts are to be worn by all occupants in a moving motor vehicle while on company business, in company provided vehicles or while on company property, including parking lots inside the company property security fence.
2. **Lock, Tag, Clear and Try.** This procedure identifies the means to prevent unexpected release of hazardous energy when such release could cause injury to personnel or damage to systems, equipment, or machines. It includes the minimum requirements for locking, tagging, clearing, trying, and testing systems, equipment, or machines to allow work without injuring personnel or damaging property.
3. **Line Break**. This procedure establishes the minimum safety requirements for isolating and opening potentially hazardous piping systems, piping components, equipment, and vessels to prevent possible injury or environmental release. It specifies the minimum safety requirements for making line breaks into piping systems, piping components, equipment and vessels in a way that prevents injury or adverse environmental releases.
4. **Permit-Required Confined Spaces**. This procedure identifies the minimum safety requirements necessary to enter a permit-required confined space and, where appropriate, to protect employees from the hazards of entry into permit-required confined spaces, and for regulating employee entry.
5. **Fall Prevention**. The procedure identifies the minimum requirements for protecting people when working at heights including ladders. Each employee working on a walking/working surface with an unprotected side or edge which is six feet or more above a lower level shall be protected at the walking/working surface.

Company G

Inviolable Standards

These standards are established to ensure a clear understanding of the site management's high expectations regarding prevention and elimination of injuries and environmental incidents. These standards will provide supervisors, employees and contractor personnel with examples of what may be considered a flagrant violation of Safety, Health & Environmental Standards, in which immediate disciplinary action (with removal from the site) is initiated. Contractor companies may also be removed.

Intentionally Directing or Condoning Violations of Safety or Environmental Standards - Supervision will not direct or condone an employee violating any safety or environmental standard.
Lock, Tag and Try - No person shall work on any piece of equipment for which he or she has not personally inspected and insured is de-energized, isolated and affixed his or her approved locking device and tag.
Confined Space Entry - No person shall enter a confined space unless all routes from which hazardous material can enter are disconnected and blanked, the atmosphere in the confined space is checked, prescribed PPE is worn, a lifeline is present, a harness is worn, and a standby person is in place. Entry occurs at the time when any part of the body breaks the plane of a point of entry.

Line Break - Prescribed PPE must be worn when exposed to open systems containing hazardous materials.
Fall Protection - Prescribed fall protection must be in place when working from a position where there is a potential for falling or slipping, and in places where lifting of disabled personnel may be necessary (such as in a confined space entry).
High Energy - Protective equipment and isolation requirements for energized electrical equipment and proximity to energized electrical must be followed.
Barricade Access - Entry to any barricade is through the identified gate/gates. Any unauthorized entry, anyone ducking under, stepping over, or going through a barricade will be subject to removal from the site.

(16b) Alarm Management for Improved Process Safety, Reliability, and Product Quality

Ankur Pariyani[1], Jeffrey Arbogast[2], Warren D. Seider[1], Ulku Oktem[3], Masoud Soroush[4] and Olivier Cadet[2], (1)Chemical and Biomolecular Engineering, University of Pennsylvania, Philadelphia, PA, (2) Process Control & Logistics, Air Liquide, Newark, DE, (3)Risk Management and Decision Center, Wharton School,University of Pennsylvania, Philadelphia, PA, (4)Department of Chemical and Biological Engineering, Drexel University, Philadelphia, PA

Most chemical plants have hundreds of variables that monitor the dynamics of their processes. When the variables move out of their normal operating ranges, alarms are triggered to notify the human operators. The frequency and selectivity of alarms influence the performance of operators, which in turn, impacts the plant performance and reliability. This work introduces techniques to perform *alarm management* that improves the reliability of plants. To ensure the relevance and specificity of alarms, particularly during abnormal situations, the work focused on providing operators with actionable alarm information. *Bad Actors* (BAs), which are alarms that deteriorate the performance of alarm systems, are identified and resolved. Typically, BAs are poorly-configured alarms that have the highest alarm counts, often contributing to periods of high alarm frequency (referred to as *alarm floods*). In this work, they include alarms that experience the *most-critical abnormal events* (when variables cross their Emergency Shutdown thresholds) and/or lengthy departures from their normal operating ranges - jeopardizing plant safety, reliability, and product quality. The latter are important for many plants operated unmanned over nights/weekends.

Several months of alarm data were analyzed for an industrial air-separation unit operated by Air Liquide. The BAs were identified and assigned different grade levels. Several Key Performance Indicators (KPIs) were developed to evaluate the performance of alarm systems. The methodologies developed in this work improve upon the existing industry-standard techniques, which focus only on the alarm counts and not on the criticality of abnormal situations. Results show that the average frequency of alarms was well under manageable limits, with the alarm system in the flood condition just 1% of the total time. However, 45% of the total alarms during the study period occurred during alarm floods. These alarm floods are referred to as *near-misses*, which often distract operators, but don't necessarily result in trips or accidents.

See more of this Session: Application of Process Safety Principles
See more of this Group/Topical: Topical 1: Global Congress on Process Safety

Incident Investigation – Addressing the Worst That Could Have Happened

S. Gill Sigmon
Honeywell
15801 Woods Edge Road, Colonial Heights, VA 23834
Steve.Sigmon@honeywell.com
Tel 804-520-3205

Samuel A. Rodgers
Honeywell
15801 Woods Edge Road, Colonial Heights, VA 23834
Samuel.Rodgers@honeywell.com
Tel 804-520-3706

Keywords: Worst Credible Consequences, Actual Consequences, PHA, RCA, Recommendations, Solutions, Incident Investigations, Human Factors

Abstract

Incident Investigation techniques typically start with a detailed description of the actual consequence and work backward toward the root causes. A lot of time is spent to qualify and quantify the actual consequence so as to appropriately apply resources to prevent recurrence. A frequent outcome of such an approach is to develop recommendations that mitigate the actual experienced effects on the person, possibly through personal protective equipment or improved operating procedures. This approach has a basic flaw in that it does not address what could have happened, specifically, the worst that could have happened. Thus, the recommendations might not address preventative safeguards or passive design necessary to eliminate the scenario altogether. By formally requiring an assessment of the worst credible consequence of the initiating event, the investigation team will then have a better perspective when considering alternative recommendations to prevent recurrence. This broader perspective assists the investigation team in culling the many potential recommendations to those which not only address the actual consequence, but also are effective against the worst credible consequence. Additionally, by having identified the worst credible consequences, the investigation team can then select those recommendations sufficient to achieve the company's risk criteria.

1. Introduction

A number of examples of incident investigations are presented to illustrate specific situations where the difference between what actually occurred and what could have happened did or could have resulted in vastly different number and types of recommended mitigations. Root Cause Analysis (RCA) tools generally allow solutions for each branch of the RCA tree. This paper proposes that evaluating recommendations relative to the worst credible consequences and company risk criteria results in the minimum effective set.

2. Why focus on Worst Credible Consequences

2.1 RCA #1: Crucible failure and loss of containment

Quartz is melted in an induction-heated crucible, which is centered in a multi-layer insulating containment. A crucible fabrication error promoted localized heating and led to a large hole in the bottom of the crucible, releasing molten material. The molten material ignited a gasket on the insulating container, the only combustible component in the design. The gasket was consumed in a very short time and no other combustibles were in the vicinity so the fire extinguished itself. The crucible held on the order of two gallons of molten material, which solidified as soon as it breached the insulating containment. The insulating containment and induction-heating coils were damaged beyond repair, resulting in a multi-week loss of production. The RCA recommended sixteen solutions, including hiring an engineer with induction-heating expertise. Figure 1 shows the crucible installation and illustrates the equipment scale..

Given the volume of molten material, the initial insulating containment and physical separation between operations, the safety consequences of the incident were limited. However the financial consequences were significant. The worst credible consequences for this incident were identical to the actual consequences, so consideration of worst credible consequences should not in theory change the outcome of the incident investigation. In this case, the RCA team did not consider if the incident was likely to re-occur and focused only on the severity of the business loss. There were nine "Required" recommendations and seven "For Consideration" recommendations.

Figure 1 Induction-Heated Crucible

In many companies a risk criteria is used to differentiate which recommendations are required versus those that are to be considered. When both safety and financial consequences exist, the risk should be evaluated separately for each and appropriate recommendations made based on the degree of risk. In many RCA tools this distinction is not made, which can lead to generating excessive recommendations. After including an evaluation of risk, a majority of the "Required" recommendations were changed to "For Consideration".

2.2 *RCA #2: Compressor excess vibration and pipe rupture*

A compressor out-of-balance condition caused rupture of an instrument connection and a significant release of compressed gas. The RCA determined that the connection needed to be braced according to site standards for installations with possible severe vibration. The installation had a history of out-of-balance events and of minor piping leaks, which had been repaired. None of the prior leaks resulted in a total rupture of the line and the actual consequences were minimal. Prior incident investigations were conducted with recommendations limited to increased inspection frequency to detect the out-of-balance condition.

In the prior investigations a change in a procedural safeguard, equipment inspection, was considered adequate for the actual minor consequences. Had the worst credible consequences of a pipe rupture been considered when addressing these earlier minor leaks, the passive solution of improved pipe supports might have been identified and implemented prior to the pipe rupture.

2.3 *RCA #3: Human error during instrument return-to-service*

An operator opened a manual valve to pressurize an instrument that had a "zero-energy" drain valve mistakenly left open. When a hazardous material was observed flowing from the open valve, the operator immediately closed the manual valve to mitigate the release and suffered transient effects. Due to the quick operator response, there was no reportable release. An incident investigation addressing only the actual consequence of operator exposure might focus on post-release protections for the operator and miss the potentially more severe remote effects.

Operator local response toward a hazardous material leak can not be required by procedure and should not be expected. Had the operator moved away from the manual valve and been unable to return to isolate the leak, the release could have continued for many minutes until more remotely located manual valves could be accessed and closed. In the case where the leaking material is a flammable or toxic, the worst credible consequence from an extended leak could well be multiple fatalities. Recommendations should more appropriately address preventing return-to-service leaks by verifying system integrity prior to charging with hazardous material. Alternatives would include automatically actuated isolation valves. Such changes would address not only the actual consequences of operator exposure but also the worst credible consequences.

2.4 *RCA #4: Manual mitigation using a spray mitigation system*

A toxic gas leak was visually detected by a field operator who then manually activated a nearby water spray mitigation system. As the leak continued it rapidly grew in size. The water spray mitigation contained the leak until the system could be safely shut down. As a result of the operator actions there were no off-site effects. The actual consequences were limited to reportability and minor equipment repair.

Had the leak not been detected at the incipient stage by the operator, the water spray would not have been active as the leak rate increased. The worst credible consequence could have been an unmitigated leak from the ultimate hole size with potential off-site exposures. The magnitude of potential consequences led to a passive equipment redesign.

When there are no actual consequences it is sometimes easier to focus on the worst credible consequences. When there are actual consequences involving personnel, environment, or significant property damage, resolution of these is a necessary part of the investigation. While an actual injury to a single person rightfully requires attention, the RCA team should also consider potentially much worse consequences. The goal should be to develop recommendations that address both the actual and the worst case consequences.

3. When should Worst Credible Consequences be defined?

3.1 Type of Incident

We must all be stewards of our companies' resources so we must carefully balance how many resources to apply to incident investigations versus the potential to identify significant safety improvements. Some incidents offer important broadly-applicable insights. For example incidents occurring when procedural steps are not defined for a hazardous process might deserve a formal investigation involving multiple disciplines to consider why the procedural steps were not defined. Some incidents have more narrow application, potentially limited to single individuals. For example an operator forgetting to perform a procedural step might require a simple investigation limited to the operators' supervisor.

Under Process Safety Management regulations [1] an incident investigation is to be performed whenever a catastrophic release did or could have occurred. In order to fulfill the second part of the requirement, to understand if a catastrophic release could have occurred, it is necessary to evaluate worst credible consequences whenever a covered process is involved.

When the hazard is one that could cause effects remote from the event, such as a toxic or flammable material, evaluation of worst credible consequence should be included.

3.2 Decision Criteria for Formal RCA

Many companies have established decision criteria to determine how many resources to apply to an incident investigation based on actual consequences. If actual consequences are deemed significant, then often a formal RCA is required. If actual consequences are less severe, then less formal methods may be allowed, including investigation limited to the first line of management. If these decision criteria are limited to actual consequences instead of the worst credible consequences, then opportunity for significant safety improvements might be missed.

Repeat or similar incidents often are investigated formally, even if there were no consequences. This might occur because the incidents are perceived as related to failure of an underlying management system. Such a failure could become an initiating event for a more severe incident.

The focus on actual events is related to historical reliance on lagging metrics. Regulations require reporting of actual effects, such as OSHA recordable injuries, Lost Work Days, and recordable releases of hazardous materials. However, the most advantageous learnings are those that occur before an event, instead of after it. A company's "Near-Miss" reporting process is a good indicator of the potential to identify safety significant improvements by considering worst credible consequences. Many companies recognize and report near-miss incidents when preventative safeguards function as intended and there is no actual consequence. If the near-miss process documents sufficient information to evaluate worst credible consequences, there is higher likelihood that an appropriate level of investigation will be performed.

Although a formal RCA resource commitment is not practical for all incidents, an estimate of worst credible release is important to include in all investigations, including those for a near-miss.

4. How to determine Worst Credible Consequences

4.*1 Identifying Safeguards that Worked*

Incidents occur from a failure followed by a sequence of events leading to a consequence. The RCA process identifies the management system that failed so that the RCA team can suggest systemic improvements within their scope-of-control. One of the events between the management system failure and the consequence is an immediate cause, termed the Initiating Event according to Process Hazard Analysis (PHA) terminology. The Initiating Event is a failure in an engineered device or procedural step, and is an action cause according to RCA terminology.

The connection between RCA and PHA is somewhat obfuscated by the use of different terms for the same thing, and by the time sequence of steps in the two processes. The RCA Primary Effect is equivalent to the PHA Consequence. The RCA process starts with Primary Effect/Consequence then proceeds backwards in time to the preceding events while the PHA starts with an Initiating Event then proceeds forward in time to the resulting Primary Effect/Consequence. In a PHA the sequence of events between the Initiating Event and the Consequence/Primary Effect may be termed the Intermediate Events. The RCA continues until management system failures are identified, which may be identified as Root Causes. When the Initiating Event is a human error, the PHA may investigate the causes via a special study called the Human Factors Assessment. Figure 2 is a typical event tree derived from a synthesis gas heater failure [2], expressed in RCA terminology. Figure 3 shows the same incident expressed as a PHA scenario. Both PHA and RCA seek to reduce future risk by making improvements, which typically are called "Recommendations" in the PHA but may be called "Solutions" in the RCA.

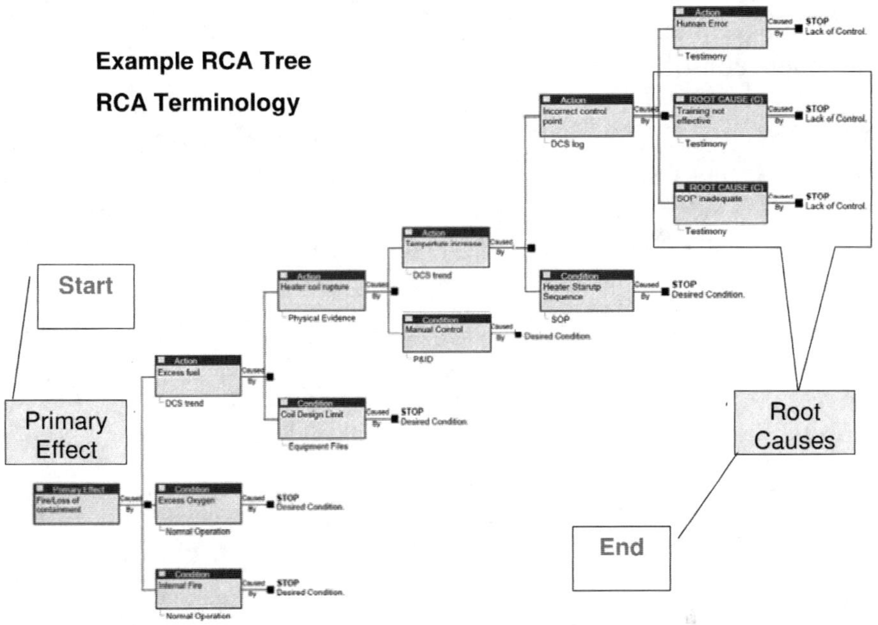

Figure 2 Event Tree with RCA Terminology

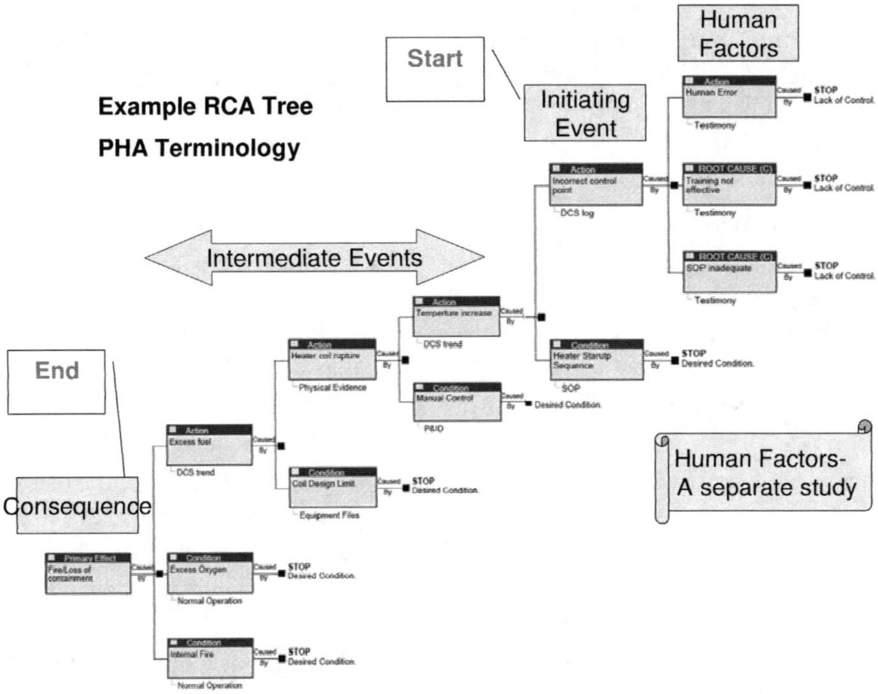

Figure 3 Event Tree with PHA Terminology

The sequence of events in the RCA is constrained only by the scope of the investigation, and includes management systems at the operational, site or even industry level. The sequence of events in the PHA begins with an equipment or administrative failure, and typically excludes the management systems that may have contributed to the failure.

Between the Initiating Event and the Consequence the sequence of events includes preventative and mitigating protective systems, termed Safeguards according to PHA terminology. For near-miss incidents at least one of these preventative protective systems worked. For other incidents the preventative protection systems failed but commonly a mitigating system worked to some extent. For these incidents the worst credible consequences are the consequences that would have occurred assuming that the working protective systems did not in fact work as intended. The RCA team should take care that the assumed failures are consistent with the PHA consequence assessment. For worst case incidents, all of the preventative and mitigation protective systems failed and the actual consequences are equal to the worst credible consequences.

4.2 Integration of Worst Credible Consequence with Formal RCA

Usually a formal RCA process uses a team to bring multiple skill sets to the assessment and might include members qualified to judge worst credible consequences. If not, then the RCA process should allow the team access to persons experienced in consequence analysis. Commonly the consequence analysis skill set is defined in the company's PHA process.

One of the first steps in the RCA process is to document the event consequences. This is a good time for the team to discuss what could have happened. The team could then include the potential consequences in consequence description along with the actual consequences. For example, if the actual event included mitigation the team would discuss what might have happened if the mitigation did not occur.

Later as the RCA team develops the RCA tree, certain events will be identified as having lessened the consequences. Often these can be identified by team members noting "we were lucky". For example the incident might have occurred during shift change when field operators are in the control room. If these circumstances are unlikely should the incident re-occur, the worst credible consequences should reflect the more likely outcome. In the example it is more likely that field operators will be performing their field work instead of attending shift break.

The authors suggest that the worst credible consequences assessment used to support incident investigations should be consistent with the company's PHA practices. Typically PHAs assume that active protections such as human intervention are not considered when establishing worst credible consequences.

Before the RCA report is issued it is beneficial to review it for this consistency. For OSHA-regulated processes the worst credible consequence assessment occurs when incidents are incorporated into the revalidated PHA. A contemporaneous review of the RCA report assures that the future revalidation PHA team will agree with the RCA team assessment and reduces the PHA team work when they incorporate the incident into the revalidation PHA.

4.3 Other Benefits

The beneficial relationship between the RCA and PHA processes has been identified by recognized RCA authorities, but the impact on solution/recommendation selection can be overlooked. One paper that linked RCA and PHA [3] listed the following criteria for solution/recommendation selection:

- Management buy-in
- Cost-value
- Addresses multiple causes, common causes and systemic causes
- Reliability
- Implementation speed

Note that risk-reduction potential is not included in this list.

If the company has objective risk criteria, the number and type of solutions/recommendations should be assessed against the risk criteria. Are they sufficient to mitigate the risk, considering the worst credible consequence? Do some reduce risk while others do not reduce risk? For example, a one-time, company-wide alert makes the company aware of the potential lessons-learned but does not by itself assure that the lessons will be retained in the corporate memory and would not normally be assigned a risk-reduction credit by a PHA team.

5. Risk Criteria with Worst Credible Consequences

A RCA team will often develop a time line as a basis for the RCA tree. - Here is a time line of a spill that occurred during routine sampling.

1. Operator opens needle-type valve to obtain a routine sample
2. Operator opens valve further due to blockage
3. Blockage clears with sudden increase in flow
4. Operator drops sample container and evacuates area
5. Release continues unmitigated for fifteen minutes.
6. Leak mitigated by Emergency Response team

What are the worst credible consequences for this event?

If the event were to re-occur should we assume that the operator would again safely evacuate? Here the operator evacuated safely, although in fact his PPE was not effective against the abnormally high localized concentrations from the spill. Should we consider only the actual flow that occurred, or a higher flow based on a full open valve? Here the operator opened the valve more than normal, but less than full open so a higher flow could have occurred. Should we consider only the actual down wind consequences or could they have been more severe? Here the wind direction was favorable and the down wind corridor was not occupied, had the flow been greater and the wind direction unfavorable multiple personnel injuries might have occurred.

The RCA team recommendations considering only actual consequences were to (1) install a sample bottle holding device and (2) install a pressure regulator to reduce the sampling pressure The RCA team did not recommend improved PPE because in the actual event the operator evacuated safely, nor did it fundamentally change the sampling operation because the event was safely mitigated without injury.

If the RCA team had recognized the potential harm to the operator and to other personnel, it is likely that their recommendations would have either considered improved PPE or fundamentally changed the sampling operation, or possibly both. Incorporating objective risk criteria can help the RCA team determine if one or both such recommendations are required. Typical risk criteria when a simple operator error can cause serious injury would require multiple protections so improved PPE and an enclosed sample container might both be required.

Similarly when RCA studies are incorporated into PHAs, the PHA team should consider what could have happened instead of actual consequences. The example RCA tree was based on a synthesis gas fire that occurred during a heater startup and where there were no safety consequences. If a PHA team assesses this event based on actual safety consequences only, the PHA team may not include the event in the PHA or, if included, may not evaluate safeguards or develop recommendations. If worst credible consequences are considered the PHA team might determine that trouble-shooting during startup could place an operator near the heater and that worst credible consequences are one or more injuries.

6. Conclusion

The authors' believe that requiring evaluation of worst credible consequences has great potential for focusing the RCA process on those recommendations that give the most risk-reduction impact. In this scheme the minimum RCA recommendations selected for implementation are those that assure that the company's risk-criteria are met. Additional recommendations would be implemented based on the risk-benefit assessment.

7. References

[1] Process Safety Management of Highly Hazardous Chemicals, 29CFR1910.119 (2002)

[2] Perrow, Charles, "Normal accidents : living with high-risk technologies", Princeton University Press (pub), 1999, p. 116

[3] Hall, M., Hughes, B., and Rygaard, D. "Using Root Cause Analysis to Improve Risk Management," Professional Safety Journal, 2010

PSM Auditing Dilemmas

Michael J. Hazzan, P.E.
AcuTech Consulting Group, Inc.
mhazzan@acutech-consulting.com

David Heller, CSP
AcuTech Consulting Group, Inc.

Lisa Morrison
BP International

Laurie Brown
Eastman Chemical Company

John Traynor
Evonik Degussa Corporation

Keywords: PSM Audits, PSM Auditing

Abstract

In any given PSM audit, the auditors will usually face one or more situations that represent a dilemma because the situation has not happened before or no thought has been given on how to resolve it. These dilemmas usually require that the individual auditor and/or the audit team leader resolve the situation in the field. These "on the fly" resolutions require both astute judgment and practical solutions that fit not only the governing regulations or company/facility standards for PSM, but also how those mandatory requirements should be interpreted and applied to the specific design, operations, and PSM program of the facility being audited. The PSM audit dilemmas presented in this paper are based on the experiences of seasoned auditors during actual PSM audits.

1. Introduction

Examples of common PSM audit dilemmas are presented below. In general, there are no absolute right or wrong answers to solving PSM dilemmas, although in some cases a resolution might seem obvious. Often, there is a good resolution, but many times such resolution would require adoption of a policy or practice that is not a compliance requirement. This should be carefully considered before a final decision is made. Possible resolutions or conclusions to these dilemmas have been offered to readers for consideration. Even when the dilemma seems to represent a compliance issue, and its resolution seems clear-cut, there may be substantial flexibility to craft the resolution and correct the problem. The resolution of these dilemmas, should they exist at a given facility or company, should be carefully determined on an individual basis by each user/reader of this paper and the facility or company they represent. Therefore, a single, common resolution or conclusion to these situations is not appropriate for all users, and the resolutions/conclusions presented below should be carefully considered before being adopted.

In describing the dilemmas, the PSM element nomenclature from the CCPS *Guidelines for Risk Based Process Safety* is used. For example, Hazard Identification and Risk Analysis (HIRA) is used instead of Process Hazard Analysis (PHA); Asset Integrity (AI) is used in lieu of Mechanical Integrity (MI), etc. Additionally, references to "related" findings means a finding that does not represent a compliance issue but is a finding against an audit question or criteria that is related to good/successful/common practices in PSM, written clarifications by regulators, citations, and other sources that are not compliance related.

The topic of PSM audit dilemmas is also addressed in the recently published CCPS *Guidelines for Auditing Process Safety Management Systems, 2nd Ed.,* and several of the following dilemmas are PSM adaptations of those published by Cahill et al., in *Environmental Health and Safety Audits.*

2. PSM Audit Dilemmas

2.1 Dilemma: Evidence of Systemic Problems

You are performing the Asset Integrity portion of a PSM audit at a large petrochemical facility. While reviewing the operations of the spare parts warehouse, you notice several parts in the warehouse that have expiration dates. There are no labels on the storage locations warning of expiration dates, nor does the inventory management software provide any information about parts with expiration dates. During the interview with the warehouse supervisor, he states that it is not necessary to track or manage expiration dates for the spare parts because the usage rate always results in the part being issued for use before the expiration date is reached. However, during the warehouse walk around you notice that the expiration date on a spare chemical hose has been reached. As an auditor, what is the nature of your finding(s) and recommendation(s)?

Possible Resolution or Conclusion: The finding, which is a compliance one, is that the warehouse contains spare parts that have exceeded their shelf life. This situation certainly is a deficiency with respect to the requirement that spare parts be suitable for the process application for which they are intended. However, based on the interview with the warehouse supervisor, this situation does not appear to be a simple oversight for one chemical hose, but represents a lack of a proper management system and internal controls associated with expired spare parts. Historical usage rates may change over time as facility throughput or operations change, and reliance on them is generally not adequate to ensure that parts do not expire. Since the governing regulations (i.e., the PSM Standard) do not explicitly require a management system for spare parts, the compliance finding is simply that an expired spare part was found in the warehouse. However, the systemic finding regarding the lack of a management system and internal controls to deal with expired spare parts would be a related finding.

2.2 Dilemma: The VPP Defense

You are conducting a PSM audit of a chemical manufacturing facility. There are several PSM-covered processes at the facility. While auditing the Asset Integrity element, you discover that the internal inspections and wall thickness measurements for five pressure vessels, out of a total of two dozen pressure vessels in PSM service on-site, are overdue, in some cases by a few years,. The same recurring maintenance tasks for 3 of 12 low-pressure storage tanks in PSM service are also overdue, again by a few years. The Maintenance Manager tells you that the plant is considered a safety model for the region and the facility has never been cited or threatened to be cited for overdue vessel and tank preventive maintenance. He also states that the site has been an OSHA VPP Star site for nearly 10 years, and the relationship with the local OSHA field office is excellent. Also, the facility is not in a state that regulates unfired pressure vessels. The time and effort to quickly perform the overdue vessel and tank inspections will be substantial and will result in some unscheduled down time and late product shipments because of the need for this equipment to make and store the products. The Maintenance Manager and Plant Manager are definitely opposed to incurring these production upsets on what they believe to already be a "best in class" operation. What is the nature of your findings?

Possible Resolution or Conclusion: If there are overdue ITPM tasks on important components such as pressure vessels and storage tanks, these are clear compliance findings and should always be reported as such. Pressure vessel ITPM tasks and frequencies may also be specified by state law or regulations, depending on the state, and these local laws or regulations may specify different RAGAGEPs other than API-510, for example, those published by the National Board of Boiler and Pressure Vessel Inspectors (i.e., NB-23). The status of the facility as a VPP site and its relationship with the local OSHA field office are not relevant with respect to the validity of this or any other finding. Auditors must be very careful when relating the VPP program to the status of the PSM program. Facilities that have achieved VPP status are usually very proud of this accomplishment; however, it should not result in a "free pass" on any valid PSM audit findings.

2.3 Dilemma: Practices Not Institutionalized (#1)

You are conducting the Contractor Management element of a PSM audit at a chemical plant. The PSM Coordinator tells you that a purchasing supervisor screens and approves the safety performance of prospective contractors. He has developed and implemented a pre-qualification questionnaire that includes detailed information about the contractor's safety program and performance. The PSM Coordinator is able to show you an example of one of the completed questionnaires for a new prospective inspection contractor. When you ask to interview the purchasing supervisor so that you can learn more about the contractor pre-hire screening process and sample additional records, you are informed that the supervisor won the state lottery two months ago and is on an extended vacation period. To verify whether the contractor pre-screenings are being conducted properly, you request to review the computer records of them and also to review the procedure that governs contractor management. Unfortunately, facility staff cannot gain access to the records because they do not have the supervisor's password, and a contractor management procedure reflecting how the purchasing department manages contractors has not been developed yet. Calls to the supervisor are not successful; he has turned off his cell phone. How do you handle this situation as an auditor?

Possible Resolution or Conclusion: A finding that the records confirming that contractor pre-qualification reviews were not available and could not be reviewed should be created because the sampling and testing for this aspect of the contractor management program could not be completed. Because the governing regulation does not explicitly require a management system procedure for this PSM element, the finding that a procedure does not exist and that the purchasing supervisor's good practices are not institutionalized is a related finding and not a compliance finding. Alternatively, the auditor can include a recommendation to develop a management system procedure to help correct the systemic problem.

2.4 Dilemma: Practices Not Institutionalized (#2)

You are conducting an audit of the Process Knowledge Management element of a specialty batch chemical plant that manufactures a diverse set of products, and functions as a toller for several different industries. New products involving new chemicals and conditions that have not been experienced by the facility are not an unusual situation. The facility has a reputation for quickly incorporating new chemical products into their processes with very high quality. During a review of the relief device design and design basis records, you notice that many of them are not complete and do not reflect the properties and conditions imposed on the reactor relief device (rupture disks) by the current products being manufactured. The Engineering Manager states that because of the quick turnaround on product incorporations required by their customers, a full quantitative analysis of the relief device design and design basis is not possible. She also states that the pressure vessels, piping, and relief devices are all over-designed and can handle the full range of pressures and temperatures that their processes impose, and there has never been any process leak that has been traced back to an overpressure or over-temperature transient. Technical interviews with the Engineering Manager and several engineering and operations personnel involved in the introduction of new products into the facility indicate that relief device issues are studied qualitatively during MOC safety reviews and HIRAs. This includes a review of the properties of the materials and some simple lab tests that are performed on-site. Interviews with facility engineering and operations personnel involved indicate that they seem to understand the ramifications of the prospective new chemicals on reactor pressure. Reviews of operational records confirm that there have been no pressure transients that resulted in any leakage or releases. How do you handle this situation as an auditor?

Possible Resolution or Conclusion: A compliance finding that relief device design and design basis process safety knowledge is missing for the currently installed reactor relief rupture disks should be created. Clearly, from the interviews and record reviews, the process for evaluating new and modified products against the relief design of the reactors is flawed; however, the governing regulations do not explicitly require a management system and internal controls for this - just that the relief design basis is performed and documented properly. Therefore, the auditor can include a recommendation to include such a management system to help correct the systemic problem, or alternatively, a related finding can be created to address the lack of a management system for evaluating reactor relief capability when products are changed.

2.5 Dilemma: The Boundaries of PSM Programs (#1)

You are conducting a PSM audit of a petrochemical plant with large inventories of flammable materials. All the processes using, storing, or manufacturing these materials are included in the PSM program as defined by the facility PSM manual. The PSM manual also states that the fixed and mobile fire protection system is included in the PSM program. When reviewing the HIRAs, you notice that a HIRA of the fire protection system has not been performed, nor have any of the process HIRAs included an analysis of the fire protection system, except to list it as a safeguard. Also, the Asset Integrity program procedures do not include any ITPM information for the fire protection system. The Safety Manager states in an interview that they test the fire protection system, but a review of the test records reveals that the fire pumps have not been tested in over two years and the fire monitors are lubricated but not flow tested. The only records you can find are monthly external inspections of the sprinkler systems, some testing records by a contractor for the fire alarm system, and a lengthy list of fire extinguishers that shows the dates and technician who inspected them. You prepare two findings that state: 1) the HIRAs do not include analysis of the fire protection system failures, and 2) the AI program does not include all the ITPM tasks specified in NFPA 25 for water-based fire protection equipment. Are these appropriate findings? If yes, what are the appropriate recommendations?

Possible Resolution or Conclusion: The two compliance findings regarding the lack of HIRAs for the fire protection system and missing ITPM tasks required by the governing RAGAGEP for water-based fire protection systems are correct. Fire protection systems are not explicitly required to be included in a PSM program by the governing regulations, and water is not a highly hazardous chemical per those regulations. Therefore, the inclusion of these systems and equipment in the PSM program of the facility in question is voluntary. However, because the facility has defined the PSM program to include them, the other elements of PSM could be considered compliance requirements. Also, taking credit for the fire protection system in a HIRA means that it has to be functional.

2.6 Dilemma: The Boundaries of PSM Programs (#2)

While auditing the AI program at a chemical plant that manufactures toxic materials covered by OSHA's PSM Standard, you determine from document reviews and interviews that the area toxic gas detectors are not included in the AI program. The fixed detectors are used to detect the same highly hazardous chemicals covered by the PSM Standard (in this case chlorine), the detectors are located inside the battery limits of the PSM-covered process and provide indications of chlorine concentration levels and alarms if the concentrations reach pre-set limits. Therefore, they fit the definition of a control, indication, and alarm in paragraph (j)(1)(v) of the PSM Standard. Do you have a finding?

Possible Resolution or Conclusion: Since the fixed chlorine detectors are within the process area covered by the PSM Standard, and are specifically intended to alarm when a highly hazardous chemical is released, a compliance finding should be written.

2.7 Dilemma: PSM Program Boundaries (#3)

During the Asset Integrity portion of a chemical plant PSM audit, you discover that the facility is very sensitive to the loss of cooling water to a set of batch reactors, and rapid exothermic reactions will occur if cooling or power is lost. The facility has SISs that measure temperature and pressure and initiate an emergency shutdown on high temperature and pressure. These devices are tested regularly and the records are complete. No ITPM is performed on the cooling water system or the electrical power system that supplies it. The Maintenance Manager says it is not necessary because the facility has the SISs installed and they are tested. Is this a finding?

Possible Resolution or Conclusion: Some utility systems are very important to process safety because their failure could directly contribute to a possible catastrophic release. When such is the case, there are usually multiple layers of protection for the hazard scenario in question. However, the utility system, which is the first line of defense, is one of those layers and should not be ignored when defining the PSM boundaries (and in this case the boundaries of ITPM activities within the PSM program) because there are other layers. To do so would be to consciously sacrifice the first layer of protection because by itself, it does not contain the chemicals or materials required for formal regulatory coverage. These important utilities would likely be included in HIRAs as safeguards, and may be included in the SOPs and PSK as safety systems. While this applicability philosophy is desired, it is not a mandatory requirement, and any findings generated because these critical utilities are not included in the PSM program would be written against related rather than compliance criteria.

2.8 Dilemma: Repeat Findings

You are auditing the HIRA element during a PSM audit. You have observed that the facility has not yet resolved 10 recommendations from a PHA performed on the Reactor #1 process that was completed two years prior to the audit. The compliance audit report from three years ago includes the following finding: "Fifteen recommendations from the most recent HIRA on the Reactor #3 process are overdue for resolution." You have determined that the 10 current unresolved recommendations are overdue, but they were not overdue three years ago, and all 15 recommendations that were noted as overdue during the previous audit were resolved in the intervening three years. Is this a repeat finding?

Possible Resolution or Conclusion: The question in this situation is whether overdue HIRA recommendations in successive PSM audit cycles, and not just a particular HIRA recommendation represents a repeated finding. The OSHA Field Operations Manual (OSHA, 2009b) states that an employer may be cited for a repeated violation if that employer has been cited previously for the same *or substantially similar* conditions or hazards. Also, if an originally cited violation has at one point been abated but subsequently recurs, a citation for a repeated violation may be appropriate (state plan citations cannot be used as a basis for repeat violations). Therefore, if overdue HIRA recommendations are findings in audits, even if they are not consecutive audits, and even if the finding is not being written for the same exact overdue recommendations, it should be treated as a repeat finding. Of course, the definition of "timely" is also relevant when determining if a recommendation is overdue. Also, the management system for the HIRA recommendations should also be reviewed to determine if the management of PHA recommendations is addressed.

2.9 Dilemma: Just-in-Time Compliance (#1)

You are assigned to audit the hot work permit (HWP) part of the Safe Work Practices element of a PSM program. You first request a copy of the site's hot work procedure, which is provided to you. The procedure tracks the PSM and referenced regulations exactly. However, the effective date of the procedure is three days prior to the audit you are now conducting. The facility Safety Manager admits that the procedure is brand new, that it replaced a more simplistic and informal hot work permitting process that had existed before, and that the facility had worked very hard to complete it and get it approved in anticipation of the PSM audit.

When you ask about the required training for operations and maintenance employees on the new procedure, which is one of the provisions of the new procedure, the Safety Manager tells you it is scheduled for next week. They had to wait until approval and issuance of the final procedure and could not conduct it prior to the audit. You also ask if you can review any completed hot work permits and are told that, since the procedure is so new, no completed permits are available. Next, you ask if you can observe some hot work while you are on-site. The Safety Manager says that there is no hot work scheduled for the week of the audit.

The bottom line is that they seem to be knowledgeable of the requirements and are putting an excellent program in place. It is just not fully implemented. Is there a finding here? If so, what would the finding be? Other than reviewing the program again on the next audit scheduled in three years, is there a way you can be sure that the program is real and not just "paper"?

Possible Resolution or Conclusion: The governing regulations do not explicitly require a hot work procedure, only that a hot work permit be issued and that it address the issues in the referenced regulations—OSHA Standard 1910.252(a), Fire Protection for Welding and Brazing. Therefore, the existence of a hot work procedure is not, strictly speaking, a compliance requirement. However, since the facility has issued such a procedure its contents are compliance requirements. The finding in this case should be that not all the provisions of the approved hot work procedure have been implemented. The fact that no hot work took place during the week of the audit is not relevant.

2.10 Dilemma: Just-in-Time Compliance (#2)

Another auditor during the same PSM audit who is reviewing the MOC program observes that the forms for 12 active and recently completed MOCs are incomplete. Signatures are missing and various data fields on the form requiring information be entered are blank. The auditor prepares a draft finding that the facility MOC procedure is not being followed, citing the specific deficient MOCs. The PSM Coordinator acknowledges the problems and proceeds to correct them by re-routing the forms to get the incomplete information and signatures inserted. Before the closing meeting the PSM Coordinator returns to the MOC auditor with copies of the 12 corrected MOCs forms and requests that the finding be deleted. Are these corrected MOCs still findings?

Possible Resolution or Conclusion: A fairly common situation in PSM audits is that the facility attempts to correct the findings as the on-site portion of the audit is progressing. There are two schools of thought regarding this practice:

- A simple finding that reports that the records are incomplete in a particular PSM element point only requires that the missing data be inserted for the subject records to be complete and satisfy the compliance requirement. Such findings should be able to be corrected at any time after they are discovered, and if they are corrected before the closing meeting or the issuance of the final audit report, they should not be mentioned in these forums because they are moot.
- An audit finding is a report of the conditions as the auditor found them, and the findings should be described and published as the auditor found them because that was the status of that aspect of the PSM program on the date found. The dilemma facing auditors in this situation is that a facility that strives to correct findings as they occur may be more concerned with the existence of the findings (or the number of them in a specific audit) rather than what the findings are telling them about their PSM procedures and practices.

The other issue associated with this dilemma is that facilities that work hard to close findings during the audit usually feel that they are being unfairly "punished" if the audit team advocates the second school of thought. Incomplete records can be a simple oversight or they may represent a systemic problem with the procedure or practice that is generating the records in question. Strictly speaking, if the procedure is inferred rather than explicitly written and the governing regulations (if any) do not require a procedure or specify a documentation method for the activity, then the incomplete records could be related rather than compliance requirements. Therefore, auditors should attempt as much as possible to determine if there is a systemic problem so that the finding can be written to fully describe not just the evidence discovered but any possible problems with the way the PSM procedure or activity related to the evidence in question is being practiced. It is not likely that a facility will be able to correct systemic problems during the on-site portion of an audit, given that these corrections will generally require changes to procedures, additional training, revised recordkeeping procedures, additional administrative steps, or other activities before the finding can be permanently corrected. By ensuring that all parties fully understand and agree in advance on the audit ground rules about allowing or not allowing the correction of findings during the audit, the dilemma can be resolved.

2.11 Dilemma: When PSM Elements Require Training or Informing Employees

In several elements of the PSM Standard, e.g., HIRA, MOC, and Incident Investigation, employers are required to inform or train employees whose jobs are affected by the outcome of the activity. You are interviewing non-management employees during a PSM audit and covering this aspect of the PSM program, that is, how they are informed about the results of HIRA recommendation resolution, approved changes, and the lessons learned from incident investigations. The procedures for each of these three elements contain provisions for providing copies of the HIRA and incident reports, as well as MOCs to the operators in the control room building and requiring that they sign forms stating they have reviewed the documents and understand them. Four operators separately state that they have not been informed of the outcomes of these activities, even though they were fairly recent and each of them signed the appropriate forms. Do you have a finding here?

Possible Resolution or Conclusion: The meaning of the terms "communicate" (HIRA recommendations), "informed in and trained" (MOCs), and "reviewed" (incident investigation lessons learned) in the PSM Standard has been a matter of continuing debate among PSM practitioners for years. While some practitioners interpret these terms to require face-to-face training or briefings, others have interpreted them to mean that providing various forms of written information is sufficient. The governing regulations and written forms of interpretation are silent on this issue, and the actual practice of informing employees of the results of the subject PSM activities has spanned the spectrum of practitioner's interpretations. There is no one-size-fits-all solution to this requirement and a mix of both face-to-face and written communication or training may be appropriate. In addition, the governing regulations do not require confirmation that the employees understood the information being presented to them. This is different from the Training element of the PSM Standard, where the demonstration of understanding is required. Therefore, auditors should not generate findings if face-to-face presentations of this information are not being conducted, unless the facility or company procedures require them. Auditors should review the methods used to provide written information to employees to determine that they are proactive and contain some confirmation that each employee received the information.

2.12 Dilemma: Stretched Definition of "Annual"

In November 2008 you are conducting the Asset Integrity portion of a PSM audit at an oil refinery, which has 350 employees and 50 maintenance personnel. In assessing the SWP training for the maintenance department, you note that an excellent needs assessment matrix—computerized and pretty impressive at first glance—has been developed for all applicable training modules and for all job classes of maintenance personnel. You check on 10 employee records and note that 3 of them have missed their required annual training for hot work permits or line breaking/process opening for 2008. The last recorded training on these topics for the 10 employees was in January–February 2007. The EHS training coordinator, who did not seem aware of these deficiencies, says not to worry; this was probably due to summer vacations and a short turnaround in early 2008. The employees will make up the training in December, which means that some employees will have an interval of almost two years between sessions. The EHS training coordinator argues that this meets the annual requirement for this training. Is this a finding?

Possible Resolution or Conclusion: In this context, "annual" is interpreted to mean a rolling 365-day period and not one occurrence of an activity in successive calendar years. Therefore, competing SWP training anytime in 2007 and anytime in 2008 for a given person does not meet this definition. However, some facility PSM personnel argue that "annual" means once in the calendar year. The most common practice is to observe the rolling 365-day definition. Therefore, findings that describe exceeding this limit typically survive because it makes common sense to most people. This issue also applies to any requirement in the PSM Standard with a time limit, i.e., HIRA revalidations, compliance audits, certification of SOPs, etc.

2.13 Dilemma: What is "Timely"?

While performing a PSM audit at an oil refinery, the auditors for the HIRA, audit, and incident investigation elements notice that none of the recommendations for these elements are overdue. However, many of them have very long due dates, even for the simple recommendations (e.g., modify a procedure). Also, it appears that the due dates for many of the recommendations have been extended multiple times, and that many of the new dates were changed very shortly before the audit you are conducting. Is there a finding here? What does "timely" mean with respect to resolving and implementing recommendations in a PSM program?

Possible Resolution or Conclusion: Very long due dates for administrative changes such as changing the wording of a procedure fails the test of "timely." These would be compliance findings because the facility's schedule was not being followed. However, each situation must be examined on its own merits before the compliance finding can be written. For example, a wholesale change to a procedure that implements new or modified documentation methods requiring new software, substantial training, etc. may reasonably take more than multiple months to accomplish, whereas the changing of the wording of one warning statement in a SOP should be reasonably completed within a few months, or even less. Also, the multiple extensions indicate a breakdown in the application of "timely" in practice at the facility. Since the term "timely" has no single uniform definition that covers every situation, auditors should apply a reasonable definition on a case-by-case basis, seeking consensus from within the audit team for each situation. The last-minute extensions just before the audit are probably attempts to avoid findings without addressing the underlying issues.

2.14 Dilemma: Referenced Documents in SOPs

In the same audit, you are informed that the SOPs are maintained electronically rather than as hard copies. In the shift to electronically maintained SOPs, the facility has included many links and references to other documents to meet the regulatory requirements for SOP content, e.g., the MSDSs for safety/health and exposure information, and to other engineering documents for information about safety systems. Are these documents that are incorporated by reference or are linked in the SOP subject to the annual certification requirement?

Possible Resolution or Conclusion: A strict interpretation of the certification requirement might conclude that any document incorporated by reference or linked in a SOP would be subject to the same annual certification requirement as the host SOP itself. However, the regulators have been silent on this issue, and no level of acceptable practice has emerged. Some PSM practitioners treat incorporated references and linked documents separately and apply a different review and verification process to them. This seems to be a sufficient practice for now. The governing regulations, particularly the PSM Standard, were adopted when current document publishing capabilities and techniques were in their infancy and not widely used (particularly document hyperlinks). The governing regulations were written upon the assumption that the SOPs were stand-alone documents that contained all the required information in one place. Document management practices have evolved well beyond that assumption. Therefore, auditors should apply a common sense approach to this complicated issue until regulatory interpretation or industry common practice indicates a more uniform approach. If the referenced or linked documents have a periodic and recorded review process associated with them, the periodicity of that process is reasonable and that process is being observed by the facility, then findings should not be necessary in this area.

2.15 Dilemma: RAGAGEPs—Safety Instrumented Systems

You are auditing the Asset Integrity element of a chemical facility PSM program. The facility has not implemented ANSI/ISA S84.01 for safety instrumented systems (SISs). Despite the clear indication of interlocks, trips, and other automatic controls on the P&IDs and in other PSK, the facility lead control systems engineer states that the facility has no SISs or ESDs. Is this a finding? This dilemma could also be faced by the auditor of the Process Knowledge Management element.

Possible Resolution or Conclusion: ANSI/ISA S84.01 is a RAGAGEP for safety SISs. But like any RAGAGEP, it can be substituted with an equivalent written practice, even a company-designed practice, if it accomplishes the same goals and objectives. Several large companies in the chemical/processing sector have long-standing engineering standards in place that specify how control systems are to be defined, specified, designed, installed, and tested. These company procedures offer an alternative approach to that provided in ANSI/ISA S84.01 and have been found to be acceptable, as demonstrated by long and successful practice, although acceptance has not been made in writing by regulatory authorities. Therefore, not using ANSI/ISA S84.01 is not a compliance finding if an equivalent practice is in place. However, simply declaring that there are no SISs/ESDs at the facility without a documented risk-based analytical process in place that confirms it would be a compliance finding.

2.16 Dilemma: RAGAGEPs—Is Vibration Monitoring Required?

You are auditing the Asset Integrity element of an oil refinery PSM program. The facility does not perform vibration monitoring of rotating equipment. The Maintenance Manager argues that vibration monitoring would not yield useful data about the rotating equipment and that the careful engineering and maintenance the refinery performs, e.g., laser alignment of all rotating equipment, frequent oil change and analysis, frequent draining of water from lube oil systems, have virtually eliminated seal and bearing failures. Is this a finding? During the same audit you notice later that some, but not all of the OEM manuals for the rotating equipment recommend vibration monitoring. Does this change the finding? Also, if the great majority of facilities in the chemical/processing industry have adopted vibration monitoring, does that create a de facto mandatory requirement?

Possible Resolution or Conclusion: Vibration monitoring of rotating equipment has proven to be a successful asset integrity activity, and it has also become a very common, time-tested industry practice. Hence, it has become a level of acceptable practice for rotating equipment in ITPM programs. However, this status within the industry does not mean that vibration monitoring is a mandatory or compliance requirement. It is still a voluntary practice.

2.17 Dilemma: Conflicts of Interest

You are auditing the MOC element of a chemical plant PSM program. During the review of the active and recently completed MOC packages, you notice that a great majority of them are initiated by the same person, and the same person signs the MOC form indicating that the safety review has been completed and the handwriting in the space to record the results of the safety review appears to be the same person. The same person is also one of the approvers of the MOC, and on some of the MOCs is the only approver. Is this a finding?

Possible Resolution or Conclusion: The MOC element of the governing regulations does not stipulate that conflicts of interest be avoided in the MOC procedure. While not advisable, it is sometimes unavoidable given the staffing, especially in small-to-medium-sized facilities. The MOC procedure should require that the MOC initiator not be an approver, or at least not the only approver, and the initiator should not be the only person who assesses the impact on safety and health of the proposed change if at all possible. However, this is a related rather than a compliance finding.

2.18 Dilemma: The Value of Near Misses

You are auditing operating procedures and are reviewing the inside operator's log and see an entry that indicates that a safety instrumented system was activated. The control loop functioned as designed and the process shutdown safely. A review of the incident investigation files reveals that no incident report was generated. Do you have a finding?

Possible Resolution or Conclusion: The definition of a near miss in the governing regulations is a general one, i.e., an incident that could reasonably have resulted in a catastrophic release. Any chain of events that results in the activation of the safety instrumented systems, which are usually the last line of defense for automatically preventing catastrophic releases, should be classified as a near miss and investigated properly. There is often a hesitance to do this because there is a belief by some that if the SISs worked as designed then that is what is supposed to happen and therefore a catastrophic release was not possible. This ignores the fact that one of the failures in the hazard scenario in question did occur and that the SISs were called upon to "save the day." This meets the definition of a near miss and therefore a compliance finding could be generated for not investigating the scenario in question. Beyond the issue of whether uninvestigated near misses should be audit findings, they represent important opportunities to learn about possible flaws in policies, practices, or procedures without suffering any of the adverse consequences of an actual incident and this is another reason to thoroughly investigate them and disseminate the lessons learned to all whose jobs are affected by these lessons.

3. Conclusion

PSM audit dilemmas can occur during individual audit interviews, during the daily debriefs with the facility, while evaluating collected information, during the closing meetings, in the audit report, or any other forum where the audit results are being discussed, evaluated, or communicated. When they occur – and they will occur - the audit team must be able to explain their findings clearly and thoroughly with supporting facts and a convincing interpretation of the underlying requirements. If the dilemma results in objections from the facility the auditors must be sure that they correctly understand these objections. The evidence that the auditors are using to draw their conclusions must be both sufficient and adequate. If the audit team has its facts straight and has correctly interpreted how those facts result in a finding given the governing requirements, challenges, when and if they occur, can be addressed professionally, calmly, and rationally.

4. References

[1] Center for Chemical Process Safety, *Guidelines for Auditing Process Safety Management Systems, 2nd Ed.*, 2011.

[2] Center for Chemical Process Safety, *Guidelines for Risk Based Process Safety*, 2007.

[3] Cahill, et. al., Government Institutes, *Environmental Health and Safety Audits, 8th Ed.*, 2001.

The UK Experience in Managing Risks Arising from Human Error in the Oil and Gas Sector

David Embrey
Human Reliability Associates
1, School House
Higher lane
Dalton
Wigan
Lancs WN87RP
UK
dembrey@humanreliability.com

Jamie Henderson
Human Reliability Associates
Henderson@humanreliability.com

Keywords: Human factors risk analysis UK regulatory requirements process safety

Abstract

Following the Piper Alpha incident in the North Sea, both offshore and onshore Oil and Gas operators in the UK were required under the COMAH (Control of Major Accident Hazards) regulations to submit a Safety Report setting out how the risks in their installations were to be managed. After many years, it was realised that few Safety Cases considered the risk potential of human activities in these installations, even though the majority of major accidents could be clearly traced to errors either at the operational, design or safety management level. A new regime was implemented, in which there was a clear requirement to address human caused risks in a systematic and auditable manner. This paper sets out the regulatory requirements and processes for assessing human caused risks.

1. Introduction

Following severe incidents in the oil, gas and chemical processing sectors in the eighties, such as Piper Alpha and Seveso, a regulatory framework was developed in the UK by the regulating authority, the Health and Safety Executive (HSE), to address the control of major accident hazards in these industries. The COMAH (Control of Major Accident Hazard) regulations were introduced in 1984 and amended in 2005 to reflect the changes in the Seveso II directive. Sites with high inventories of hazardous materials, referred to as Top Tier sites, were required to develop a comprehensive Safety Report. A Safety Report is a document prepared by the site operator and provides information to demonstrate to the HSE that all measures necessary for the prevention and mitigation of major accidents have been taken.

The safety report must include:

- a policy on how to prevent and mitigate major accidents;
- a management system for implementing that policy;
- an effective method for identifying any major accidents that might occur;
- measures (such as safe plant and safe operating procedures) to prevent and mitigate major accidents;
- information on the safety precautions built into the plant and equipment when it was designed and constructed;
- details of measures (such as fire-fighting, relief systems and filters) to limit the consequences of any major accident that might occur; and
- information about the emergency plan for the site, which is also used by the local authority in drawing up an off-site emergency plan.

After several years of reviewing Safety Reports, it became apparent that there was rarely a consistent and explicit treatment of the risks arising from human error. This was even though incident investigation reports suggested that human error was a more frequent contributor to major accidents than hardware or equipment failures. Accident investigations in the USA by organisations such as the Chemical Safety Board, for example the Texas City incident, also suggest that human failures in design, operation and process plant management are significant contributors to accident causation in the process industries.

Guidance from the HSE indicates that sites subject to the COMAH regulations should use predictive qualitative assessment techniques to identify potential human factors issues [1]. The guidance acknowledges that 'this will be a relatively new area for many duty holders' and explains that 'our expectation is that they conduct qualitative analyses of human performance – identifying what can go wrong and putting remedial measures in place.' Largely as a result of this regulatory interest, sites subject to these regulations are starting to undertake human reliability risk assessments as part of their overall COMAH Safety Reports. Although there is no formal regulatory requirement in the USA, OSHA's PSM standard explicitly states that PHAs must address Human Factors. However, the standard does not specify exactly how this is to be done, although the appendices provide some guidance in this area. Nevertheless, this falls some way short of the UK requirement that that the safety report submitted to the regulator should explicitly address risks arising from human errors as part of a demonstration that this area has received the same level of attention as hardware initiated process hazards. Apart from regulatory requirements, there is compelling evidence that effort applied to this area has the potential to make significant impact on reducing risk and financial losses.
This paper presents practical guidance in addressing risks arising from human error in process plants, based on several years of experience in supporting organisations to satisfy HSE requirements in this area.

In general, once the commitment has been made to undertake this type of work, its ultimate success depends on how certain organizational issues are managed. This paper is designed to provide some insights into these issues for organisations taking their first steps in this field. To this end, a case study is used to demonstrate one type of qualitative human reliability risk assessment and illustrate potential pitfalls when carrying out these types of analyses. Guidance is provided regarding the planning, resources and timescale issues that need to be managed in order to successfully complete such analyses.

.

2. Process description

HSE guidance suggests that sites subject to the COMAH regulations should use predictive qualitative assessment techniques to identify and manage potential human factors issues (HSE, 2007). They provide an outline description of one such technique, describing a 7 stage process:

Step 1: consider main site hazards;
Step 2: identify manual activities that affect these hazards;
Step 3: outline the key steps in these activities;
Step 4: identify potential human failures in these steps;
Step 5: identify factors that make these failures more likely;
Step 6: manage the failures using methods that reflect hierarchy of control principles;
Step 7: manage error recovery.

Over the last few years we have applied an analysis technique that is consistent with this structure, and this has been used by many organisations as part of their response to satisfying regulatory requirements in their Safety Reports. The following sections provide an overview of the process and further information is available at www.humanreliability.com.

2.1 Steps 1 and 2 – Consider main site hazards and identify human activities that affect these hazards

Tasks related to hazard areas where human failures have the potential to lead to significant consequences should be identified. HSE guidance [1] suggests that the following types of activity, in particular, should be examined:

- tasks that have the potential to initiate a major accident sequence (e.g. inappropriate valve operation causing a loss of containment);
- tasks designed to mitigate the consequences of failures (such as activation of Emergency Shut Down (ESD) systems);

tasks associated with systems designed to prevent or mitigate an incident (e.g. maintenance of safety systems). One of the most commonly advanced arguments against addressing human factors issues in safety reports is that analysing every human activity on a site would be an impossibly time consuming and expensive procedure. To address this criticism, the systematic process that we will describe begins with the development of a Task Inventory, to identify the major areas where human activities may impact on risk, together with a systematic risk ranking procedure to prioritise which of these activities will be evaluated in detail. Using the existing site COMAH site safety report and a list of site procedures as a guide, a workshop involving people with good operational knowledge of the site and process can be used to develop a list of tasks that may initiate MAH (Major Accident Hazard) scenarios. Tasks designed to mitigate the consequences of, or prevent such accidents are also considered. By considering the scale of hazard involved and factors that may affect the likelihood of failure, such as task complexity, distractions and fatigue, the task list can be prioritised in terms of criticality to enable analysis effort to be targeted at the most important activities.

2.2 Step 3 - Outline key steps (Analyze tasks)

Developing a clear understanding of how these critical tasks are performed is an important stage of the process, since an incorrect understanding will significantly undermine the validity of the subsequent analysis.

A formal analysis process, such as Hierarchical Task Analysis (HTA) can be used to lend structure to the analysis (see, for example [2], and [3]. Experienced operators often find it easier to describe what they do rather than why they do it. This technique groups tasks into their constituent subtasks and task steps. Organizing tasks in this way helps participants to describe the reasons for their actions. This often leads to suggestions for better task performance and makes subsequent stages of the analysis easier to manage. HTA is particularly effective for analyzing tasks of a step-by-step nature, such as preparation for maintenance. Figure 1 illustrates the top level of a formal task analysis for the task of unloading a tanker at a jetty. The subtasks at this level are then broken down further to their constituent steps to document how they are achieved. This is illustrated in Figures 2 and 3 for subtasks 1 and 3. Note that subtask 2 is not broken down any further, as it already provides sufficient detail for the analyst to identify any significant risks. This is an important aspect of HTA: it does not try to analyze all subtasks to the same level of detail. Instead, those tasks that are scored as being of high risk potential are selected on the basis of the severity of the consequences if they fail, the likelihood of failure and the likelihood of recovering the error before the consequences arise. The product of these assessments is used to rank order tasks in terms of risk, and only the most highly ranked tasks are analyzed in detail.

Figure 1: Top level of a task analysis for a tanker unloading operation

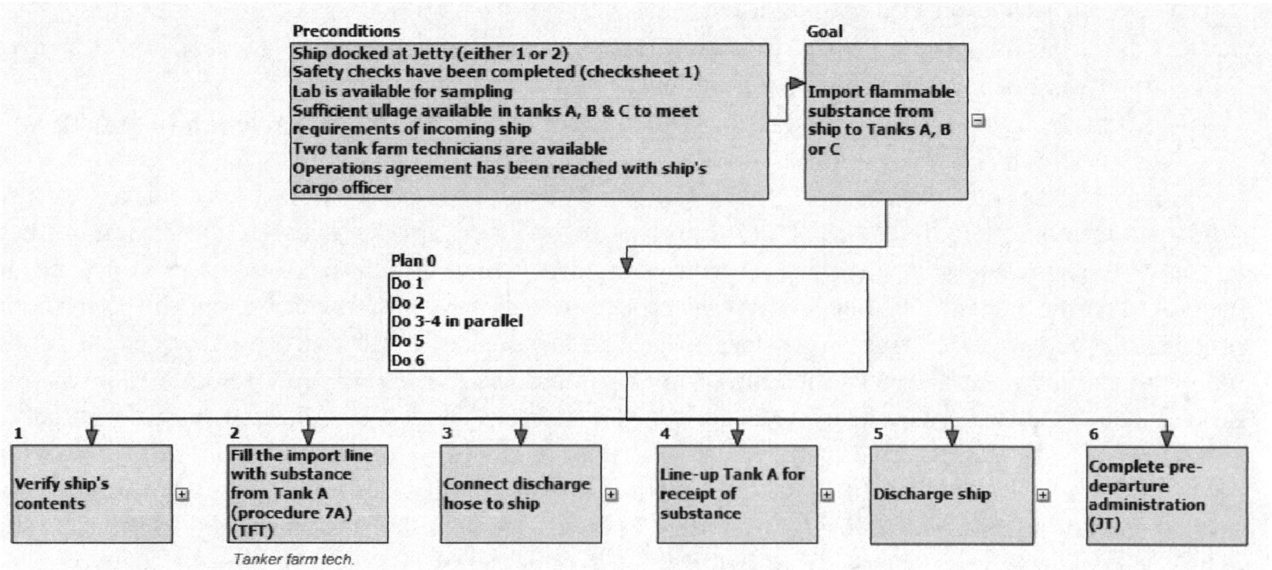

Figure 2 breakdown of subtask 1: Verify ship's contents

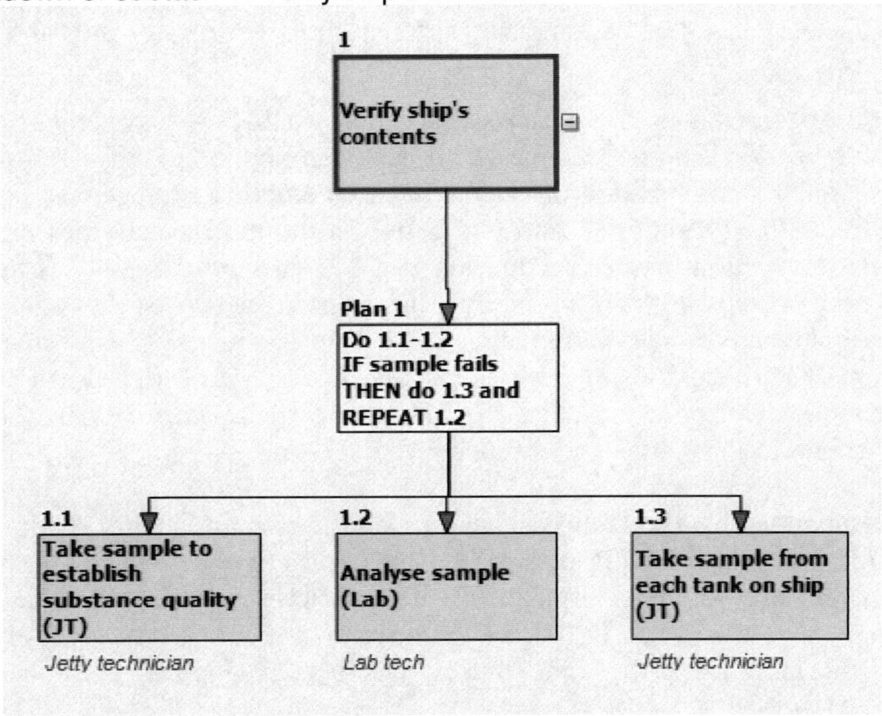

Figure 3 breakdown of subtask 2: Connect discharge hose to ship

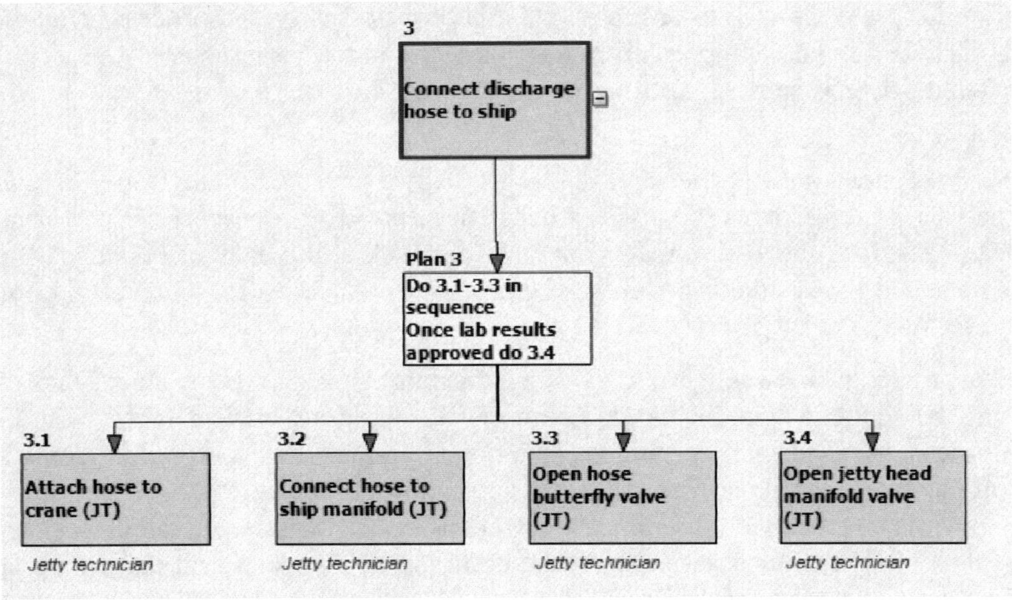

2.3 Step 4 – Identify potential human failures

Workshops are also required for this step. Ideally, the same people who attended the task analysis workshop would attend these sessions. However, as a minimum, at least one individual should have an understanding of the task practicalities and one should have an understanding of the potential consequences of failures. If procedures are out of date or do not reflect actual working practices, much time in the workshop can be spent addressing these issues. The outputs of this process make an excellent basis for new reference procedures, job aids and training specifications.

Human Reliability Associates use a process called SHERPA (Systematic Human Error Reduction and Prevention Approach), [4] which has been applied extensively over the past 25 years. It is essentially a combination of HAZard and OPerability Studies (HAZOP) and FMEA (Failure Mode and Effects Analysis), and is focused on the human activities in the system. It uses guidewords (e.g. action omitted, action too early) to prompt the analysis team to consider failures in subtasks or steps in the task that may have serious safety or economic consequences. It then considers alternative interventions that could lead to the risks being managed in a cost effective manner. The process also identifies existing Risk Control Measures (RCMs) designed to prevent human failures from contributing to a Major Accident Hazard event, these include, for example, alarms, interlocks, deluge systems and relief valves.

These tools are supported by a software package called the Human Factors Workbench which is described in more detail in a previous paper [5]. This allows the graphical format of the analysis to be developed, and then automatically converts it to a Microsoft Word table that can also be used to document the results of the FMEA/PHEA error identification process. An example of the output from this analysis for the tanker unloading subtask 1 (Figure 2) is shown in Figure 4.

2.3.1 Practical considerations in performing the analyses

The primary focus of these types of analyses should be on MAHs. However, as tasks are scrutinized, other issues related to occupational and process safety and process efficiency may be identified. These are also important issues, and are useful to pursue if time is available. However, where time is limited, facilitators must be careful to keep the focus on MAH issues during workshops.

This process is systematic, rigorous and thorough. However, as with any tool, there is a danger that the analysis itself becomes the goal, rather than improved process safety performance. The analysis team therefore needs to remain vigilant to ensure that the balance between effort put into the assessment and the resultant benefits is appropriate. In order to achieve this objective we suggest the following strategies:

Transfer lessons learnt between tasks. During the initial prioritization stage, tasks of similar types should be grouped together. For example, the top four critical tasks identified by the screening process may all involve process isolations. Rather than repeat the complete analysis for four similar tasks, the most critical process isolation can be used to identify general issues regarding the management of isolations, as well as specific issues related to that particular isolation. It is likely that these issues will be common to all isolation operations, saving time when analyzing further tasks of this type.

Keep the level of detail proportionate to risk. We recommend the use of structured task analysis, since we have found that reliance on written procedures as the sole source of information on how jobs are performed can lead to significant issues being missed. One of the strengths of HTA is that it allows the analyst to focus on the most hazardous parts of the task. With experience, the analyst can use this feature of the analysis technique to reduce the effort required

Figure 4 Tabular version of SHERPA analysis generated by Human Factors Workbench software

ID	Description	Additional Info	Failure Mode	Potential to Recover from the Failure	Consequences	Risk Control Measures	Performance Influencing Factors	Actions
1	**Verify ship's contents**							
Plan 1	Do 1.1-1.2 IF sample fails THEN do 1.3 and REPEAT 1.2							
1.1	Take sample to establish substance quality (JT)	This is an aggregated sample from all ship tanks to be discharged. If there is a problem then individual tanks are sampled to establish where the problem is.	A9 Operation omitted	May be picked up at a later stage of the process	Fail to take sample - import sea water tainted substance OR import substance other than intended - possible damage to reactors/reaction	Ships only carry one substance, certificate of quality (Load port).	ACCESS: Access and visibility at sampling points can be difficult. LABELLING: Tanks are generally well labelled.	Implement access and visibility audit at key sampling points
			A10 Operation incomplete		Fail to take sample from all tanks - see above		TIME PRESSURE/ HAZARD PERCEPTION: Field operators may not appreciate the consequence of sea water contamination.	TRAINING POINT: Ensure that all techs understand the risks associated with sea water contamination.
1.2	Analyse sample (Lab)				See above			
1.3	Take sample from each tank on ship (JT)				See above			

2.4 *Step 5 – Identify factors that make failures more likely*

Performance Influencing Factors (PIFs) are the characteristics of people, tasks and organisations that influence human performance and therefore the likelihood of human failure. PIFs include time pressure, fatigue, design of controls/displays and the quality of procedures. Evaluating and improving PIFs is one approach for maximizing human reliability. Some PIFs influence individual task steps, for example ease of access to an emergency valve. Other PIFs have a broader influence, such as quality of training or level of task experience. The potential impact of these factors for the task in question should be considered.

In our experience, operators accept problems that have existed for a long time as part of the job. Having an independent facilitator can help operators to challenge these accepted conditions and practices. Checklists and software tools such as the Human Factors Workbench [5] can be used to ensure coverage of different types of issues.

2.5 *Steps 6 and 7- Manage the failures using the hierarchy of control and manage error recovery*

The principal outputs from the risk assessment will be an analysis document summarizing the failure modes, potential consequences and existing risk control measures and a list of potential improvement actions.

There can be a danger, when responding to issues identified by the analysis, to rely on training and procedures as solutions. These are an important part of a site's risk management systems. However, a proper review of hierarchy of control and cost-benefit principles must be undertaken for all identified issues before deciding on appropriate responses. This can be assisted by a employing a multidisciplinary team (e.g. operators, maintenance engineers, managers, human factors specialists) working together to identify the best response.

As described previously, the nature of the process means that task specific issues are readily identified. The facilitator should consider whether the issues raised are indicative of wider problems (e.g. a problem with a difficult to operate valve may indicate issues with the valve maintenance regime). In particular, management factors such as allocation of resources, determining priorities and management of change should be considered as important potential issues.

In many cases, the existing RCMs will be adequate. If this is the case, then they should be documented in the analysis output to provide an audit trail demonstrating that risks are being assessed and managed. If the RCMs are not adequate, or some aspects of the process need to be altered to make the task safer, then these decisions should be made with reference to cost-effectiveness and hierarchy of control considerations.

3 Guidance for planning human factors risk assessments

From our experience in a range of organisations, the success of the programme will depend to a large extent on how well it is planned, supported and resourced. The following issues should all be considered:

- **Set objectives.** For example, an initial pilot phase may introduce the concepts to relevant staff and deliver some visible benefits to encourage buy-in to the process.

- **Define scope and duration of the initial project**. Set some achievable goals, for example number of critical tasks to be looked at.

- **Select site facilitators**. Many sites do not have internal human factors expertise and buy this in. However, they will require the close support of at least one individual on site to act as a process facilitator. The role of this facilitator includes organising workshops, supplying supporting documentation, reviewing analyses and providing feedback to participants. They should have a good understanding of the plant and have contacts across the site. It is particularly important that they are respected by, and able to communicate effectively with, both engineering and operating level staff.

- **Select participants.** This is a bottom-up process, in that the quality of the output relies heavily on having people with the right knowledge and skills participating in the workshops. For the task analysis workshops, this means including people with extensive practical task experience (i.e. operators). This is so they can describe real working practices and task constraints arising from local working conditions, as well as any differences that exist between shifts

  Ensure adequate follow-up and implementation. If resources are not available to implement recommendations the process will be seen as a time consuming paper exercise with little benefit. In particular, there will be poor buy in from staff which inevitably undermines the quality of the process and makes it difficult to sustain. It is vital that all proposed actions are managed using an action-tracking process. Most organisations have these systems. At a minimum, the system should prioritize the actions required, allocate responsibilities and provide a clear timeframe in which they should be completed. In all cases, the reasons for choosing a particular option, even where the decision has been to take no action, should be noted and fed back, along with the outputs, to participants in the process and the wider organisation. It also helps if the modifications recommended during the process compete on an equal footing with other site priorities.

4 Conclusions

This task-based approach to human factors risk assessments has been implemented successfully in both large and small sites. One of the reasons for its success is that it quickly draws attention to task specific MAH safety issues (e.g. task design, valve labeling, access, unreliable instrumentation, poor interface design, alarm issues). These issues are easy to communicate, clearly linked to risk management and specific enough to address. For a facility new to the topic of human factors, this clear link between human factors issues, MAH and solutions can be the easiest introduction to human reliability and human factors. Other approaches, for example human factors audits of site systems (e.g. shift handover, procedures, training, change management), whilst important, can result in general recommendations that can be hard for a site to address. In addition, when first introduced to human factors, some people find it difficult to make the link between these general issues and process safety. Starting with a task-based risk assessment helps a site identify specific issues as well as providing an insight into strengths and weaknesses of more general site systems. Having established weaker areas in such systems, these can then be subjected to more detailed human factors reviews.

The primary application of the human factors risk assessment approach described earlier has been in onshore oil and gas processing plants to address potential failures in routine operations. However, it has a significant value in assisting organisations to analyze tasks that arise in emergencies. In Appendix 1, we have provided a short analysis of one of the human failures that contributed to the Deepwater Horizon incident, and suggest how the methodology described in this paper could have been applied to reduce its likelihood of occurrence.

5 References

1. Health & Safety Executive Core Topic 3: Identifying Human Failures http://www.hse.gov.uk/humanfactors/comah/core3.pdf (accessed January 2010)

2. Embrey, D. E., Green, M, and Kontogiannis, T. Guidelines for Preventing Human Error in Process Safety Center for Chemical Process Safety New York 1994

3. Kirwan, B., & Ainsworth, L. K. A Guide to Task Analysis. London: Taylor & Francis. 1992

4. Embrey, D.E. SHERPA: A systematic human error reduction and prediction approach. Proceedings of the International Topical Meeting on Advances in Human Factors in Nuclear Power Systems, Knoxville, Tennessee. American Nuclear Society La Grange Park, Illinois 60525 USA. 1986

5. Embrey, D. E., and Zaed, S A set of computer based tools for identifying and preventing human error in plant operations *Proceedings of the 6th Global Conference on Process Safety* San Antonio, 2010

6. Internal BP investigation report on the Deepwater Horizon incident 8th September 2010 http://www.bp.com/liveassets/bp_internet/globalbp/globalbp_uk_english/incident_response/STAGING/local_assets/downloads_pdfs/Deepwater_Horizon_Accident_Investigation_Report.pdf (accessed January 2010)

Appendix 1

Human factors issues in the Deepwater Horizon incident

The Deepwater Horizon incident in the Gulf of Mexico has shown that human factors issues at the operational and safety management system levels had a major impact on the failure to manage the event.

The initial analysis of the incident causes [6], indicated failures of the following barriers:

1. Annulus cement barrier did not isolate the hydrocarbons
2. The shoe track barriers did not isolate the hydrocarbons
3. The negative pressure test was accepted even though well integrity had not been established
4. Hydrocarbon Influx not recognized until it was in the riser
5 Well control response activities fail to gain control
6. Diversion to mud gas separator resulted in gas venting on to rig
7 Fire and Gas system did not prevent hydrocarbon ignition
8 The BOP emergency modes did not seal the well

Human failures had an impact on a number of these barriers, but for the purpose of illustration we will focus on Barrier 3, the acceptance of a negative pressure test even though well integrity had not been established.

A positive pressure test was carried out successfully, followed by a negative pressure test. The objective of the negative pressure test is to test the ability of mechanical barriers to withstand the pressure differentials during subsequent operations; the reduction of hydrostatic head to seawater and disconnection of the Blow Out Preventer (BOP) and riser.

During the negative testing, the BP Macondo team did not recognise the significance of certain events, e.g. high fluid returns (15 bbls taken rather than 3.5 that the analysis team felt should have been expected). This excess flow through the drill pipe should have indicated to the rig crew that there was a flow path to the reservoir through failed barriers. The accident investigation states that:

- There were 'broad operational guidelines' for the test.
- The rig crew and well site leader were 'expected to know' how to perform the test.

However, witnesses state that the (Transocean) rig crew's 'preferred method' was different to the written procedure provided by the BP Macondo well team, in that they monitored the drill pipe line rather than the kill line. The well site leader noticed the discrepancy and they proceeded with the BP Macondo method. The rig crew was therefore likely to be unfamiliar with the procedure they were now using, and both their ability to carry out the operation successfully and to correctly interpret the information sources may have been severely compromised.

During the test the rig crew and the well site leaders both misinterpreted the drill line pressure of 1400psi. Witnesses state that the toolpusher (rig crew) suggested that the pressure on the drill pipe was due to a phenomenon they (toolpusher and driller) had seen before called 'annular compression' or the 'bladder effect'. The well site leaders and rig crew accepted this and carried on.

It is clear that this test is extremely safety critical, and hence should have been subjected to a prospective human factors risk analysis, using the task analysis methods described earlier in this paper. It would then have been clearly defined, standardized and documented, and potential errors could have been anticipated. This would have meant that both the Macondo and the Transocean team would have used the same procedure, and both would have been trained in and familiar with its use. This would have greatly facilitated the communication between the rig crew and the well site leaders and enhanced the likelihood of the correct interpretation of the situation.

Specifying Reliability Performance that Meets Process Safety Management Expectations

Kenneth Bloch
Flint Hills Resources, LP
PO Box 64596
St. Paul, MN 55164
ken.bloch@fhr.com

Doug Dunmire
Western ROPE, LLC

Jeremy Bertsch
Flint Hills Resources, LP

Keywords: Safety, Safeguards, Reliability, PHA, HAZOP, LOPA, MI, Pump, PSM, Failure, Fire, Leak, Risk, Loss Control, Refinery, Petroleum, Chemical

Abstract

Process Hazard Analysis (PHA) and Mechanical Integrity (MI) programs are two of the essential elements that the Process Safety Management (PSM) Standard uses to prevent or minimize the consequences of catastrophic toxic, reactive, flammable, or explosive chemical releases. In cases where process containment is essential for maintaining process safety, equipment that does not meet reliability expectations is more likely to be involved in a PSM failure. When a PSM failure occurs, the equipment maintenance history often exposes the failure as an accident that was waiting to happen.

The MI program and safeguards must control the consequences of equipment failure process safety hazards identified during a PHA to an acceptable level. A quantitative approach to evaluating equipment failure risk can be used to determine the reliability needed to adequately prevent or minimize potential process safety consequences. Additional safeguards are needed when the MI program alone cannot realistically achieve an acceptable level of equipment performance. This paper explains how one refinery uses a quantitative approach to satisfy PSM objectives on potential releases represented by process pump mechanical seal failures.

1. How Equipment Reliability Impacts Process Safety

OSHA's PSM Standard (CFR 1910.119) details the requirements for preventing or minimizing the consequences of catastrophic releases of toxic, reactive, flammable, or explosive chemicals. Process Hazard Analysis (PHA) teams assemble to identify and evaluate hazards that represent the potential release of dangerous materials as described in subpart e of the PSM Standard (CFR

1910.119(e)). As part of the PHA process, action items are assigned to manage the consequences of identified hazards to an acceptable level. For this reason, it is not uncommon for historical PHAs to be evaluated in response to an accidental release of potentially hazardous process materials. This review is triggered to understand how previous PHA teams assessed the hazard. It is therefore possible to interpret any failure that results in the accidental release of potentially hazardous process materials as a PHA team failure. More precisely, it indicates that the previous PHA teams may have failed to adequately identify, evaluate, and control the process safety hazard. Its potential consequences were therefore left for discovery by the process safety failure that exposed them.

Many, if not most, process safety failures are preceded by lower consequence repeat failures. These low consequence failures can form a failure "rap sheet" documented in the equipment work order history of a plant's Computerized Maintenance Management System (CMMS). When a process safety failure is investigated, a persistent history of failures stands out as an obvious warning of greater risks that should have been identified and controlled. However, prior to the process safety failure, these lower consequence failures can easily be seen as a normal part of equipment operation and maintenance. This type of "normalization of deviance" received much attention within NASA after the loss of fourteen astronauts and two space shuttles. Evidence of it can also be found in the manufacturing industry. Consider the following examples of catastrophic releases caused by centrifugal pump failures.

2. Case Histories in Responsive Risk Mitigation

2.1 Alkylate Pump Fire and Explosion [1]

A fire and explosion occurred at a large refinery operating in New Mexico, USA on April 8, 2004. The fire ignited upon the catastrophic release of flammable process liquid following a centrifugal pump mechanical seal failure. The failure resulted in six injuries as well as extensive property damage and business interruption.

Figure 1
Alkylate Pump Fire (CSB photo)

The fire resulted from loss of process containment from the alkylate recirculation pump shown in Figure 1. Alkylate is a mixture of light hydrocarbons typically in the C4 (butane) to C8 (octane) range. Process material leaking from the pump at 350 °F autoignited upon contacting air. The failure resulted from the misapplication of energy control after the pump was removed from service to address a process fouling problem. The release occurred when mechanics that were scheduled to replace a failed mechanical seal began disassembling the pump on location.

The pump involved in the release had a recurring history of seal failures leading up to the fire. In a formal report, the United States Chemical Safety Board (CSB) designates this incident as an MI program failure. In the report, the CSB cites 23 work orders issued to address seal failures in the year leading up to the catastrophic process release (Table 1). The CSB makes the valid argument that "an effective mechanical integrity program would have investigated and resolved the problems that were repeatedly causing the [alkylate] pumps to fail." Instead, the investigation showed that any opportunity to prevent the failure through reliability improvement was substituted with "breakdown maintenance." In other words, maintenance was used to address the problems caused by pump failures rather than addressing the underlying causes that would have resulted in satisfactory pump performance, sufficient to reduce the risk for a process safety failure.

The potential for a process safety failure increases during shutdowns for maintenance [2]. This case history illustrates how energy control defects experienced during routine maintenance activities can interfere with safe work execution. For this reason, multiple safe work practices like lockout-tagout and confined space entry policies are often used to mitigate additional risk when equipment is shut down for maintenance. But it is not unusual for process safety failures to occur on equipment in continuous operation when these additional safety precautions typically do not apply. The next case history illustrates how similar MI defects can be involved in process safety failures regardless of the equipment's operating mode.

Table 1. Alkylate Pump Failures

Failure	Date	Problem
1	April 17, 2003	Seal leak
2	May 9, 2003	Pump spraying from seal
3	May 23, 2003	Repair seal
4	June 9, 2003	Repair seal
5	June 9, 2003	Repair seal
6	June 18, 2003	Repair seal
7	June 20, 2003	Replaced seal
8	July 31, 2003	Replaced seal
9	August 22, 2003	Seal leak
10	August 25, 2003	Replaced seal
11	September 26, 2003	Replaced seal
12	September 26, 2003	Replaced seal
13	October 14, 2003	Seal leak
14	December 6, 2003	Replaced seal
15	December 9, 2003	Seal leak
16	December 9, 2003	Seal leak
17	December 15, 2003	Replaced seal
18	December 15 2003	Seal leak
19	January 28, 2004	Seal leak
20	March 22, 2004	Seal leak
21	April 1, 2004	Pump seal leaking
22	April 3, 2004	Pump seal leaking
23	April 7, 2004	Repair pump seal

2.2 No. 2 Fuel Oil Pump Fire

A fire occurred in a Distillate Hydrotreating Unit operated by a large refinery in the Upper Midwest, USA on November 15, 2004 (Figure 2). The fire ignited upon the catastrophic release of flammable process liquid following a centrifugal pump mechanical seal failure (Figure 3). The failure resulted in one OSHA-Recordable injury as well as extensive property damage and significant business interruption.

Figure 2
No. 2 Fuel Oil Pump Emergency Response

The fire ignited when No. 2 Fuel Oil, a diesel-range hydrocarbon mixture with carbon chain lengths ranging between C10 and C20, leaked out of the pump through a failed mechanical seal. Similar to the pump failure described in the first case history, the 600 °F process material leaked out above its autoignition temperature (about 500 °F) and therefore caught fire immediately upon contacting oxygen.

The injury occurred during the emergency response to the fire. The first responder began applying water to the pump fire without first increasing his personal protective equipment (PPE) level. At some time during the response an injury resulted from smoke inhalation. However, within two hours the fire was extinguished and the emergency situation was brought under control without any further safety consequences.

Figure 3
No. 2 Fuel Oil Pump Failure

Reviewing the maintenance record of the pumps involved in the loss of No. 2 Fuel Oil containment revealed a long history of thrust bearing failures and seal leaks (Table 2) similar to the pumps examined in the first case history. Likewise, the cause for the fire was determined to be a catastrophic mechanical seal failure. However, the physical evidence collected in the unit after the fire indicated that seal damage was the secondary in response to a catastrophic thrust bearing failure. This primary failure caused uncontrolled shaft movement in the axial (thrust) direction that destroyed the mechanical seal.

Eventually, the investigation team was able to link together the probable cause for unstable hydraulics at the pump installation. The failure mechanism was introduced by operating the No. 2 Fuel Oil pumps in continuous parallel service. Originally the pumps were designed for single-spare operation. But through years of growth and unit debottlenecking efforts, the pumps had started continuously operating in parallel to overcome rundown piping pressure constraints. In parallel operation, the pump's rotating elements came under constant stress. Before the fire, this failure mechanism was adequately managed by condition monitoring and frequent repairs. But eventually a fire in an operating unit and an OSHA Recordable Injury settled any debate over the potential consequences for accepting poor pump reliability in this service. Although this particular installation had been examined twice previously by PHA teams in accordance with OSHA regulations, the hazard remained hidden until the process safety failure exposed it.

Table 2: No. 2 Fuel Oil Pump Failures

Failure	Date	Problem
1	August 18, 1993	Leaking head
2	September 3, 1993	Failed coupling
3	January 27, 1994	Failed coupling
4	February 22, 1994	Leaking head
5	November 22,1994	Cavitating noise
6	April 27, 1998	Leaking seal
7	May 29, 1998	Failed coupling
8	November 20, 1998	Failed thrust bearings
9	April 6, 1999	Leaking seal
10	March 30, 2000	Leaking seal
11	August 3, 2000	Leaking seal
12	February 20, 2002	Failed coupling
13	October 21, 2002	Failed thrust bearings
14	September 9, 2003	Failed thrust bearings

Although these two separate failures occurred in different facilities, in different services, at different times, and under different process operating conditions, the common thread of below-expectations reliability runs between them. In both cases it is easy to look back on them as an accident waiting to happen. Unfortunately, in neither case was the MI program able to prevent repeat failures that eventually resulted in an unacceptable, non-discretionary process safety failure. In both cases, the owner-operators of the unreliable equipment were in full compliance with OSHA 1910.119(e) governing the use of the PHA program to detect hazards that could result in the potential release of the processes they contain. However, in both cases the PHA program failed to identify and adequately control the hazards that ultimately resulted in the failure. Additionally, the MI program (OSHA 1910.119(j)) was unable to achieve a level of equipment reliability sufficient to offset any PHA defect. Indeed, the MI program is just as important for the PSM standard to achieve its objective as any of its other elements.

3. Conservative, but Reasonable?

Typical PHA team members do not take their responsibility to identify hazards lightly. Rendering their services in PHA meetings requires a considerable amount of time away from their normal responsibilities. They participate with the intention of adding value by detecting, assessing, and controlling any hidden hazards to protect themselves and their fellow workers. Therefore, they take their PHA performance very seriously. For this reason, sincere PHA team members are committed to learning from their mistakes so that they can add more value in future PHA meetings. In cases where PHA teams are criticized (publicly, privately, or interpersonally) for failing to avoid an incident after it occurs, it is typical for them to "err on the safe side" in future PHA meetings.

In some cases this conservative response may be appropriate. For example, it is both reasonable and important to expect drastic changes when a facility learns of operation that is contrary to industry policies or standards. On the other hand, it would not be realistic to view all potential hydrocarbon releases equally. Yet this is what some teams do upon recognizing their failure to generate action items sufficient to mitigate the potential risk of a process safety failure in previous PHA meetings. This approach has the tendency to generate excessive action items that may or may not add value. Addressing them may create an image of improving workplace safety without making progress on hazards that truly represent unacceptable risk.

Recent events in process safety failure show how dangerous it can be to develop initiatives around safety items that represent little incremental value. This situation was brought to British Petroleum's (BP) attention after the refinery explosion in Texas City on March 23, 2005 [3]. It is not that BP was not concerned about, nor investing in, process safety improvements. The unfair truth about process safety is that there is no reward for hard work. To avoid a process safety failure the effort must be properly directed. A safety program will fail if it focuses employee attention on the wrong things. The illusion of a safe workplace is destroyed when a catastrophic failure exposes a persistent, unacceptable risk as an accident waiting to happen. Working on the wrong things creates a distraction from the greater and more realistic process safety threats that should be resolved first.

4. Risk Basics

Although the argument could be made that safer pump operation results from upgrading them with more robust seals, bearings, and monitoring systems, doing so is probably not the most deliberate way to achieve process safety. In many cases upgrades offer no incremental improvement unless they address a deficiency that causes the pump to perform below life cycle expectations [4]. Indiscriminately upgrading pumps can consume a considerable amount of resources with the intention of making a system safer, while creating a distraction from other process safety hazards that represent a greater risk.

Risk by definition is a function of frequency and consequence. Not all centrifugal process pump failures represent the same risk. For example, a hydrocarbon pump operating in the middle of a congested process unit may not represent the same potential consequences as a pump moving similar process liquids in a remote location away from an operating unit. Likewise, the high temperature gasoil fraction that leaked in the second case history may not represent the same potential consequences as a leaking gasoil fraction cooled below 300 °F, downstream from a rundown cooler.

The point here is that assessing the potential consequence of a catastrophic pump failure is not a binary process. A risk assessment is not performed by simply asking whether or not the pump contains hydrocarbon. The consequences of a catastrophic pump failure are a function of a number of critical factors. Some of these factors include the type of process material, leak rate, failure location, and temperature. Additionally, assessing the failure frequency can be aided by determining what is in the CMMS before a process safety failure triggers an investigation. This information makes it possible to detect an "accident waiting to happen" *before* it happens.

5. What Does "Good" Look Like?

The two case histories illustrate scenarios where a high frequency of seal failures preceded a catastrophic chemical release that defeated PSM objectives. These are considered MI program failures because the MI program did not drive the equipment failure frequency sufficiently low to mitigate the risk for a process safety failure. Remember that risk is a function of frequency and consequence. Therefore, driving the risk for a safety process failure down to zero (the goal of a "zero-injury" workplace) simply involves reducing the equipment failure frequency to zero. Unfortunately, this can only be achieved by shutting down the equipment whose failure may result in PSM consequences. Even the most reliable equipment represents risk as long as it is operating.

Although most companies would immediately shut down equipment that was determined to be operating unsafely, few of them would voluntarily shut down a machine to guarantee their "zero-injury" workplace goal. After everything is shut down, nobody gets hurt at work because nobody goes to work. It is more satisfying to set an acceptable risk tolerance and understand what exactly needs to be done to achieve it. By assigning risk and consequence, it becomes possible to establish equipment reliability targets based on the relationship between risk, consequence, and failure frequency. This is a much more rewarding alternative to achieving safe equipment operation. This approach helps facilities manage their MI program with performance expectations that relate to equipment failure risk tolerance.

6. A Standardized Approach to Assessing Pump Failure Risk

A risk assessment tool was constructed to evaluate the risk represented by process releases resulting from catastrophic pump failures. The guideline was developed to be consistent with, and borrows heavily from, the approach defined in API Publication 581 Risk-Based Inspection (RBI) Base Resource Document [5]. RBI is a widely accepted method currently practiced across the refining industry. Although API 581 applies primarily to fixed equipment, the approach has many parallels to assessing the risk for a rotating machinery failure. The standard RBI components are supplemented with data, methods, and tools more specific to centrifugal pumps when needed.

A standardized risk assessment approach reduces the inconsistency that different PHA teams may encounter at different times. More importantly, it adds value by connecting the reliability of a specific pump installation to process safety risk tolerance. The benefit comes from determining a realistic target for the MI program to achieve, instead of motivating reliability professionals to achieve their safety goals with nonspecific targets like "work harder," or "do better," or "fail less." Setting a tangible reliability target allows a responsible decision to be made as to whether or not risk tolerance can be achieved through the MI program alone. If the MI program cannot realistically achieve the desired level of risk control, then additional layers of protection must be added to manage the risk to an acceptable level.

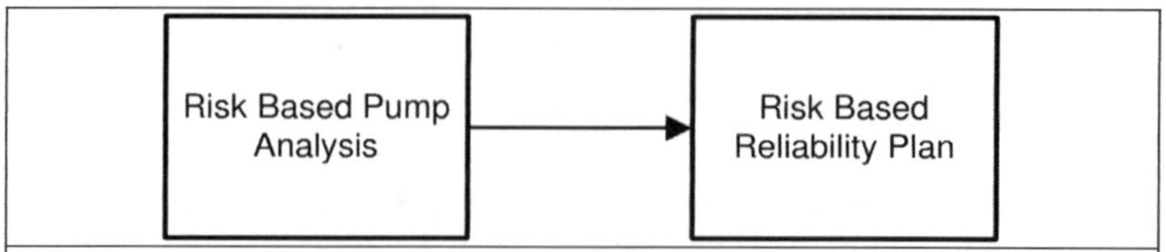

Figure 4
Catastrophic Pump Seal Failure Risk Assessment Method Overview

In some cases the MI program may adequately drive risk to an acceptable level without requiring any additional safeguards or improvements. At such a time the PHA team has a basis to conclude that no further actions are needed to mitigate the potential hazards associated with a catastrophic pump failure. In other words, the existing safeguards have been evaluated and are considered adequate. The process of evaluating the potential risk associated with catastrophic pump failures begins with determining an acceptable level of risk. This prevents the PHA process from defeating its purpose by creating action items that consume available resources that should be working on resolving more important process safety risks.

6.1 Method Overview

Figure 4 shows the basic process used to evaluate the risk represented by a catastrophic centrifugal pump seal failure. The analysis begins with a technical pump risk assessment (Risk Based Pump Analysis). This step is performed according to the consequence analysis and likelihood analysis methods described in API 581 Sections 7 and 8. Afterwards, a quantitative Layer of Protection Analysis (LOPA) is used to compare the specific pump risk against an acceptable risk tolerance. This makes it possible to develop a reliability plan to operate the pumps within risk tolerance.

6.2 Catastrophic Pump Seal Failure Frequency and Consequence

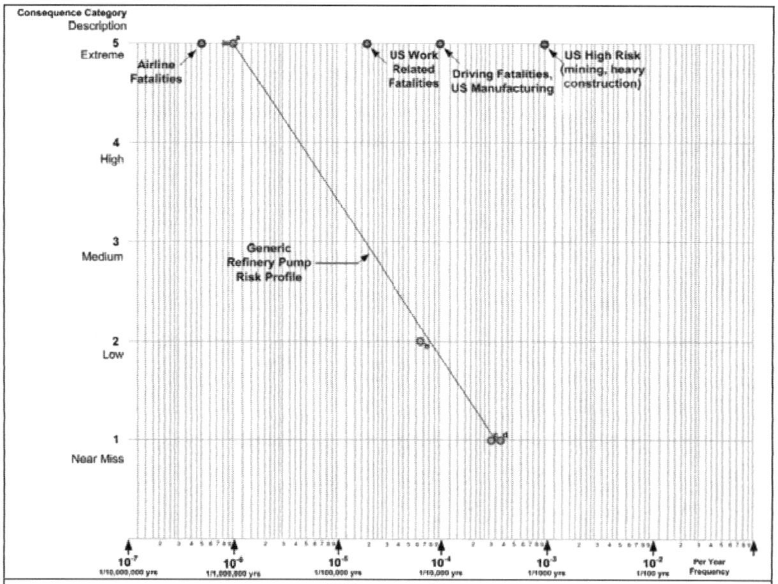

Figure 5
Catastrophic Pump Failure Frequency/Consequence Plot

OSHA data from 1992 to 2009 [6] contains a record of 36 catastrophic releases of highly hazardous chemicals that resulted in fatalities. These incidents are responsible for 52 fatalities and 250 employee injuries. Ninety-eight of these injuries were severe enough to require hospitalization. One of these incidents involved a process release that occurred while steaming-out a pump casing. The pump casing split open, resulting in a hot oil release that immediately exploded (Jan 19, 2005 Kern Oil Refinery, Bakersfield, CA). The conditions present during this failure are similar to those that the CSB documents in the first case history. However, none of the fatal incidents contained in the OSHA database resulted from a pump reliability issue.

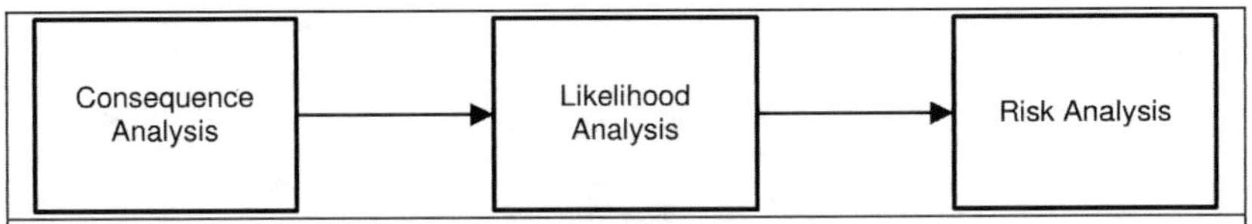

Figure 6
Risk Based Pump Analysis

It would not be responsible to conclude that a catastrophic pump seal failure could not result in a fatality based on these historical statistics. The second case history illustrates the potential for pump failure mechanisms to be directly involved in a process safety incident capable of causing severe consequences. Although there is insufficient data for a straightforward fatality frequency calculation, enough statistical information exists to estimate a minimum frequency based on site-specific data and industry averages. A frequency/consequence diagram like the one shown in Figure 5 can be constructed using this information along with the following facts and assumptions:

- A total estimated 2009 refining capacity of 17,671,550 BBL/Day [7].
- The relationship of approximately one fire for every one thousand repairs, as cited by an industry reliability authority [8].

According to this analysis, the frequency for a fatality (highest severity consequence) is estimated to be lower than $1x10^{-6}$ (1/1,000,000) years. This frequency suggests that a fatality caused by a pump reliability issue is probably more likely than an airline fatality but less likely than other typical US workplace fatality causes [9]. Based on the industry workplace fatality statistics contained in the OSHA database, this relative ranking seems reasonable.

This information makes it possible to define risk tolerance. Risk tolerance (or literally "Tolerance to Risk") implies that the choice has been made to operate equipment in a responsible manner rather than shutting it down to mitigate a process safety failure risk. Risk tolerance will vary between different organizations. It is a decision that should be made under the direction of legal counsel and supported by industry statistics.

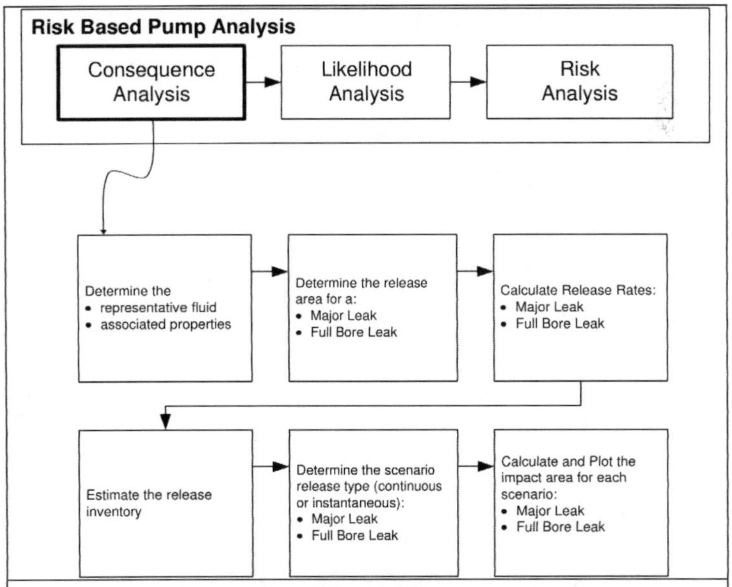

Figure 7
Consequence Analysis Steps

6.3 Risk Based Pump Analysis

Figure 6 shows the primary steps involved in the Risk Based Pump Analysis. In the Risk Based Pump Analysis, results from the consequence analysis are combined with the likelihood analysis to determine the risk associated with a catastrophic pump failure. Comparing actual operating risk against a designated risk tolerance makes it possible to assess risk reduction options that may adequately control the process safety hazard. To be effective, the risk reduction options must directly address the factors governing process safety.

6.3.1 Consequence Analysis

The consequence analysis is covered extensively in API 581 RBI Base Resource Document Section 7. It is used to calculate the release area that would develop upon a loss of process containment caused by a catastrophic equipment failure. In this case, the RBI principles of API 581 Section 7 are being applied to potential releases caused by a catastrophic pump failure. Figure 7 outlines the recommended approach for working through the consequence analysis using the methods described in API 581 Section 7.

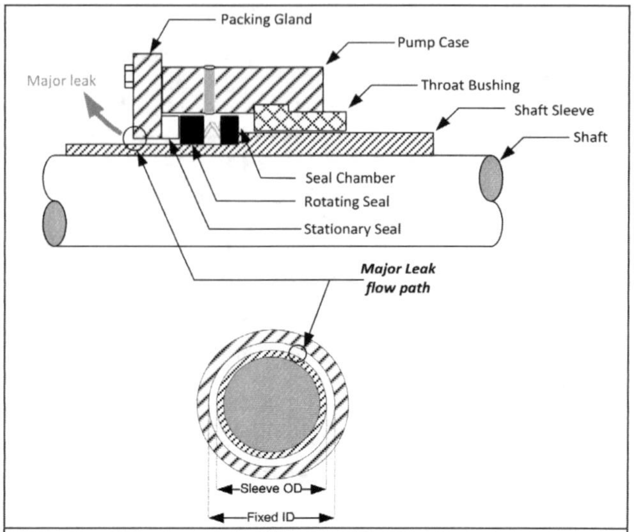

Figure 8
Simplified Seal Sketch – Major Leak Flow Path

The analysis should be based on a representative fluid, which assumes that typical refinery pump service is constantly changing and the process material properties being evaluated may be best described as an estimate of average operating conditions over a period of time. API 581 breaks process fluids down to a discrete number of representative fluids. This level of detail is sufficient for the consequence analysis.

The flow area for a major leak is represented by an annular area between the shaft sleeve and the closest fixed dimension of the pump casing or packing gland. The OD of the shaft sleeve and the ID of the closest fixed dimension of the pump casing or packing gland are determined from the seal manufacturer's detailed drawing as illustrated in Figure 8. These dimensions are then used to calculate a major seal failure leak rate.

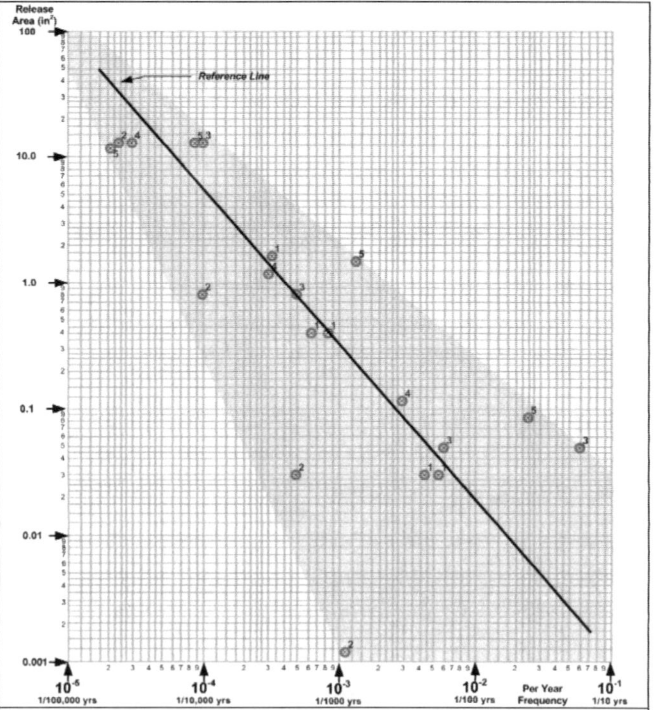

Figure 9
Generic Centrifugal Pump Leak Area / Frequency

6.3.2 Likelihood Analysis

The likelihood analysis is described in detail by API 581 RBI Base Resource Document Section 8. Its purpose is to generate an initiating event frequency for both the major and full bore leak scenarios. The likelihood analysis described in this study makes use of generic initiating event frequencies (IEF_g) that are based on the empirical data shown in

Figure 9. Figure 9 is based on catastrophic pump failure data placed into the public domain by multiple sources [10]. This information covers a wide range of leak rates from minor leaks (low severity) to full bore leaks (high severity.) The middle area of the chart represents the major leak range.

The likelihood analysis is performed by (1) selecting an appropriate IEF_g based on the analysis represented in Figure 9 then (2) adjusting the IEF_g based on the specific pump's actual reliability history ($MTBF_a$ – see Equation 1) compared with the standard reliability of a generic refinery process pump ($MTBF_g$). This adjustment is made according to Equation 2, which produces the initiating event frequency for a specific pump installation (IEF_a).

$$\mathbf{MTBF_a} = \frac{\textbf{Years of Pump Repair History}}{\sum \textbf{Repairs for All Pumps in the Functional Group}} \qquad \text{[Eq. 1]}$$

$$\mathbf{IEF_a} = \mathbf{IEF_g} \times \frac{\mathbf{MTBF_g}}{\mathbf{MTBF_a}} \qquad \text{[Eq. 2]}$$

6.3.3 Risk Analysis

The risk analysis takes place as a Layer of Protection Analysis (LOPA) that assesses the pump operating risk against the designated risk tolerance. Its purpose is to determine if a pump installation meets its reliability expectations. This is true if the frequency of mitigated consequences is less than the designated risk tolerance. If the frequency of mitigated consequences is more than the designated risk tolerance, then guidance should be suggested to improve performance to meet process safety objectives.

6.3.3.1 Probability of Personnel in Affected Area

The probability of personnel in the affected area (P_p) is a function of the size of the affected area (A_a) and the amount of time personnel are likely to be in this area. There are causes of catastrophic pump failures that increase the probability of personnel being in the affected area at the time of the event. An example may be an abnormal process condition (such as loss of flow) where the console operator calls for the outside personnel to respond. There are also causes that are random in nature where the probability of personnel in the affected area is based on the average amount of time that people are in the area on any given day.

Failure cause distribution estimates for centrifugal pumps in US Process Plants indicate that approximately 12% of failures are caused by improper operation [11]. Some of these causes result from chronic poor operating practices that reduce pump reliability. They may have been normalized over time and do not result in an operator response. An example may include cavitation noises caused by low $NPSH_a$ operation or long term flow outside of recommended reliability limits. An estimate of 10% of the causes of major releases which result in increased occupancy of the affected area is assumed for this analysis. The remaining random occupancy that does not increase the probability of personnel in the affected area would therefore be 90%.

An estimated random occupancy of 1 hr/day/1000 ft^2 is assumed for normal process areas. This estimate should be modified if there is evidence of higher or lower occupancy. Remote areas that are not frequented with multiple rounds a shift will be less. Affected areas that include known high occupancy zones will be greater. Any basis for choosing a different random occupancy should be documented. This random occupancy is further simplified to a probability of 0.04/1000ft^2. By combining Cause Generated Occupancy with Random Occupancy an overall probability of personnel in the affected area can be determined by Equation 3.

$$\mathbf{P_p = 0.10 + A_a\left(0.04/1000\,ft^2\right)} \qquad \text{[Eq. 3]}$$

6.3.3.2 Probability of Ignition

API 581 reports probability of ignition (P_i) for five potential outcomes in tables. The proper table in API 581 Section 7 should be selected based upon the process leak assessment made during the consequence analysis.

6.4 Risk Based Reliability Plan

The output from the Risk Based Pump Analysis feeds into a process for managing the risk of a process safety failure. The frequency of mitigated consequences (F_m) is the product of the frequency of the specific pump's initiating event frequency (IEF_a), the total probability of failure on demand for each independent layer of protection (PFD_t), the probability of personnel in the affected area, and the probability for ignition as calculated in Equation 4. If the frequency of mitigated consequences (F_m) is higher than the designated risk tolerance, then a Risk Based Reliability Plan must be developed to manage the risk for a process safety failure. This can be accomplished by either increasing the pump's reliability (MTBF) or by applying safeguards sufficient to mitigate the consequences of a catastrophic pump failure to an acceptable level. The basic process used to develop a Risk Based Reliability Plan is shown in Figure 10.

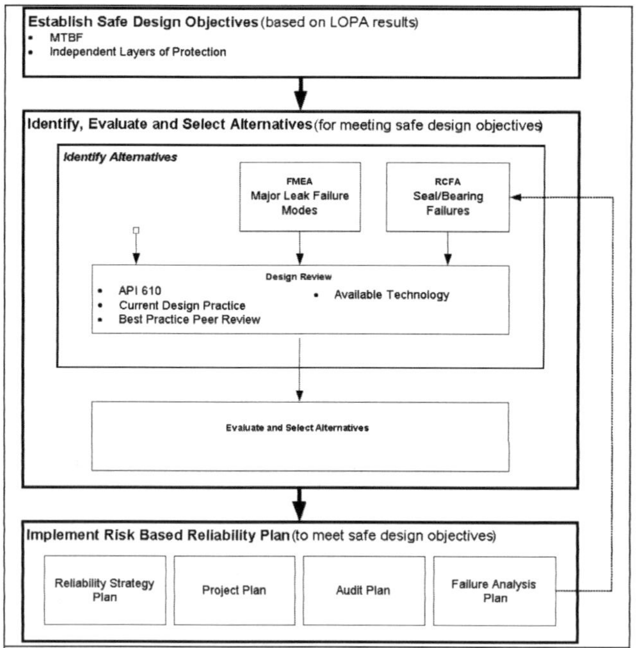

Figure 10
Creating a Risk Based Reliability Plan

$$\mathbf{F_m = IEF_a\left(PFD_t\right)\left(P_p\right)\left(P_i\right)} \qquad \text{[Eq. 4]}$$

The LOPA results designate a target MTBF for meeting a designated risk tolerance. MTBF improvements have a number of advantages. For example, they reduce both maintenance costs and the potential to introduce of some major leak failure modes during repairs like the one described in the first case history. MTBF improvements are typically preventive instead of reactive. However, it may be difficult to quantify the expected MTBF improvement available through failure analysis and investigation. Failure analysis skills, training, and methods are involved in developing an effective set of corrective actions to increase MTBF. This depends greatly upon the failure investigator's individual capabilities.

Machinery engineers must be consulted to determine if the MTBF improvement is realistically achievable. Consideration should be given to proven technology and both industry and personal experience with the process requirements. MTBF improvements can be applied together with additional safeguards to meet the overall risk tolerance criteria. If MTBF alternatives are selected as a part of the strategy to meet the risk tolerance, MTBF becomes a part of the process safety risk management for the pump group under consideration. It should therefore be managed with the same diligence and priority as defined by safe operating limits.

7. Case History in Preventive Risk Mitigation

An investigation was used to determine the cause for a series of recurring seal and thrust bearing failures on a pair of Heavy Vacuum Gas Oil (HVGO) service pumps operating in a refinery Vacuum Crude Unit. The maintenance history of these pumps is shown in Table 3. The investigation determined that high frequency vibration caused by vortex cavitation suction recirculation (VCSR) [12] was resulting in the low MTBF. Based on this diagnosis, an action item was created to increase the pumps' NPSH margin ratio to reduce the cavitation forces responsible for excessive stress on the thrust bearings [13].

Addressing this action item would require either redesigning or replacing the pumps at considerable expense. Based on the resulting maintenance expenses, other competing reliability improvement projects offered a greater return on investment. Therefore, it was determined that the risk for a catastrophic HVGO pump failure would continue to be managed by repairs triggered by condition monitoring until the higher priority reliability improvement projects could be completed.

The potential consequences of HVGO leaks in Vacuum Crude Unit service are not comforting (Figure 11). A leak of sufficient size would likely autoignite upon contacting air. The consequence for a catastrophic pump failure represents a potential PSM incident in addition to property damage and business interruption. But condition monitoring seemed to be an acceptable approach to managing the risk for a catastrophic HVGO pump failure based on previous operating history.

Figure 11
Catastrophic HVGO Pump Seal Failure Consequences

Upon developing the Risk Based Pump Analysis guidance, the HVGO pumps were reevaluated, to verify that the reliability strategy was in agreement with refinery risk tolerance. The analysis showed that the pump group was one protective layer short at its present MTBF and its reliability would have to be increased to at least six years MTBF to operate within refinery risk tolerance. This immediately changed the basis for the project from a reliability improvement opportunity to a process safety risk mitigation project. The priority of the HVGO pump project was therefore elevated and an execution date was scheduled.

Table 3. HVGO Pump Work Order History

Date	Problem	Cost
December 13, 2006	Need to replace inboard and outboard bearings	$1,445.31
April 9, 2007	Replace inboard and outboard pump bearings	$1,974.29
March 31, 2008	Inboard seal leak	$4,771.32
November 1, 2008	Inboard pump seal leaking	$3,852.63
April 7, 2009	Outboard bearing leak	$3,486.53
		$15,530.08

7. Disclaimer

The guideline and methodology discussed in this document attempts to be generally applicable to all centrifugal pumps. However, good engineering judgment must prevail while applying this guideline. The approach can be modified as appropriate following a recommended peer-review and documenting the technical basis for deviations.

8. Conclusions

Tolerating repeat failures on machinery that contains potentially hazardous process materials can have disappointing consequences. However, it is not uncommon for equipment failures to be accepted without comparing actual reliability performance against a designated risk tolerance. In cases where breakdown maintenance is the option selected to manage the risk for catastrophic process releases, a definitive and objective basis is needed to expose a potentially unacceptable process safety hazard before an incident occurs.

A standard risk assessment method can be developed to evaluate pump reliability on the basis of managing its failure frequency sufficiently low to realistically avoid a process safety incident. However, the MI program by itself may not sufficiently elevate equipment reliability to a level where process safety consequences can confidently be prevented. Risk tolerance ultimately determines the complete plan needed to fully address a process safety risk. In many cases, a complete plan represents a combination of reliability (MI program) improvements and safeguards (Layers of Protection).

The RBI method described in API 581 Sections 7 and 8 provides a sound engineering basis for performing a risk analysis on equipment whose failure may represent process safety consequences. This information can be supplemented with site specific failure data and industry statistics to develop risk assessment criteria for centrifugal pumps operating in the petroleum and chemical processing industries. Practicing this approach to process safety is expected to provide more satisfying and effective results than dedicating resources to random safety improvements that may ultimately fall short in avoiding a process safety failure.

8. References

[1] "Oil Refinery Fire and Explosion," 2004-08-I-NM, U.S. Chemical Safety and Hazard Investigation Board, October 2005

[2] *Safe Ups and Downs*, 1st ed. (booklet), Standard Oil Company, 1960

[3] "The Report of the BP U.S. Refineries Independent Safety Review Panel," January 2007

[4] Kenneth P. Bloch, "Centrifugal Compressor Failure Investigation Report," July 29, 2010 (unpublished)

[5] "Risk Based Inspection (RBI) Base Resource Document," API Publication 581 1st ed., May 2000

[6] "OSHA National Emphasis Program Directive," CPL 03-00-004, June 7, 2007, http://www.osha.gov/ (accessed December 26, 2010)

[7] "Refining Capacity Report, January 1, 2009" National Petrochemical & Refiners Association, August 2009

[8] H. P. Bloch, "Understanding Canned Motor Pumps," *Lubrication Management and Technology*, September/October 2008

[9] *Layer of Protection Analysis Simplified Risk Assessment*, Center for Chemical Process Safety (CCPS) of the American Institute of Chemical Engineers, 2001

[10] OREDA, "Offshore Reliability Data," 4th and earlier eds.," SINTEF, 2002

HSE, "Offshore Hydrocarbon Releases Statistics and Analysis, 2002," HSR 2002 02, 2003

S. Mannan, *Lees' Loss Prevention in the Process Industries*," 3rd ed., 2005

Cox, A.W., Lees, F.P., and Ang, M.L., "Classification of Hazardous Locations," Institute of Chemical Engineers, 1990

DNV (for DTI in UK), "White Rose DA Volume 5 Part Two (Concept Safety Analysis)," July 2000

HSE, "Offshore Technology Report – OTO 1999 079," January 2000

J. Spouge, "New Generic Leak Frequencies for Process Equipment", *Process Safety Progress*, Vol. 24, No.4, December 2005

"Guidelines for Quantitative Risk Assessment," *Purple Book*, CPR18E, SDU, Committee for the Prevention of Disasters (CPR), The Hague, 1999

Flemish Government, LNE Department, "Handbook Failure Frequencies 2009 for drawing up a Safety Report," May 5, 2009

[11] Bloch, H. P. and Geitner, F. G., *Machinery Failure Analysis and Troubleshooting*, Volume 2, 1999

[12] Kenneth P. Bloch, "Centrifugal Pump Failure Investigation Report," December 17, 2009 (unpublished)

[13] Bruno Schiavello, "Cavitation and Recirculation Troubleshooting Methodology," *Proceedings of the 10th International Pump Users Symposium*, 1993

An Approach to Integrate Plant Reliability Efforts With a Mechanical Integrity Program

Bruce K. Vaughen
Cabot Corporation
700 E US Highway 36
Tuscola, IL 61953
bruce_vaughen@cabot-corp.com

John F. Nagel
Cabot Corporation
Tuscola, IL
john_nagel@cabot-corp.com

Matthew J. Allen
Cabot Corporation
Billerica, MA
matthew_allen@cabot-corp.com

Keywords: maintenance, planned, preventive, predictive, process safety management, six sigma, mechanical integrity, quality assurance, process safety information, operational discipline,

Abstract

Maintenance is expected to ensure reliable operation of the equipment they maintain to meet operation's demands. Operations is expected to ensure sufficient, quality product is made to meet the business demands. Engineering is expected to design, fabricate, and install equipment that minimizes risk. Safety is expected to ensure that the engineering and administrative controls are in place to minimize that risk. Each of these groups is expected to work without causing safety, health or environmental harm while focusing on their individual group demands. This paper presents a preliminary concept that combines these reliability and safety efforts, such that the maintenance, operations, engineering and safety groups speak the same equipment reliability and operational integrity language to minimize overall business risk.

This proposed approach combines existing plant reliability efforts with several Process Safety Management (PSM) elements. The basic steps include using Process Hazards Analysis (PHA) risk-assessments to identify safety-critical equipment, ensuring accurate and up-to-date Process Safety Information (PSI), establishing the Mechanical Integrity (MI) Program structure, establishing Planned Maintenance (PM) programs for inspections and tests, and creating equipment-specific maintenance procedures.

The resultant MI Quality Assurance (QA) Program sustains these combined equipment reliability and operational integrity efforts, ensuring that the equipment is properly designed (via PSI), fabricated as designed (via Management of Change, MOC), and installed as designed (via Pre-Startup Safety Reviews, PSSR). The MI QA Program ensures that the equipment is operated and maintained as designed by trained personnel, as well. The paper concludes with an example of this approach on a process safety-critical piece of equipment. Some benefits include being able to: 1) share and document best MI and PM practices between different sites; 2) focus sometimes limited PM-related resources on the process safety-critical equipment; and 3) help monitor the essential reliability efforts to ensure continued safe operation to meet each group's demands.

1. Introduction

This paper is designed to share an approach combining a site's existing Plant Reliability (PR) programs with its Process Safety Management (PSM) efforts, with a focus on its Mechanical Integrity (MI) program. The background section provides frameworks for the overall PSM system and how current PR and MI efforts can be combined to keep the plant processes running safely and meet its other often-conflicting business risk reduction efforts – including its personnel and environmental safety, quality, customer-focused and financial-profitability goals.

An approach that can be used to combine PR and MI efforts is described next. Examples of the steps that need to be taken and an example MI Quality Assurance (QA) gap analysis tool are provided. The paper concludes with a summary of potential benefits when the MIQA gaps have been identified and addressed.

2. Background

A Process Safety Management (PSM) Model showing 12 key PSM System Elements is shown in Figure 1. These twelve elements essentially capture the 20 key pillars of the PSM Model described by the Center of Chemical Process Safety (CCPS) [1, 2]. A business' PSM Model, whether it has "elements" or "pillars" cannot be designed without accurate and up-to-date Process Safety Information (PSI) – the PSM system's foundation. In addition, the success of any PSM program is predicated on a framework of Information Systems. These systems are used to document and communicate essential process safety information across the organization [3]. And, successful risk reduction interactions within this complex framework are dependent upon strong Operational Discipline and Conduct of Operations [4].

As is shown in Figure 2, the Process Safety Information (PSI) is used and updated by the site within its Process Hazards Analysis (PHA), Management of Change (MOC), Pre-Startup Safety Review (PSSR) and Mechanical Integrity (MI) protocols. This PSI flow chart depicts how information is used to ensure safe and reliable operation while meeting customer demands for receive quality product on time, as well as the business' financial expectations to meet the product as well as the business' financial expectations. Hence, plant reliability – the expectation that equipment will run safely and correctly when needed – is the "life-blood" of a site's performance. As is shown in Figure 2, these MI efforts can be envisioned as the "heart" of the information flow, pumping the site's life-blood required for successful site risk minimization.

The efforts combining a site's Plant Reliability (PR) and Mechanical Integrity (MI) programs can be shown in Figure 3, a MI Quality Assurance (QA) program framework. A successful MI Program includes the design, fabrication, inspection, operation and maintenance of a site's equipment and systems to prevent unexpected shutdowns and adverse safety, health, and environmental (SHE) consequences to the operation of the business. Planned Maintenance (PM) programs include equipment that must be maintained to ensure safe operation, as well as equipment required to ensure quality product and satisfied customers. These PM programs include several approaches, such as run-to-failure of equipment for systems that have little consequence when they fail, inventorying critical spare parts or operating with in-line spares, and implementing both preventive and predictive maintenance programs. The inspection, replacement or refurbish frequencies for these different PM programs are based on the combination of vendor recommended programs, actual site experience and equipment reliability data (such as the Process Equipment Reliability Database, PERD, through the Center for Chemical Process Safety, CCPS).

As is shown in Figure 3, as well, additional PSM Elements are monitored within a site's MIQA Program. These include Training of the operating and maintenance personnel with accurate and up-to-date Operating Procedures (OPs) and equipment-specific Maintenance Operating Procedures (MOP).

However, if a site lacks Operational Discipline (OD), the PM efforts will suffer due to inadequate or sporadic support from all levels of the organization. To help quantify and describe the impact of poor PM Program Operational Discipline, the following simplified risk equation is used with the equipment's reliability risk represented as a function of the equipment's failure frequency, the equipment's failure consequence and the site's safety culture and subsequent adherence to its PM Program (also known as its PM-related OD) [1]:

$$\text{Risk} = \frac{(\text{Frequency})\ (\text{Consequence})}{(\text{Operational Discipline})} \qquad \text{[Eq. 1]}$$

No matter how well designed a site's plant reliability effort is, poor PM program Operational Discipline will result in poor equipment reliability, increasing the downtime risk and potentially increasing the site's overall process safety risk. The managed risk assumes that everyone in the organization performs their PM-related task correctly 100% of the time (i.e., the value of OD equals 1). If, however, individuals perform their PM tasks correctly only half of the time (at 50% OD, or OD equals 0.5), the actual risk designated by Equation 1 may be twice as high as the perceived risk. The authors recognize that the terms in this simple risk equation are not independent. Poor adherence to recommended PMs on engineering controls, for example, will increase the frequency of failure which then increases the overall risk, as well. The effect of poor OD on the perceived risk is shown on the risk matrix in Figure 4. Hence, the better the PM Program Operational Discipline is, the more closely the actual equipment reliability risk matches its managed or perceived risk [5, 6, and 7].

3. An Approach Combining Efforts

The purpose of this paper is to describe an approach that combines the plant reliability and mechanical integrity efforts to ensure reliable and safe equipment operation. The steps a business can take to develop a combined PR/MI program are shown in Figure 5. These steps are based on the Six Sigma DMAIC approach: Define, Measure, Analyze, Improve and Control (Sustain) [8]. The business must establish its overall scope (its framework), discuss and obtain agreement between the individual sites and groups, perform its gap analysis, implement changes as needed, and then sustain (control) its gains. The business must define its terms, such as "mechanical integrity" and establish its risk criteria (its "tolerable" risk) before performing the gap analysis. Once the business framework is established, the sites can then use the framework to identify their critical site-specific equipment and ensure that their PMs are adequate and being followed to help reduce both their reliability- and safety-related risks.

This proposed approach combines existing plant reliability efforts with several PSM elements. The basic steps are: 1) refer to the PHA "tolerable risk" assessments to identify safety-critical equipment; 2) ensure accurate and up-to-date PSI; 3) establish an MI Program structure; 4) establish Planned Maintenance (PM) programs for inspections and tests; and 5) create maintenance-specific procedures. The next step is to establish a MI Quality Assurance (QA) Program to sustain these combined equipment reliability and integrity efforts. An effective MIQA Program "audits" to ensures that the equipment is properly: 1) designed (via the PSI); 2) fabricated as designed (via MOC); and 2) installed as designed (via the PSSR). In addition, the MIQA Program ensures that the equipment is operated and maintained as designed by trained personnel, PMs are performed on schedule, and that equipment that does not meet its inspection criteria is repaired or replaced (via OPs, MOPs, and Training). The MIQA Program is essentially a site's internal auditing system that identifies gaps in its equipment reliability efforts – an overarching view of many of the PSM Elements, as is shown in Figure 3.

A proposed MIQA Program auditing approach uses an "MIQA Gap Analysis" template shown in Figure 6. The "aspects" of a successful MIQA program include a thorough understanding of the PSI, consistent training for the operators, electricians and mechanics with current and accurate OPs, properly designed and implemented the implemented inspection and testing PMs, and an effective QA program for monitoring and tracking issues. The different subjects for each of these aspects are detailed in Figure 7. Once the site has evaluated its program (filling out the audit columns shown in Figure 6), management can determine if the gaps are minor or major and assign appropriate resources and timing to close the gaps.

4. Application and Benefits

As the reader may know, sometimes proposed changes are easily implemented, whereas other times they are not. To date, applying this approach at five different sites across the US, Europe and China has fallen in the latter category, with each site embracing the MIQA auditing process and enhancing the template shown in Figure 7 with additional site-specific items (rows) to address their own local operational reliability requirements and needs. The simple structure and framework of the MIQA Gap Analysis template is built off the strong process-safety risk reduction foundation that has evolved over many years, after the business struggled with its definition of tolerable risk, established its specific tolerable risk matrix, established standards for its Process Safety Information (PSI) and Process Hazards Analyses (PHAs), and created a variety of site-specific safety, health and environmental risk reduction efforts.

After training each site on the MIQA Gap Analysis approach and template, each site completed the analysis on the same piece of critical process equipment and its control system, noting gaps between PSM Elements and actual equipment reliability PM program implementation. One example of such a finding is as follows:

> An interlock was defined by a PHA Team as a safeguard to prevent damaging the critical piece of equipment and causing a loss of containment. Implementing the interlock ensured that the equipment was operated reliability at its tolerable risk level. However, the scheduled PM to test this critical interlock was inadvertently left out of the maintenance work order system, was not performed with all the other interlock tests during the scheduled outage and, hence, was overdue when discovered at the time of the gap analysis. Metrics reporting overdue PMs did not capture the missing PM since it was not in the system. Note that this scenario is shown schematically in Figure 4, with the perceived, tolerable risk location potentially shifting to a higher actual and undesirable risk location. Responses to this finding included ensuring prompt completion of the interlock PM and adding its PM schedule into the work order system. A subsequent system-related fix included updating the PHA and MI procedures to ensure that critical equipment identified as risk reduction safeguards be formally communicated to the PM program manager to ensure entry into the work order scheduling system.

Another benefit for this finding included improving the site's "communication" culture: all groups involved now understood the interlock's important role in helping reduce the specific process safety risk, the technical basis for the interlock, its role for sustaining equipment reliability. The maintenance, operations, engineering and safety groups each understood their important role, the reason for the interlock and why its PM schedule must be adhered to. And, each of these groups continues to meet their own expectations to work without causing safety, health or environmental harm. The results also helped upper management prioritize its resources to address gaps in its process-safety risk reduction efforts.

Future plans (and benefits!) include sharing PMs between sites on common safety- and business-critical process equipment and systems. By leveraging common critical equipment-related process safety risk reduction PM programs, the global process safety-related risk will be reduced by ensuring that the best available technologies are understood and applied consistently across the organization. And, monitoring and sustaining both the MI-related and the essential operational- and reliability-related efforts ensures continued safe operation to meet the often conflicting demands of the operations, maintenance, engineering and safety groups.

5. Conclusion

This paper presents an approach that has been used to successfully integrate current plant reliability efforts with its mechanical integrity efforts. Inherent in the approach is a common understanding of the organization's "acceptable risk," adequate understanding and sharing of critical process safety information (PSI) between the PSM System's Elements (PHA, MOC, PSSR, OP, etc.), and adherence to the Operational Discipline (OD) Characteristics. Closure of identified MIQA gaps reduces a site's overall risk and helps to ensure reliable and safe operation. It is hoped that by combining these reliability and safety efforts, the maintenance, operations, engineering and safety groups will speak the same equipment reliability and integrity language, thus minimizing the overall business risk while meeting the business' personal safety, environmental safety, product quality, customer-focused and financial-profitability goals. Keep in mind, however, all of these efforts will suffer if people from the top to those touching the process materials and equipment do not have strong OD.

7. References

[1] Center for Chemical Process Safety, "Guidelines for Risk Based Process Safety," American Institute of Chemical Engineers, New York, NY (2007).

[2] Sepeda, A. L., Understanding Process Safety Management, *Chemical Engineering Progress*, pp. 26-33, August 2010.

[3] Alloco, M., "Safety Analyses of Complex Systems: Considerations of Software, Firmware, Hardware, Human, and the Environment," Wiley, 2010.

[4] Center for Chemical Process Safety, "Essential Practices for Conduct of Operations and Operating Discipline," American Institute of Chemical Engineers, New York, NY (in press 2010).

[5] Klein, J. A., Operational Discipline in the Workplace, *Process Safety Progress*, 24(4), pp. 228-235, 2005.

[6] Klein, J. A., and B. K. Vaughen, A Revised Model for Operational Discipline, *Process Safety Progress*, 27(1), pp. 58-65, 2008.

[7] B. K. Vaughen, and J. A. Klein, J. A., Improving Operational Discipline to Prevent Loss of Containment Incidents, *Process Safety Progress*, in press, 2010.

[8] G. Liebermann, Six Sigma for Chemical Engineers, Archived AIChE Webinar, July 8, 2009, AIChE ChemE On Demand.

Figure 1

A Framework for Process Safety Management (PSM)

Process Safety Information	PSI	The PSM Element Foundation
Process Hazards Analysis	PHA	Evaluates and reduces risk
Management of Change	MOC	Ensures adequate communication between groups
Pre-Startup Safety Reviews	PSSR	
Mechanical Integrity	MI	Ensures plant reliability
Operating Procedures	OP	Ensures consistency
Safe Work Practices	SWP	Ensures consistency
Training and Performance	T&P	Ensures consistency and competency
Contractors	Contractors	Ensures consistency and competency
Emergency Management Plan	EMP	Ensures adequate response
Incident Investigation	Incidents	Ensures learning and sharing
Auditing	Audits	Ensures gap analyses and improvement
Information Systems	IS	Ensures consistency
Operational Discipline	OD	The PSM System Foundation

PSM System

12 Elements

OD

Figure 2

A PSM System PSI Information Use and Updating Flow Chart

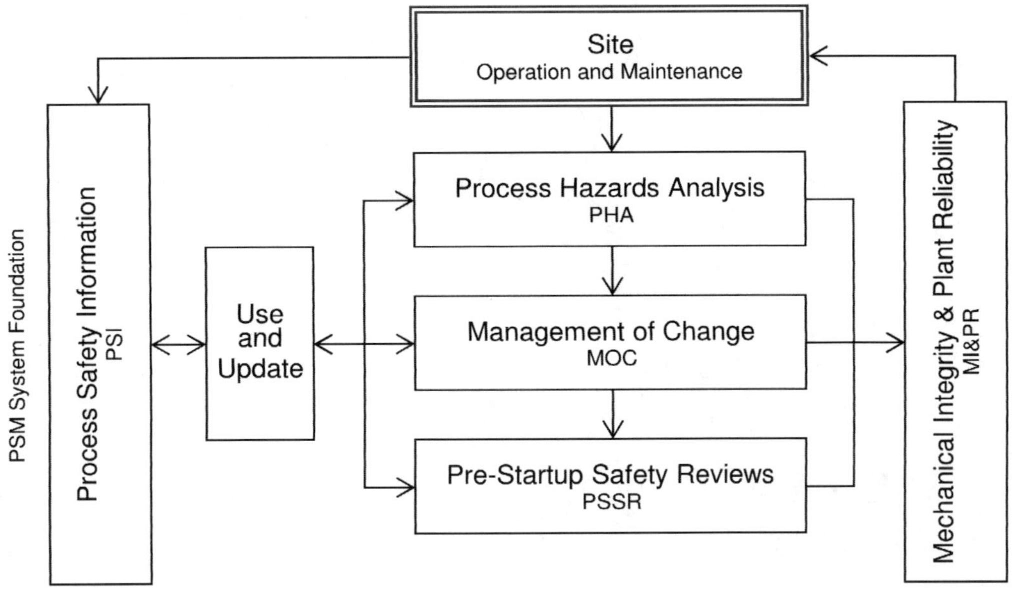

Figure 3

A Framework Combining Plant Reliability and Mechanical Integrity Efforts

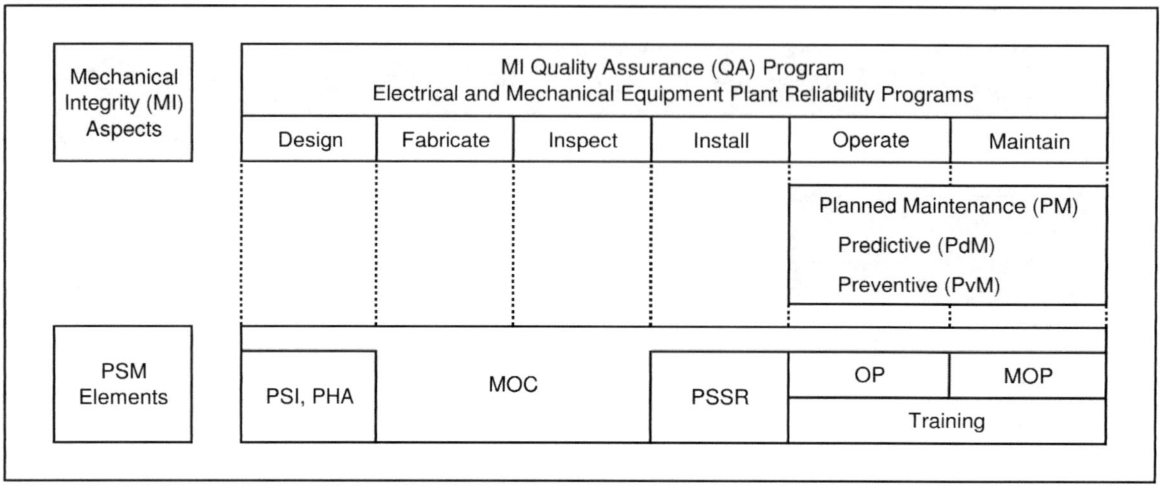

Figure 4

The Effect of Poor Operational Discipline on Process Safety Risk

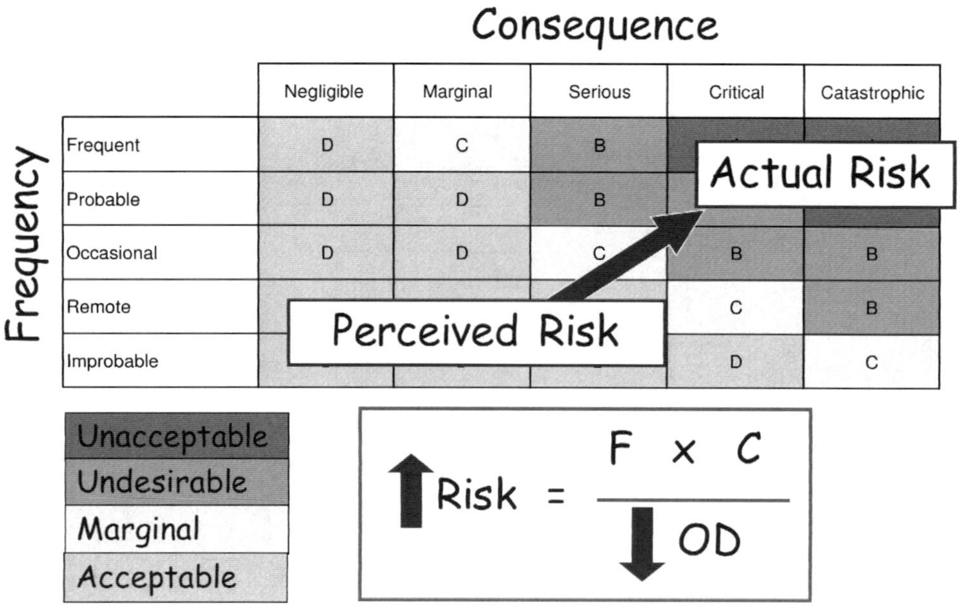

Figure 5

Steps Used to Establish a Mechanical Integrity Quality Assurance (MIQA) Program

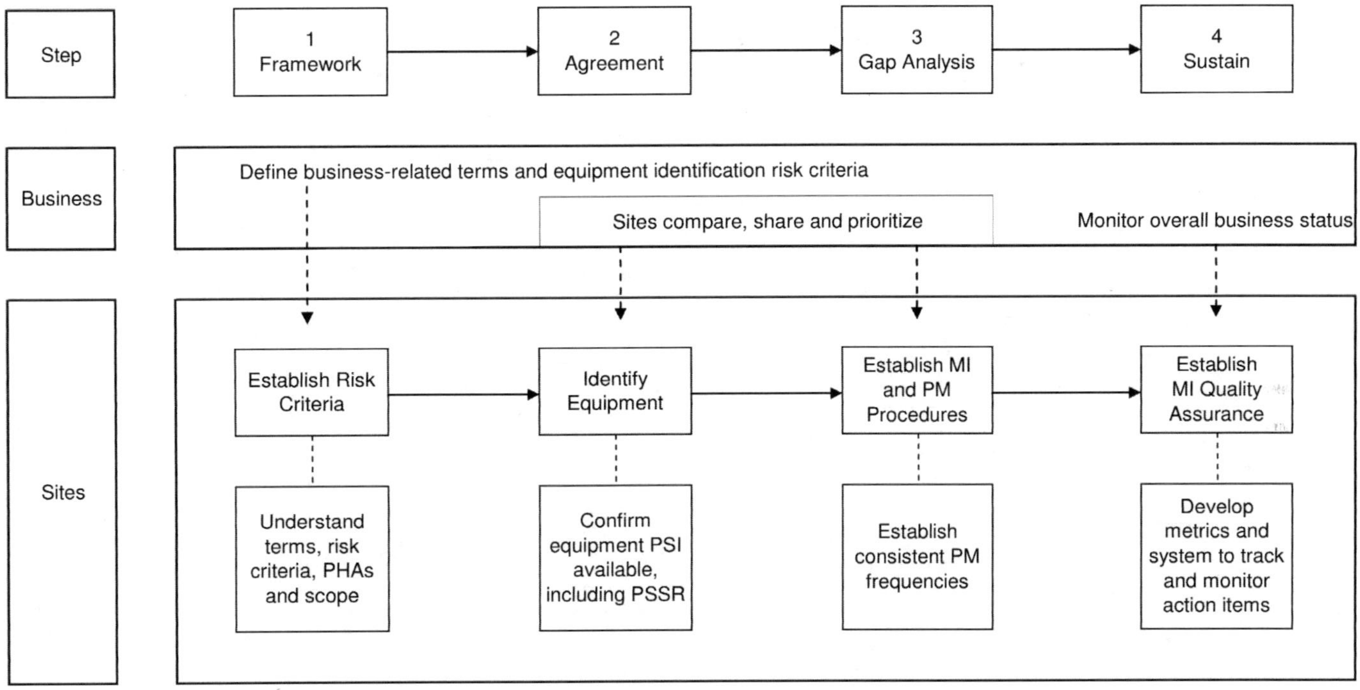

Figure 6

MIQA Gap Analysis Template – Program Aspects and Gap Assessment Columns

	MI Program Aspect	Current Location, Reference or Status	Gap? (No, Minor, or Major)	Proposed Action	Proposed Timing
1	PSI - Process Safety Information				
2	OP - Operating Procedures				
3	PM - Planned Maintenance				
4	QA - Quality Assurance				

Figure 7

MIQA Gap Analysis Template – Program Aspects and Gap Assessment Row Details

		MI Program Aspect	Examples
1		PSI - Process Safety Information	
	1.1	Hazards of Materials	MSDS
	1.2	Process Technology	FMO Process Description
	1.3	Equipment Design Basis	Equipment specifications
2		OP - Operating Procedures	(Consistent with PSI?)
	2.1	Operations (SOP)	Written and current
	2.2	Maintenance (MOP)	Written and current
3		PM - Planned Maintenance	
	3.1	Run-to-Failure (RtF)	Not Acceptable for hazardous materials
	3.2	Spares	If Applicable
		3.2.1	In-line
		3.2.2	Warehouse
	3.3	Preventive Maintenance (PvM)	
		3.3.1	Emergency response systems
		3.3.2	Relief system
		3.3.3	Instruments (Transmitters)
		3.3.4	Interlocks
	3.4	Predictive Maintenance (PdM)	
		3.4.1	Emergency response systems
		3.4.2	Internal Visual Inspection
		3.4.3	External Visual Inspection
		3.4.4	UT Thickness - NDT
		3.4.5	Dye penetrant - NDT
4		QA - Quality Assurance	
	4.1	Design	
	4.2	Fabrication	
	4.3	Installation	
	4.4	Operations Training	
	4.5	Maintenance Training	
	4.6	Inspection and Testing System	
	4.7	Inspection Response Tracking System	

Evaluation of Spring Operated Pressure Relief Valve Maintenance Intervals and Extension of Maintenance Times using a Weibull Analysis with Modified Bayesian Updating

Stephen P. Harris, Ph.D.
Savannah River National Laboratory
Aiken, South Carolina
stephen.harris@srnl.doe.gov

Emily M. Mitchell
Emory University
Decatur, Georgia
emitch8@emory.edu

Robert E. Gross, P.E.
Savannah River Nuclear Solutions
Aiken, South Carolina
robert.gross@srs.gov

Keywords: API RP 581, Weibull, Bayes, Spring Operated Pressure Relief Valves

Abstract

The Savannah River Site (SRS) spring operated pressure relief valve (SORV) maintenance intervals were evaluated using an approach provided by the American Petroleum Institute (API RP 581) for risk-based inspection technology (RBI). In addition, the impact of extending the inspection schedule was evaluated using Monte Carlo Simulation (MCS). The API RP 581 approach is characterized as a Weibull analysis with modified Bayesian updating provided by SRS SORV proof testing experience.

Initial Weibull parameter estimates were updated as per SRS's historical proof test records contained in the Center for Chemical Process Safety (CCPS) Process Equipment Reliability Database (PERD). The API RP 581 methodology was used to estimate the SORV's probability of failing on demand (PFD), and the annual expected risk.

The API RP 581 methodology indicates that the current SRS maintenance plan is conservative. Cost savings may be attained in certain mild service applications that present low PFD and overall risk.

Current practices are reviewed and recommendations are made for extending inspection intervals. The paper gives an illustration of the inspection costs versus the associated risks by using API RP 581 Risk Based Inspection (RBI) Technology. A cost effective maintenance frequency balancing both financial risk and inspection cost is demonstrated.

1. Notation

API	American Petroleum Institute
ASME	American Society of Mechanical Engineers
β	Weibull shape parameter
η	Weibull characteristic life parameter
CCPS	Center for Chemical Process Safety
$F(t)$	The probability that a PRV will fail before it acquires t years of operating time
MCS	Monte Carlo Simulation
MM	Million
MTBF	Mean Time Between Failures
PERD	Process Equipment Reliability Database
PM	Preventative Maintenance
Proof Test	The practice of pressurizing the inlet of a new or used pressure relief valve on a test stand. Popping pressure and seat tightness are tested and the as-found values are compared to the stamped set pressure.
PFD	Probability of failure on demand
PRV	Pressure relief valve(s)
R_p	Ratio of proof test pressure to set pressure
RBI	Risk-Based Inspection
RP	Recommended Practice
SORV	Spring operated pressure relief valve(s)
SP	Set Pressure
SRS	Savannah River Site
TP	Test Pressure

2. Introduction

API 580 [1] introduces principals and presents minimum general guidelines for RBI of pressurized process equipment. The quantitative methods to determine an inspection plan using RBI methodology are provided in API RP 581. As the standard says, the API Risk-Based Inspection (API RBI) methodology can be used to manage the overall risk of a plant by focusing inspection efforts on the process equipment with the highest risk.

For this paper, the U.S. Department of Energy's SRS SORV maintenance intervals were evaluated using an approach provided by API RP 581 [2] for RBI technology. In addition, the impact of extending the inspection schedule was evaluated using Monte Carlo Simulation (MCS). The API RP 581 approach is characterized as a Weibull analysis with modified Bayesian updating using SRS SORV proof testing experience. The approach uses the two-parameter Weibull distribution [3], which is a popular distribution for life data analysis.

Initial Weibull parameter estimates contained in the CCPS PERD were updated using the API RP 581 methodology. The methodology was also used to estimate the SORV's PFD, and the annual expected risk. For this purpose, risk is defined as the product of *Demand Rate*, *Vulnerability* (PFD as a function of time), and *Consequence*. *Demand Rate* is quantified by number of events per year, and is qualitatively the frequency of overpressurization on systems with in-service valves. *Consequence* is due to the on-demand failure of a valve, estimated in terms of US dollar costs due to human injury and restoration of the process to its original state.

Results of this study indicate that the current SRS maintenance plan is risk averse, and that cost savings may be attained in certain mild service applications that present low PFD and overall risk. Herein, current practices and recommendations for extending inspection intervals are reviewed. An illustration is provided that contrasts inspection cost versus the associated risk using API RP 581 RBI technology. Also demonstrated is a cost effective preventative maintenance (PM) frequency that balances financial risk with inspection cost.

3. Background

SORV's used in Newtonian fluid services perform an important safety function in the process industries by preventing system pressures from exceeding the maximum system design pressure. Routine pressure relief valve (PRV) inspection and testing are required because the device may fail to relieve overpressure events that can cause failure of the equipment protected by the device, leading to a loss of containment.

To comply with ASME Boiler and Pressure Vessel code [4, 5] Section VIII for unfired pressure vessels, SORV's are required to be inspected, tested and serviced on a periodic basis. These PM intervals are best optimized by studying past performance and proof test data. If insufficient data exist to provide a decision on optimum intervals, confidence may be improved by shortening the SORV's time in service. That, however, is not cost effective over the life cycle of valves. Additionally, if the primary failure mode is neither corrosion nor high stress in service, shortening time in service provides little valve performance improvement.

Consensus codes allow a company or facility to set and adjust maintenance and inspection intervals for PRV's, including the SORV. As such, methods must be developed internally to ensure that risk reduction and cost effectiveness are balanced while providing an adequate safety margin. The industry utilizes a trio of approaches to conduct inspections based on time-in-service, performance, and risk. The time-in-service method is used primarily at SRS but modified qualitatively by performance and risk. Risk, as previously defined, is difficult to quantify with data.

This paper provides an application of RBI methodology at SRS using the guidelines in API RP 580 [1] and the methodology in API 581 [2].

4. SRS Pressure Protection Program

When a pressure relief valve is removed from service for PM, proof testing correlates the "as found" lift pressure of the valve to its in-service "on demand" lift pressure performance. A valve whose ratio (R_p) of Test Pressure to Set Pressure (TP/SP) greater than or equal to 1.5 in the as-found condition is considered by industry and API 576 to be "stuck shut." This means that during an actual overpressure event, failure of the valve to open by 1.5 times the set pressure will challenge process piping and vessel integrity, causing failure on demand.

The confidence in the current SRS program of proof testing and inspection is considered to be highly effective in selecting parameters from API RP 581. SRS operates a local Valve Repair Shop certified by the National Board of Pressure Vessel Inspectors code NB-23 [6] where maintenance forces perform inspection, testing, and repair of approximately 1,200 SORV each year as the valves are cycled in from the field. A valve is released for field installation if the actual lift pressure (proof test) is within $\pm 3\%$ of SP (for ≥ 70 psig set). If a test is outside of this range, the valve is adjusted and then retested [6].

SRS now has over 7000 certified test records in the Computerized Maintenance Management System including both new and used valve tests. "Used" is defined as a pressure relief valve that was installed and active in a process between inspection times. A summary of used valve data is presented in this paper.

Used valves must first pass an initial external visual inspection before proof testing. If the initial proof test is within 10% SP and the average of the next three tests are within 3% SP, then the valve has passed testing and inspection without requiring further work [6]. Any used valve whose proof testing reveals higher than 1.1 times SP is disassembled for repair. The valve is subsequently reassembled, tested, tagged and returned to the field as "like new."

Guidelines [7] are provided for disposition of each valve based on the proof test results, but the final decision to disassemble and repair is made by National Board certified technicians. Currently, the average PM interval is 3.7 years for used PRV's. Intuitively, longer intervals between maintenance and proof testing should increase the likelihood of a valve being stuck shut on demand. Prediction of a failure on demand as used in API RP 581 is when $R_p \geq 1.3$.

A key assumption of the API RBI 581 methodology is that a bench test of a PRV performed in the as-found condition from a process unit will result in a true determination of the valve's performance on demand. A good inspection program will track the history of inspection and testing of each PRV.

The API RBI 581 method allows for varying degrees of confidence in the inspection. When combined, the pre-test inspection and as-found proof test provide a Pass/Fail evaluation that is given the highest degree of confidence. Conversely, if a valve is inspected and overhauled without a pre-test, a lower confidence level is associated with the validity of inspection results.

The API RP 581 standard provides default values indicating the level of confidence that proof testing will accurately represent in-service PRV performance in an overpressure demand situation. The effectiveness associated with passing the bench test means that there is a certain probability that the valve would have properly opened on demand while in service. By default, a 90% effectiveness carries a 10% probability that the valve would have failed on demand while in service. Accordingly, these probabilities are considered conditional, and reflect the confidence that an inspection result will predict the valve's performance on demand while in service.

5. Inspection Updating using API RP 581

PRV's are classified by service severity as mild, moderate, or severe. Since severe fluid service ASME Section VIII valves are not in use at SRS, the CCPS/PERD database includes proof tests based on mild and moderate Newtonian fluid service ASME Section VIII valves. Included are proof test results of new valves before installation, and over 900 used valves tested between May 2003 and January 2010. A random sample of used in-service valves was incorporated into Inspection Updating with API RP 581, providing estimates of PFD for pressure relief valves in service at SRS. The estimates were adjusted with the collection of proof test data per the API RP 581 methodology. Used valves in mild and moderate fluid services exhibited differences in proof test results that were irresolvable by frequency of failure. The severities were aggregated in given estimates.

The API RP 581 method is related to a Bayesian approach in that it uses a weighted function of current and past proof test results. When applying it to PRV's, the standard application of Bayes' Theorem resulted in PFD values that were unrealistically high. The equations were modified to include inspection effectiveness in order to produce reasonable results.

5.1. Estimating the Probability of Failure On Demand

The two parameter Weibull distribution [3, 8, 9] is used as a model for estimating PFD. In this approach, as proof test data are collected, the Weibull shape parameter β remains constant, and the characteristic life parameter η is adjusted.

The two parameter Weibull distribution for the probability of failure at t years is defined as:

$$F(t) = 1 - \exp\left[-\left(\frac{t}{\eta}\right)^{\beta}\right], \quad t \geq 0 \qquad (1)$$

where $F(t)$ is the probability that a PRV will fail before it acquires t years of operating time.

The probability density function for the Weibull distribution (1) is:

$$f(t) = \left(\beta/\eta\right)\left(t/\eta\right)^{\beta-1} \exp\left[-\left(\frac{t}{\eta}\right)^{\beta}\right], \quad t \geq 0 \qquad (2)$$

η is measured in years, while β is unitless. The probability of failure by η years is 0.632 (63.2%) regardless of the shape parameter (Equation (1) with $t = \eta$). Initial values for β and η may be determined from plant-specific data if available, or starting with default values indicated in API RP 581.

5.2. Reliability

Suspensions are valves that were tested and did not fail. Reliability (R), is defined as the probability at time *t* that failure will not have occurred by that time. Suspensions affect primarily the η parameter (η increases), with generally little effect on β [3]. Because of this, API RP 581 provides for adjustment of the η parameter to increase or decrease PFD according to proof test results.

The reliability at time t is calculated by:

$$R(t) = 1 - F(t) \qquad (3)$$

For example, when $\eta = 13$ years, $\beta = 2.3$ and $t = 6$ years, $F(6) = 0.155$ from equation (1) so there is a 0.155 probability that the PRV will fail in 6 years while the estimated probability that a failure will not occur up to $t = 6$ years in service is $R(6) = 0.845$ from equation (3).

5.3. Mean Time Between Failures

The mean time between failures (MTBF) is calculated for a Weibull distribution as:

$$MTBF = \eta \cdot \Gamma\left[1 + \frac{1}{\beta}\right] \tag{4}$$

where $\Gamma(.)$ is the gamma function. Note that MTBF=η when $\beta = 1$ (the exponential distribution) since $\Gamma[2] = 1$.

The contour plot for MTBF in Figure 1 shows that Eta (η) has a major impact on MTBF, while the impact of Beta (β) is minor over the displayed range for both parameters. This offers justification for holding β constant and updating η as data are collected.

5.4. The Weibull Hazard Function

The Weibull hazard function (5) determines the rate of failure during the characteristic life.

$$h(t) = \frac{f(t)}{R(t)} = \left(\frac{\beta}{\eta}\right)\left(\frac{t}{\eta}\right)^{\beta-1} \tag{5}$$

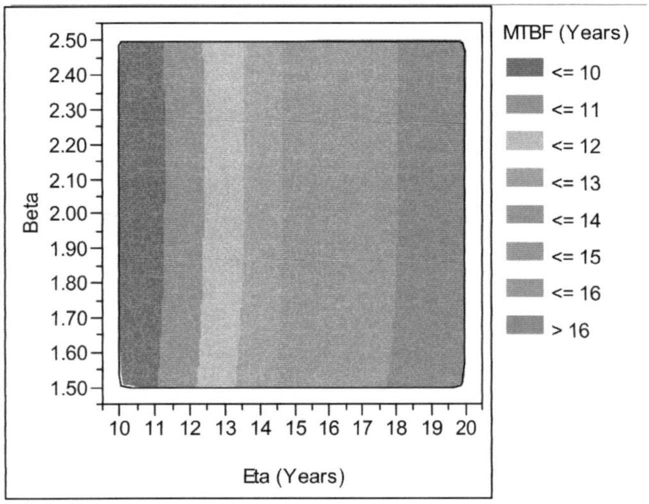

Figure 1: Contour plot of MTBF (years) assuming a two parameter Weibull distribution.

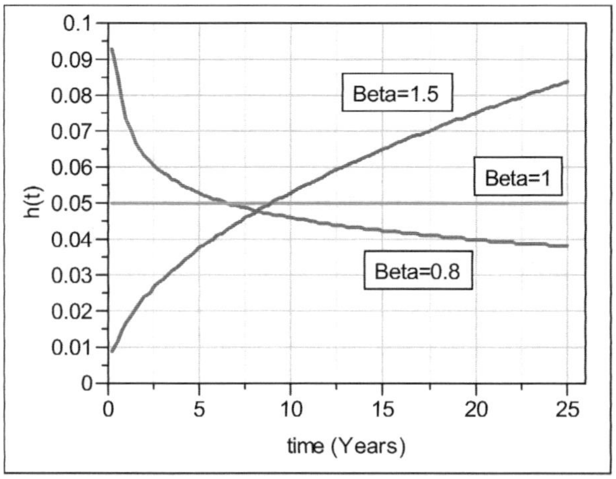

Figure 2: Hazard plot for eta=1.5 years when beta=0.8, 1, and 1.5

5.5. Conditional Probability of Failure

With the hazard plot (Figure 2), we see the following:

a) a decreasing failure rate corresponds to a burn-in period, and occurs when $0 < \beta < 1$.
b) a constant failure rate corresponds to the design life time period, when $\beta = 1$
c) an increasing failure rate corresponds to a wearing out mode, occurring when $\beta > 1$.

Initial estimates of η and β are required to start the updating process. When the PRV passes the inspection, the conditional probability of failure on demand is adjusted using the calculation below:

$$F^{cond}(t) = \left(1 - k_{pass}\right) \cdot R(t) \qquad \textit{(Passed inspection)} \qquad (6)$$

where k_{pass} is the confidence factor for the effectiveness of the proof test, specifically that the proof test determines the performance of the valve under a demand situation. As such, k_{pass} is the probability that functioning on-demand valves will pass the proof test. Typical values for highly effective pass, and for highly effective fail (k_{fail}) range from 90% to 95%. For example, for a highly effective passed inspection at six years $(t = 6)$ with 90% confidence $(k_{pass} = 0.90)$:

$$F^{cond}(6) = \left(1 - k_{pass}\right) \cdot R(6) = \left(1 - 0.90\right) \cdot 0.845 = 0.0845.$$

The $F^{cond}(6) = 0.0845$ will lead to an improvement in the weighted probability of failure (Section 5.6 below) from the previous estimate of $F(6) = 0.155$.

If the PRV fails the inspection, the conditional PFD is calculated as follows:

$$F^{cond}(t) = k_{fail} \cdot F(t) + \left(1 - k_{pass}\right) \cdot R(t) \quad \textit{(Failed inspection)} \quad (7)$$

where k_{fail} is the probability of failing on demand for valves failing the proof test. For example, with a highly effective failed proof test at $t = 6$ years, and $k_{fail} = 0.95$,

$$F^{cond}(6) = k_{fail} \cdot F(6) + \left(1 - k_{pass}\right) \cdot R(6) = 0.95 \times 0.155 + \left(1 - 0.90\right) \times 0.845 = 0.232.$$

Since the valve failed the proof test, the weighted probability of failure (Section 5.6 below) will increase from the previous estimate of $F(6) = 0.155$.

5.6. The Weighted Probability of Failure

API RP 581 weighting factors allow for variations in proof tests, giving more credit to proof tests conducted later in the characteristic life. For a highly, usually, or fairly effective proof test resulting in a pass, the weighted probability of failure can be determined by:

$$F^{wgt}(t) = F(t) - \lambda\left[F(t) - F^{cond}(t)\right]\left(\frac{t}{\eta_{\mathrm{mod}}}\right) \qquad (8)$$

where λ is the selected weighting factor and η_{mod} is the modified characteristic life. In Figure 3, the possible values for λ are 0.1, 0.2, 0.3 and 0.4 for determining the weight given to a current passed proof test. For example, with $t=6$ years, $\lambda=0.2$, and $\eta_{\text{mod}}=13$ years.

$$F^{wgt}(6)=F(6)-0.2\left[F(6)-F^{cond}(6)\right]\left(\frac{6}{13}\right)=0.155-0.2\left[0.155-0.0845\right]\left(\frac{6}{13}\right)=0.149$$

The passed proof test at $t=6$ years results in an improvement in the probability of failing within $t=6$ years, $F^{wgt}(6)=0.149$ compared with $F(6)=0.155$ previously. So passing the proof test lowered the probability of failing at $t=6$ years to 0.149.

For a highly (also usually) effective proof test resulting in a failed proof test

$$F^{wgt}(t)=F^{cond}(t) \tag{9}$$

5.7. Weibull Updating

The Weibull characteristic life is updated as:

$$\eta_{upd}(t)=\frac{t}{\left(-\ln\left[1-F^{wgt}(t)\right]\right)^{\frac{1}{\beta}}} \tag{10}$$

where $\eta_{upd}(t)$ is the updated characteristic life. For example at $t=6$ and $\beta=2.3$,

$$\eta_{upd}(6)=\frac{6}{\left(-\ln\left[1-F^{wgt}(6)\right]\right)^{\frac{1}{\beta}}}=\frac{6}{\left(-\ln\left[1-0.149\right]\right)^{\frac{1}{2.3}}}=13.27 \text{ years}$$

The $\eta_{upd}(6)=13.27$ is an improvement to the initial $\eta(6)=13$ years as a result of the valve passing a highly effective proof test. Recall that for the Weibull distribution, the probability of a valve failing by η years is 0.632 (63.2%).

Figure 3 illustrates the impact of a passed or a failed proof test on updating the estimate of η_{upd}, where the initial $\eta=20$ years, $\beta=1.5$, and $\lambda=0.1, 0.2, 0.3$, and 0.4. The y-axis is the updated characteristic life $\eta_{upd}(t)$.

During the updating process, the following must be true:
1) tests for valves less than one year in service are not included in the updating process
2) the characteristic life cannot decrease after a passed proof test, and
3) the characteristic life cannot increase after a failed proof test.

Figure 3: The impact of a single pass or fail proof test result on the current estimate of eta $(\eta = 20)$ for various weight factors:

$\lambda = 0.1, 0.2, 0.3$ and 0.4. $(\beta = 1.5)$

Note that

1) For passed inspection at 5 years, $\eta \simeq 20$ years (essentially no change)
2) For failed inspection at 5 years $\eta \simeq 13$ years

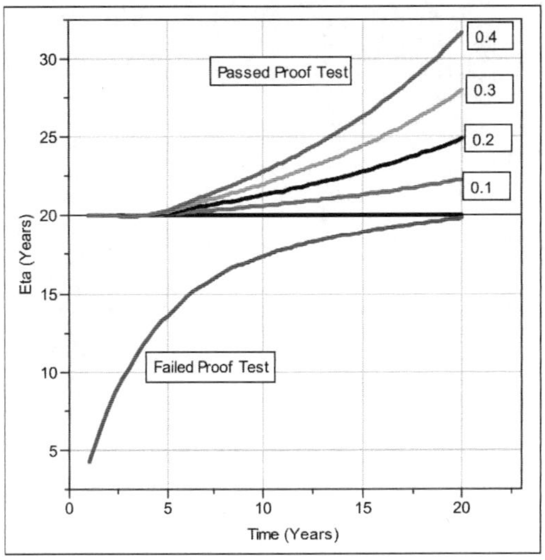

6. Statistical Review of SRS Used Pressure Relief Valve Proof Test Data

A plot of R_p versus time in service (in years) for 935 used valve proof tests does not display any obvious time trends (Figure 4). However, the time trend may not be apparent because the data are plotted according to the maintenance test times instead of failure times. For valves passing the proof test, it is given that they will fail at some future time. Figure 4, which represents approximately 12% of the used valve proof tests, does show that no failures $(R_p \geq 1.3)$ were recorded within the first 2.5 years in service.

The maintenance time distribution for SRS used valves is displayed in Figure 5. Approximately 25% of the SRS valves are tested in 3-year intervals, 50% by 3.1 years, 75% by 4.7 years. Approximately 1.4% of the valves were tested between 10 and 14 years in service.

The plots and statistics in Figure 6 do not show a statistically significant difference for passed and failed proof tests $(R_p \geq 1.3)$ for mean time in service for used valves. In reality, there may be an impact of time but it is hidden as a result of the censored nature of the data

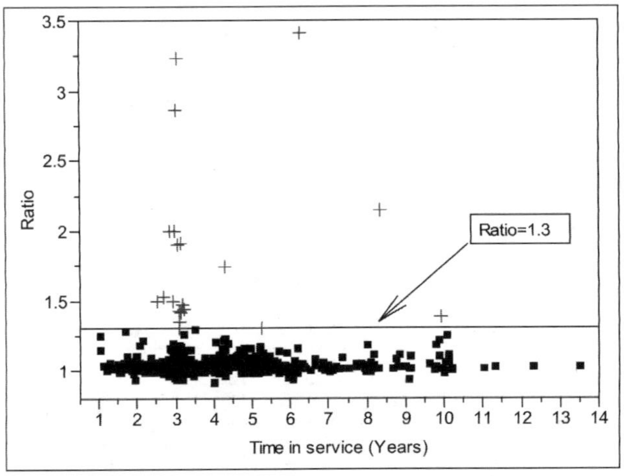

Figure 4: R_P vs. Time in service (years) for the SRS used valve proof tests.

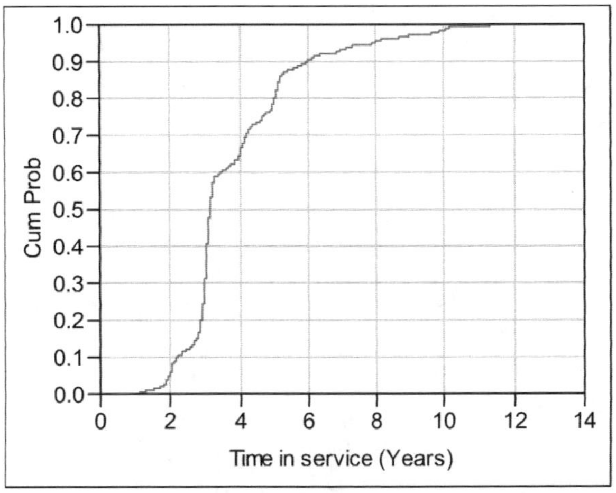

Figure 5: The cumulative maintenance time distribution for time in service for used valves

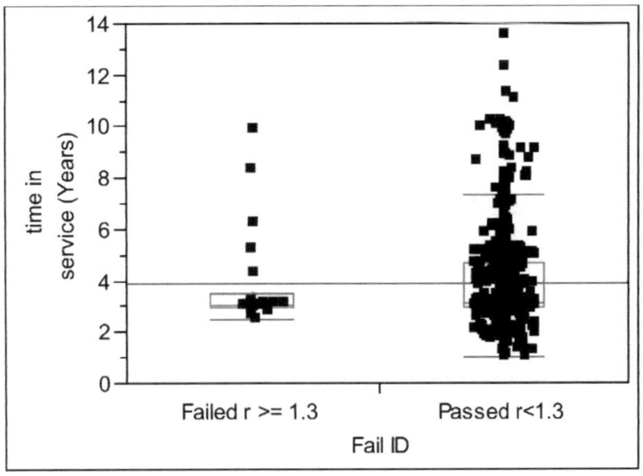

«Figure 6: Box and whisker plots for time in service of failed and passed SRS used valve proof tests

Means and Std Deviations for Time in Service

Level	Number	Mean	Std Dev
Failed $R_p \geq 1.3$	22	3.85	1.92
Passed $R_p < 1.3$	913	3.89	1.77

7. Initial Estimates for the Weibull Parameters

Initial PFD values were obtained based on in-service duration of the valves at the time of inspection, based on the maximum likelihood method [8]. The likelihood function is defined as:

$$L = \prod_{i=1}^{a} f(t_i) \prod_{j=1}^{b} (1 - F(T_j)) \qquad (11)$$

where there are $N = a + b$ proof tests in the database. Of the N valves tested, a have failed their proof test at time $t_i, i = 1,...,a$ based on engineering estimates of their estimated failure times. In addition, b of the valves tested have passed their proof test at time $T_j, j = 1,...,b$. The Weibull distribution function, $F(.)$ (1) is employed, along with $f(.)$, the corresponding density function (2) for failure times.

The usual technique for solving for the distribution parameters involves setting equal to zero the partial derivatives of the natural log of *L* with respect to the parameters, and then solving the resulting equations. For the current application, it is more convenient to use a contour plot for the estimates of η and β in the Weibull distribution.

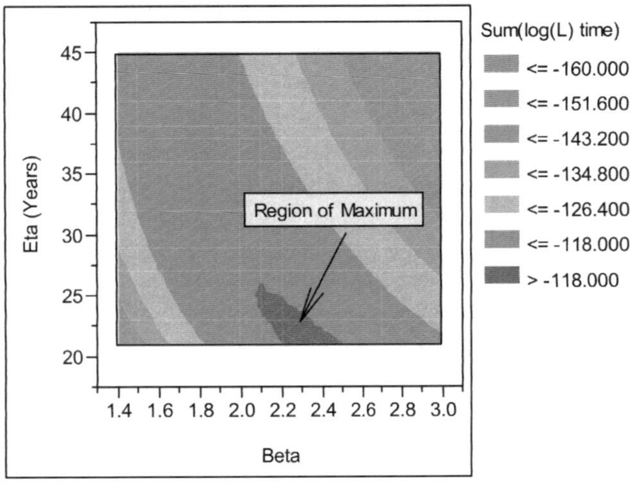

Figure 7: Contour plot of the likelihood function, L, using SRS proof test results.

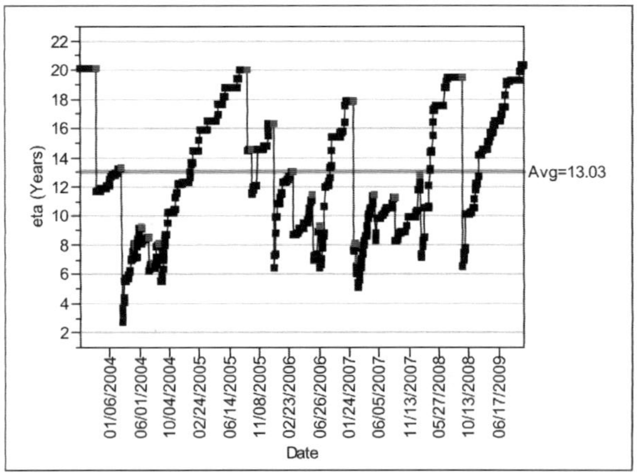

Figure 8: Updating eta based on SRS used valve proof tests

A contour plot of the likelihood function (L) is presented in Figure 7. Initial estimates of η and β were obtained from the maximum region where $\beta = 2.3$ and $\eta = 21$ years.

8. Application of API RP 581 and Monte Carlo Simulation Results

The API RP 581 methodology was applied to SRS used pressure relief valve proof test data with the initial values of $\eta = 20$ years, and $\beta = 2.3$ for the Weibull distribution. As the first failure was realized, the drop-off in eta was dramatic (Figure 8). The plot points color coded red represent valve proof test failures at $R \geq 1.3$. The maximum eta over the time period was approximately 20 years. The eta plot in Figure 8 essentially is a scaled MTBF plot since from equation (4), $\eta \simeq (1.129) \times (MTBF)$ with $\beta = 2.3$.

Four PM plans were evaluated in the MCS: the current site-wide plan, and the current site-wide plan extended by 1, 2 and then 3 years. Each of the plots in Figure 9 contains 250 of 10,000 available data points to show the impact of API RP 581 updating. As the inspection interval is increased, η, or equivalently the MTBF, increases.

API RP 581 updating was seen to be conservative, with $\eta = 21$ years for the MCS—substantially higher than the estimated mean value of eta. Specifically, the mean value for eta for the current site-wide plan was estimated to be 13.5 years, 14.1 years for the current PM plan extended by one year, and 15.0 years for the current plan extended by two years.

The average inspection time is 3.7 years for the current site-wide PM plan. The inspection cost per proof test is approximately $2,700 [10]. A demand frequency of 0.01/year was used in the simulation. The assumed population of in-service pressure relief valves was 5,000. The expected number of demands per year was calculated as $5000 \times 0.01 = 50$. The loss distribution from over-pressurization was 1 MM, 5 MM, and 20 MM with probabilities 0.50, 0.30, and 0.20, respectively.

Annual cost of testing for the current site-wide plan is approximately 3.65 MM per year for 5,000 in-service valves, while the estimated mean risk is 5.14 MM per year (Table 1). Table 1 shows that for an inspection cost decrease of 0.78 MM, risk was increased by 2 MM for the difference between the current PM intervals and a one year increase.

Figure 9: Snapshots of a 10K Monte Carlo series of eta updating for simulated time to failure ($\beta = 2.3$, and η =21 years)

right: the current valve maintenance plan
bottom left: current plan extended 1 year
bottom right: current plan extended 2 years

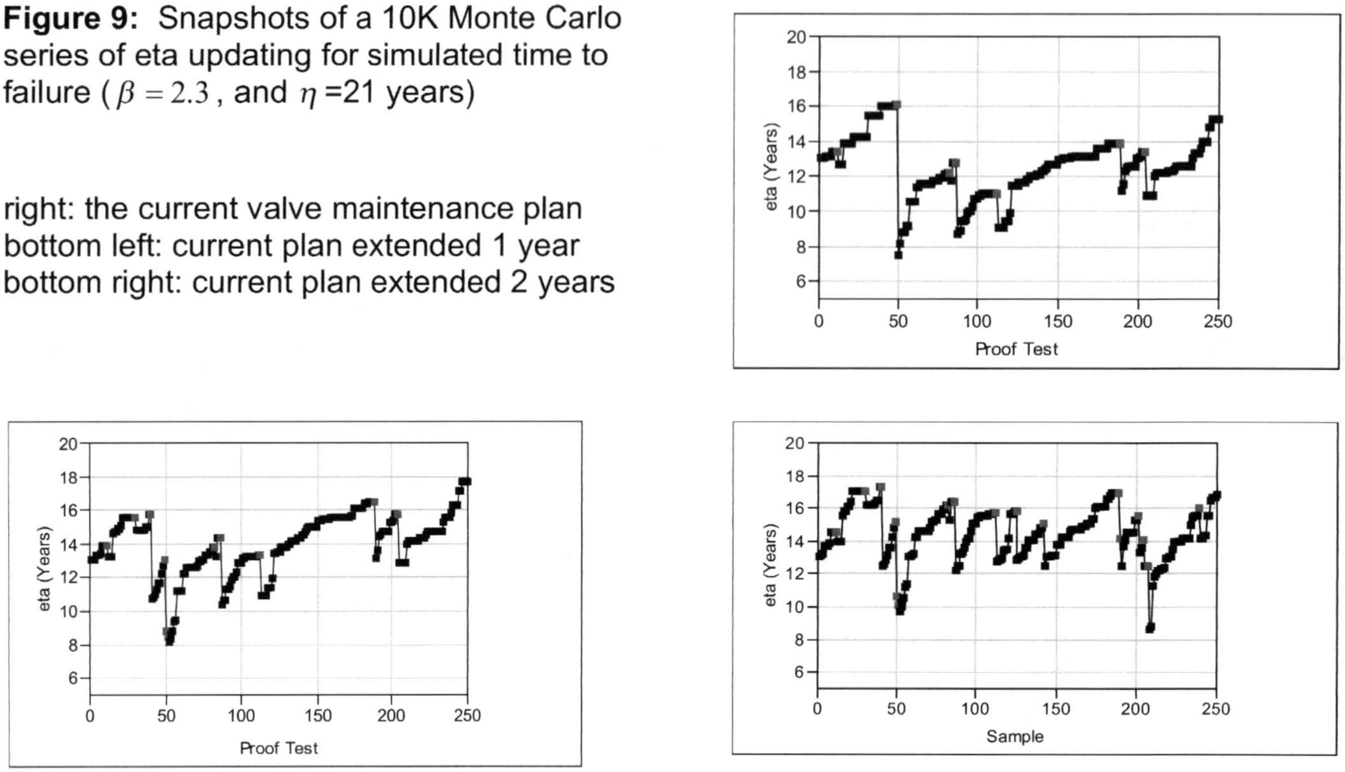

Table 1

Financial Risk of extending Preventative Maintenance Intervals

Maintenance Plan	Mean Maintenance Years	# Demands/ Year	Increase in Risk (MM $)/ Year	Total Risk (MM $)/ Year	Inspection Cost (MM $)/ Year	Total Cost (MM $)/ Year
Baseline	3.7	50	n/a	5.14	3.65	8.79
Baseline Extended 1 Year	4.7	50	2.06	7.20	2.87	10.08
Baseline Extended 2 Year	5.7	50	3.03	8.17	2.37	10.54
Baseline Extended 3 Year	6.7	50	3.41	8.55	2.01	10.57

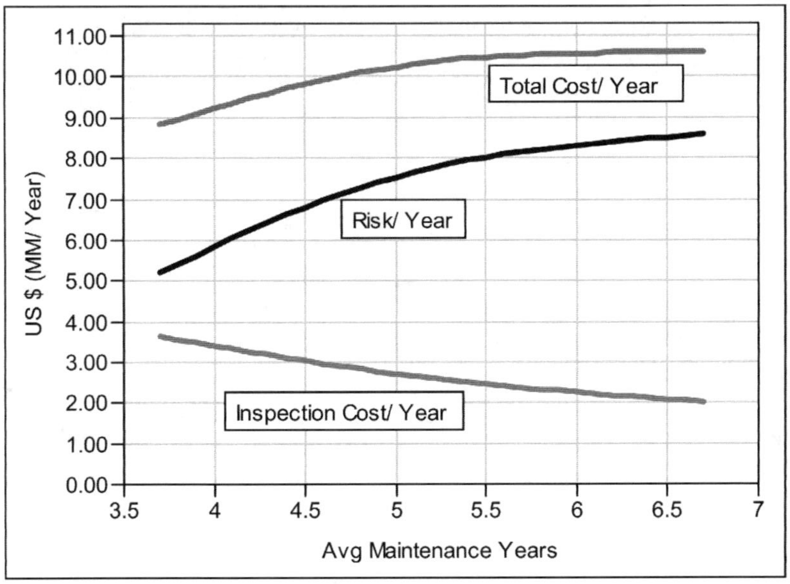

Figure 10: Risk, Inspection Cost, and Total Cost of inspection for extending maintenance intervals

A risk target is defined as the level of acceptable risk for inspection planning purposes and can be developed based on internal guidelines for risk tolerance. The level of acceptable risk can vary between companies. At SRS, current PM intervals can be extended by half a year and still fall within a target risk of $6.2 MM (Figure 10). The total cost (Risk + Inspection) increases $1.3 MM for extending the PM intervals up to one year. In implementation of these results, cost savings may be attained in certain mild service applications that present low PFD and overall Risk.

9. Conclusions

The API RP 581 methodology provides a nationally recognized process to analyze test data that not only considers inspection performance, but also includes the estimated demand rate and probability of failure on demand in the risk-based decision process. The methodology could be used at SRS to validate SORV PM intervals. Currently at SRS, a time-based approach is employed, and a risk-based approach is used for guidance in adjusting the intervals. Ideally, only zero risk is acceptable, however non-zero risks are a reality and must be minimized in production facilities. The API RP 581 method offers optimization of risk to inspection cost when making operating decisions.

For the past five years, the SRS Pressure Protection Program has endorsed individually-justified increases in PRV maintenance intervals based on an accumulation of passed/failed proof tests. Overall, the current program at SRS appears to conservatively reduce the risk of a pressure relief valve's failure on demand. Moreover, there is only a marginal difference between the reduction in the cost of testing relative to the increased risk in extending current PM intervals up one year. Intervals in specific cases—especially mild service—could be extended by one to two years and still fall below the target risk level.

The next step at SRS is to identify sub-groups of valves, like nitrogen service valves with set pressure in the 300-1,000 psig range, for applying API RP 581 with greater precision. Refinements may be made in characterizing the demand rate and the consequences ($ loss) of an actual failure on demand while in service.

10. References

[1] API RP 580 Recommended Practice for Risk-Based Inspection, American Petroleum Institute, Washington, D.C, November 2009.

[2] API RP 581 Risk-Based Inspection Technology, Section 7 Pressured Relief Devices, American Petroleum Institute (API) Recommended Practice 581, second edition, September 2008.

[3] Abernethy, R.B. (2004). “The New Weibull Handbook”, Robert B. Abernethy, North Palm Beach, Fl 33408.

[4] ASME B31.3 Pressure Piping Codes, American Society of Mechanical Engineers, New York, NY.

[5] ASME Boiler and Pressure Vessel Code (B&PV), American Society of Mechanical Engineers, New York, NY.

[6] ANSI-NB-23, National Board Inspection Code, The National Board of Boiler and Pressure Vessel Inspectors, Columbus, OH.

[7] Pressure Equipment Protection Requirements, SRS Engineering Standard 15061, Rev 6, 08/01/2009, Pressure Protection Technical Committee.

[8] Nelson, W.B. (1982), "Applied Life Data Analysis", New York, John Wiley & Sons.

[9] Lawless, J.F. (1982). “Statistical Models and Methods for Lifetime Data,” Wiley, NY.

[10] Gross, Robert E., Harris, S. P., (2008) “Analysis of Safety Relief Valve Proof Test Data to Optimize Lifecycle Maintenance Costs,” Proceedings of the 2008 Annual Reliability and Maintainability Symposium (RAMS), Las Vegas, NV. January 28-31, 2008.

Acknowledgements

Research for this paper was performed by Emily M. Mitchell while on an internship in Computational Sciences at the Savannah River National Laboratory during 2010. Ms. Mitchell received her B.A. degree in mathematics and religious studies from the University of South Carolina and is currently pursuing a PhD in Biostatistics at Emory University.

Can Functional Tests Be Replaced By Inspection After Demands?

Hui Jin
Department of Quality and Production Engineering
Norwegian University of Science and Technology
Trondheim, NO-7491, Norway
hui.jin@ntnu.no

Mary Ann Lundteigen, Professor
Department of Quality and Production Engineering
Norwegian University of Science and Technology
mary.a.lundteigen@ntnu.no

Marvin Rausand, Professor
Department of Quality and Production Engineering
Norwegian University of Science and Technology
marvin.rausand@ntnu.no

Keywords: Medium demand, Response to demands, PFD, Test coverage, Inspection

Abstract

Safety instrumented systems are used to reduce the risk of accidents in the process industry. To claim a certain risk reduction from such systems, it is necessary to perform reliability analyses, based on internationally accepted frameworks, such as IEC 61508 and IEC 61511. The main factors that influence the reliability are the failure rates of the components, the level of redundancy of hardware and software, and the frequency of periodic functional tests. The need for periodic functional testing has always been a challenge for the industry, as the tests often mandate a shorter or longer stop of the production. For those safety instrumented systems that must respond to rather frequent demands, the industry has started to ask if it is possible to replace some functional tests with thorough inspections after demands, and how information from responses to demands and the following inspections may be credited in the reliability analyses. This paper discusses the main characteristics of such a "testing strategy", and highlights some of the cautions, challenges, and conditions of use.

1. Introduction

Safety instrumented systems (SISs) are widely used as protection layers in the process industry, and a number of standards and guidelines have been published to assist the design, implementation, and follow-up of such systems. One of the most important standards is IEC 61508 [1], which outlines the key requirements in all phases of the SIS life-cycle. IEC 61508 is the basis for sector specific standards, such as IEC 61511, which applies to the process industry [2]. Both IEC 61508 and IEC 61511 have been adopted in many countries, for land-based process industry as well as for the offshore oil and gas industry. In the Unites States, IEC 61511 has been adopted with minor amendments and is referred to as ANSI/ISA 84.00.01-2004 [3].

IEC 61508 differentiates between two modes of operation, *low-demand* and *high/continuous-demand* mode of operation. In the recent edition of IEC 61508, different definitions are given to high-demand and continuous-demand mode, but throughout the rest of the standard, they are treated the same [1]. In IEC 61511, the corresponding modes are called demand mode and continuous mode of operation, but with the same interpretation as in IEC 61508 [2]. According to IEC 61508, a SIS is operated in a high/continuous-demand mode when the demand rate is greater than once per year (the test interval related criterion is removed in the recent edition of IEC 61508); otherwise it is operated in a low-demand mode [1].

The process industry employs the principle of having several independent layers of protection, in addition to the basic control system. Most SISs are therefore low-demand rather than high/continuous-demand and remain passive as long as the process is running normally. At the same time, the process industry has identified several other functions as safety-critical, and given requirements for their reliability in accordance with IEC 61508 and IEC 61511, even if the functions are needed for normal control as well as in response to hazardous events. Equipment used for well intervention and drilling operations, such as the blowout preventer (BOP) is, for example, used both during normal drilling operations and for isolation of the well in case of uncontrolled flow from the well. Such equipment is therefore operated in the high-demand mode, even if only a fraction of the demands are critical.

Recently, some manufacturers and oil companies have raised the question whether successful responses to demands, such as the non-critical demands in relation to well intervention equipment, may be credited as functional tests in reliability assessments of the safety functions [4]. So far, however, this issue has been given little attention and methods are therefore lacking. The aim of this paper is to clarify this issue and also to suggest ways to assess the reliability. More specifically, the objectives are to (i) discuss the possibility of taking credit from responses to demands in SIS reliability assessment for some specific situations, (ii) present PFD calculation formulas when responses to demands are considered, (iii) present ways to follow-up the SIS reliability when the credits from responses to demands are taken.

2. Reliability measures

A *demand* may be defined as an event or a condition that requires response by a safety related function that is performed by either a SIS or a non-SIS system. A system that uses the *same* type of components to perform safety as well as non-safety functions is exposed to a combination of (safety) critical demands and non-critical demands. In total, the demand rate may correspond to the high-demand mode, whereas the rate of the critical demands may fall in the low-demand mode. In order to avoid possible confusion and to cover systems responding to critical and non-critical demands, the expression, *safety related system* (SRS), is used in this paper. An SRS is further split into three subsystems [5]: (i) the input subsystem (e.g., sensors, transmitters), (ii) the logic solver subsystem (e.g., programmable logic controllers [PLC], relay logic systems), and (iii) the actuating subsystem (e.g., safety valves, circuit breakers). A subsystem may have one or more components. In the case of more than one component in a subsystem, the components usually work as a redundant configuration.

IEC 61508 and IEC 61511 suggest two different reliability measures for the two modes of operation: The average *probability of failure on demand* (PFD) for the low-demand mode and the average *probability of a dangerous failure per hour* (PFH) for the high-demand/continuous mode. IEC 61508 and IEC 61511 distinguish between four safety integrity levels (SILs), to each SIL, a certain PFD or PFH requirement range applies (IEC 61511 refers to IEC 61508 for the SIL 4 requirements).

When a demand for the SRS safety function occurs, the PFD is the average probability that the SRS fails to perform its safety function upon a demand. Depending on the placement of the SRS in the sequence of protection layers, the failure will have two different consequences: (i) if the SRS is the last protection layer, the failure will lead to an accident in the system. While (ii) if it is not the last protection layer, the failure will lead to a demand for the next layer. PFH is the frequency of demands for the next protection layer due to a SRS failure, or the accident frequency if the SRS is the ultimate protection layer.

The required PFD, corresponding to the required amount of risk reduction, may be determined from a hazard and risk analysis, supported by methods such as LOPA and risk graph, or by a simplified approach using, for example, the minimal SIL requirements suggested in the Norwegian guideline OLF 070 on the application of IEC 61508 and IEC 61511[6]. The required PFH is equal to or less than the acceptable frequency of a certain type of event or accident.

It is then up to the SRS manufacturer and system operator to compare the estimated reliability of each safety function with the required PFD or PFH. After the SRS has been installed, it is the responsibility of the end user to collect data, to make new reliability predictions, and make necessary improvements and adjustments to functional testing frequencies, to maintain the required PFD or PFH throughout the operational phase.

3. Classification of demands

Three categories of SRS modes of operation, instead of two as in IEC 61508, are proposed in this paper, see Table 1. In addition to low- and high/continuous-demand, a so-called *medium-demand* mode is introduced.

Table 1. IEC SIS category and proposed SRS category

IEC 61508 categorization	Low demand (<=once per year)	High/continuous demand (>once per year)	
Measure	PFD	PFH	
Proposed categorization	Low demand	Medium	High/continuous demand
Measure	PFD	PFD	PFH

The borderlines between medium-demand, low-demand, and high/continuous-demand are not explicitly defined, and will be influenced by application specific considerations. For example, in the subsea oil and gas industry, it may be reasonable to suggest the medium-demand mode as the demand rate in the range of once per year to once or twice per month. The PFD is suggested as the reliability measure for an SRS in the medium-demand mode. In this demand mode, the PFH as a reliability measure will be too conservative [7].

The demands for the SRS can be (safety) critical or non-critical. This means that the total demand rate is the sum of critical demand rate and non-critical demand rate.

$$\lambda_{\mathrm{d}} = \lambda_{\mathrm{d}}^{\mathrm{crit}} + \lambda_{\mathrm{d}}^{\mathrm{non-crit}} \quad \text{[Eq.1]}$$

where λ_{d} is the total demand rate, $\lambda_{\mathrm{d}}^{\mathrm{crit}}$ is the critical demand rate, and $\lambda_{\mathrm{d}}^{\mathrm{non-crit}}$ is the non-critical demand rate.

For an SRS operating in the medium-demand mode, different combinations of critical demand rate and non-critical demand rate may be distinguished:

1. Critical demands are dominating; non-critical demands are rare
2. Non-critical demands are dominating; critical demands are rare
3. The rates of critical demands and non-critical demands are comparable

This paper focuses on combination 2, where the total demand rate is in the medium demand mode range and the critical demand rate is in the low demand mode range. One example is the pipe RAM preventer in a BOP system that is used to hang-off drill pipes during drilling and to stop flow is case of an uncontrolled flow. Since the critical demands, that matters with respect to safety, are in the low-demand mode, the PFD is used as reliability measure.

SRSs operating in the medium-demand mode with combination 1 or combination 3 can been studied with a similar approach, but this is not done in this paper.

4. Demands as test opportunity

An SRS may suffer from both *dangerous detected* (DD) failures and *dangerous undetected* (DU) failures. Periodic functional testing is used to reveal DU-failures, while diagnostic testing is used to reveal DD-failures. In the medium-demand mode, we assume that, upon detection of a DD-failure, the *equipment under control* (EUC) [1] is taken to a safe state immediately. The contribution from DD-failures to the PFD is therefore negligible compared to DU-failures. Thus only DU-failures are considered in the paper.

Functional tests may improve the SRS reliability in two ways. First, a functional test with positive result (no failure) confirms the functionality of the SRS at the time of testing. Second, a functional test with negative result (with failure) reveals the unavailability of the SRS (components) before the next demand. When it is followed by a repair action, the reliability of the SRS is improved.

For SRSs operating in a low-demand mode, the benefit of functional tests is easily justified given the possibility of revealing a DU-failure before the next demand. In the high/continuous-demand mode, however, it may be impossible to perform regular functional testing at the frequency that is adequate to prevent an accident. However, it is still important to note that functional testing is needed to reveal DU-failures in redundant configurations. In the medium-demand mode, where the SRS responds to frequent, but non-critical demands, it is reasonable to compare the credit from this type of "random'' testing with the periodic functional tests.

A challenge with functional testing is its interference with the production. Operation of actuating devices often leads to production stops and may imply significant costs. The industry is therefore seeking for means to make functional testing more efficient. Exploiting the information gained from responses to demands may be one of the means to complement the functional testing strategy.

The response to a non-critical demand that is followed up by a thorough inspection may have a similar effect on the SRS reliability as a functional test. A successful response to a demand resembles a functional test with a positive result, and confirms the functionality. A failed response to a demand resembles a functional test with negative result. It reveals the DU-failure, and relevant repair actions may be initiated, but sometimes with additional cost.

4.1 Differences in functional test and demand characteristics

Even with the similarity between periodic functional testing and responses to (non-critical) demands (followed up by inspections), there are some significant differences that need to be considered in SRS reliability assessment.

Costs
A functional test is pre-scheduled, and the necessary equipment and resources are allocated in advance. The costs in relation to functional tests are mainly dependent on the loss of production, the resources needed to perform the test and necessary repair. The costs of the inspection after non-critical demands depend on the same factors except for most of the production loss is inevitable regardless of if an inspection is performed or not. In addition, there may be extra cost to mobilize the personnel and relevant equipment due to that the inspections cannot be planned. The total costs of inspection may increase due to the more frequent execution of non-critical demands (compared to functional testing). It may therefore be necessary to build SRSs that can automatically verify the status of all components after an activation of a system functional.

Note: Costs related to critical demands may be significant, and depend on the degree of damage caused by an SRS failure. For example, a failure to shut down upon high separator pressure will not lead to any large damage, if secondary means of protection, such as opening of pressure relief valves, is successfully achieved (assuming adequate flare capacity). For comparison, a failure of the blowout preventer (BOP) may result in fatality as well as environmental damage, such as the Deepwater Horizon accident in 2010 in the Gulf of Mexico, and the resulting costs are huge. The cost issue in relation to critical demands is, along with the safety issue, one of the reasons why taking credit from critical demands should be much more careful than taking credit from non-critical demands in SRS reliability analysis. The credit from critical demand is not considered in this paper.

Randomness

Unlike functional tests, the demands are usually random events. Non-critical demands for the pipe RAM preventer for mud control during drilling operations is one such example. The occurrence of demands may be modeled as a stochastic process. The characteristic of the process depends on the situation, but with no particular trend in the occurrence of demands, a homogeneous Poisson process (HPP) may be an adequate model.

Note: Demands may in some cases be nearly deterministic events. The pipe RAM in the BOP may also be used to hang-off drill pipes regularly. This type of event occurs at rather regular points in time. For modeling of demands, one must consider if both random and deterministic (non-critical) demands are present and if they reveal the same failure modes.

Pro-activeness

A functional test is a proactive action in terms of risk reduction. Responses to non-critical demands will have a pro-active effect as well, i.e., that the DU-failures are revealed before the next critical demand.

A response to a critical demand is, on the other hand, not a "pro-active" action in terms of risk reduction, even if the successful response may prevent a demand for the next protection layer.

"Test coverage"

For the purpose of reliability analyses, it is often assumed that a functional test has 100 % coverage, meaning that all DU-failures are revealed during the test. The realism of this assumption is sometimes debated, and some methods have introduced new parameters that account for the type of DU-failures that are not revealed by the test [8]. To achieve 100 % test coverage would require that the functional tests are performed under critical demand conditions. This is, of course, not what we would like to do in practice, for example to release hydrocarbon gas into an area to verify the performance of the gas detectors.

The "test coverage" in relation to non-critical demands may vary depending on (i) the failure modes that are covered in the activation upon non-critical demands, and (ii) whether the states of all components are verified (either by the activation or the subsequent inspection) in case of redundancy. The former issue may be exemplified by the BOP example: When the BOP pipe RAM preventer is used for hang-off of a drill pipe, it is not able to reveal the failure mode "leakage through the RAM''. The latter issue may be exemplified by a set of gas detectors working as a 2-out-of-3 (2oo3) logic. When the detectors successfully signal a gas leakage, it is not possible to known whether all three detectors detected the leakage, or only two of them, without doing any additional effort.

4.2 Taking characteristics into account in the PFD

In the PFD calculation, we usually account for the following factors (see also Figure 1): DU-failure rates, system architecture, functional test interval, and the fraction of DU-failures that are common cause failures (CCFs). DD-failures, repair rate, and test time may also be added if they are expected to influence the PFD.

In this paper, we also include the factors that may be important when non-critical demands are included in the PFD calculation. Based on the discussions in Section 4.1, these factors are:

- Non-critical demand rate
- Coverage of failure modes
- Inspections strategy

The coverage of the failure modes, i.e., the fraction of failure modes that may be revealed during a response to a demand, the inspection strategy, together with the system architecture determine what is in Section 4 referred to as the "test coverage".

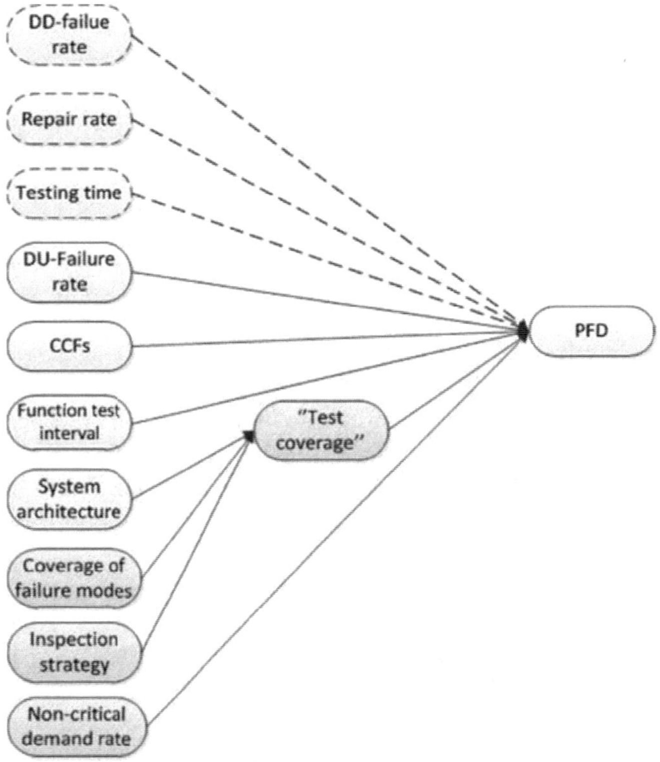

Figure 1. Influencing factors of PFD

5. Reliability modeling of demands as test opportunity

The PFD of the whole SRS can be obtained from the PFD of the subsystems. To calculate the PFD of a subsystem based on the factors in Figure 1 requires some additional detailing of the underlying assumptions. In this paper, these may be summarized as follows:

1. Non-critical demands occur according to an HPP with intensity (demand rate), $\lambda_{\mathrm{d}}^{\mathrm{non-crit}}$, within the medium-demand mode range. The impact on the SRS reliability from critical demands is assumed to be negligible compared to non-critical demands, since non-critical demands are dominating as assumed in Section 3.

2. All components have identical and constant failure rate, but it may be relevant to use the approach in ISO 13849 to account for the additional stress from frequent operations.
3. Functional testing is performed with frequency f_{FT}, where $f_{\mathrm{FT}} \ll \lambda_d$.
4. Each response to a non-critical demand (successful or failed) is followed by a thorough inspection, where the main purpose is to verify the state of all redundant components.
5. Upon detection of a failure, the EUC is taken to a safe state and the failed components are repaired immediately (This assumption means that the active repair time can be disregarded in the PFD quantification).
6. The inspection time and the following time to restore the system are negligible (might not be realistic for some systems).
7. The components are assumed to be independent, i.e., no possibility of CCFs.

Since the non-critical demands may not cover all failure modes, we introduce a ratio θ that

$$\theta = \frac{\lambda_{\mathrm{DU}}^{\mathrm{d}}}{\lambda_{\mathrm{DU}}} \quad \text{[Eq.2]}$$

where $\lambda_{\mathrm{DU}}^{\mathrm{d}}$ is the rate of failures with modes that can be revealed by non-critical demands, and λ_{DU} is the total component DU-failure rate. When all failure modes can be revealed by non-critical demands, then $\theta = 1$.

Taking credit from responses to non-critical demands is similar to using *partial stoke testing* (PST) to verify the state of valves. According to [9], the total PFD can be calculated as the sum of the PFD contribution from PST and the PFD contribution from functional testing. Therefore, a similar approach is used for response to non-critical demands. The PFD of a subsystem is approximately the sum of the PFD contribution from the responses to demands ($\mathrm{PFD_d}$) and the average PFD contribution from functional testing ($\mathrm{PFD_{FT}}$).

$$\mathrm{PFD} \approx \mathrm{PFD_d} + \mathrm{PFD_{FT}} \quad \text{[Eq.3]}$$

In order to calculate $\mathrm{PFD_d}$, two scenarios are differentiated according to whether or not the states of every component can be verified by the response to demands with respect to the failure modes that are covered by the demands.

1. **Scenario 1**: The states of every component in the subsystem can be verified by the responses to demands and the following inspections. The situation in this scenario may be that (i) the subsystem is functioning as a NooN logic or (ii) the inspection after demands can verify the state of every component.

2. **Scenario 2**: The states of all components in the subsystem cannot be verified by the responses to demands and the following inspections. The situation in this scenario may be that the subsystem is function as a MooN logic, where $\mathrm{M} < N$ and the inspection after demands cannot verify the state of every component.

For scenario 2, the voting logic of the subsystem may be changed after a demand due to the degraded operation, but this is not further investigated in this paper. We only consider scenario 1 where the $\mathrm{PFD_d}$ can be calculated as, (see also Figure 3):

$$\mathrm{PFD_d} = \int_0^{\infty} \mathrm{f}\,(\mathrm{t}) \frac{1}{t} \int_0^{t} (1 - \mathrm{R(u)})\mathrm{du}\,\mathrm{dt} \quad \text{[Eq.4]}$$

where $f(t)$ is the probability density function of the time to a non-critical demand, $R(u)$ is the reliability function of the subsystem in question, and t is the time to a non-critical demand.

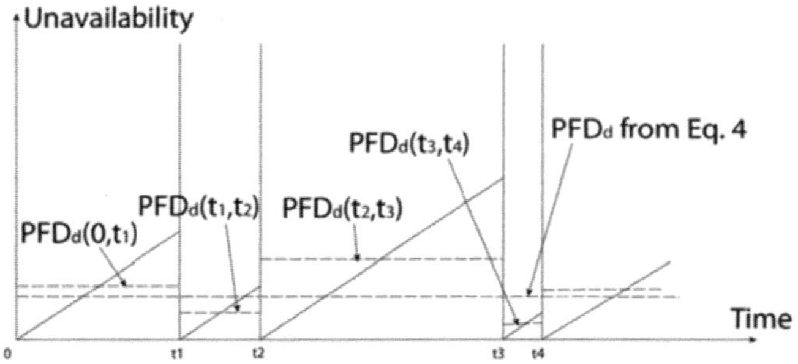

Figure 3. $\mathrm{PFD_d}$ calculated from Eq.4

The PFD can also be approximated as the average unavailability each time the SRS is demanded. This is equal to the expected $\mathrm{PFD_d}$ when a non-critical demand occurs, see Figure 4. If, for example, demands occur at times t_1, t_2 and t_3 with probabilities P_1, P_2 and P_3, respectively, the $\mathrm{PFD_d} = \sum_{i=1}^{3} \mathrm{PFD_d}(t_i)P_i$. To include all the possible times a demand may occur, the $\mathrm{PFD_d}$ is given in Eq.5.

$$\mathrm{PFD_d} = \int_0^{\infty} \mathrm{f}\,(\mathrm{t})(1 - \mathrm{R(t)})\,\mathrm{dt} \quad \text{[Eq.5]}$$

where the notation is the same as in Eq.4.

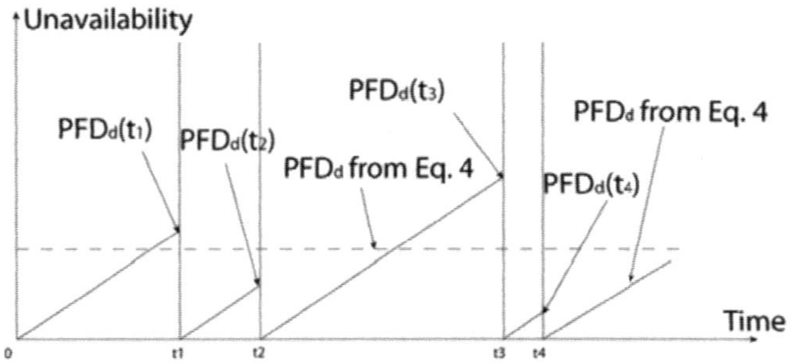

Figure 4. $\mathrm{PFD_d}$ calculated from Eq.5

Comparing the results from Eq.4 and Eq.5, it is seen that Eq.5 gives a more conservative result. The uncertainty of the $\mathrm{PFD_d}$ calculated by the approach in this paper may be higher than for the conventional approach due to the randomness of the "tests" and the "test coverage" of the demands. To compensate for this increased uncertainty, we recommend using Eq. 5 rather than Eq. 4.

5.1 Single component subsystem

For a subsystem with a single component, the average PFD of the subsystem can, according to Eq. 3 and Eq. 5, be calculated as

$$\mathrm{PFD} \approx \mathrm{PFD_d} + \mathrm{PFD_{FT}} = \int_0^{\infty} \lambda_{\mathrm{d}}^{\text{non-crit}} e^{-\lambda_{\mathrm{d}}^{\text{non-crit}} t}\left(1 - e^{-\theta\lambda_{\mathrm{DU}} t}\right) dt + \frac{(1-\theta)\lambda_{\mathrm{DU}}}{2f_{\mathrm{FT}}} = 1 - \frac{\lambda_{\mathrm{d}}^{\text{non-crit}}}{\lambda_{\mathrm{d}}^{\text{non-crit}} + \theta\lambda_{\mathrm{DU}}} + \frac{(1-\theta)\lambda_{\mathrm{DU}}}{2f_{\mathrm{FT}}} \quad \text{[Eq.6]}$$

NOTE: A condition of using Eqs. 3 and 6 is that the functional test frequency is significantly lower than the non-critical demand rate, as what has been stated in assumption 3.

When all failure modes can be revealed by demands, $\theta = 1$, thus

$$\mathrm{PFD} \approx 1 - \frac{\lambda_{\mathrm{d}}^{\text{non-crit}}}{\lambda_{\mathrm{d}}^{\text{non-crit}} + \theta\lambda_{\mathrm{DU}}} = \mathrm{PFD_d} = \frac{\lambda_{\mathrm{DU}}}{\lambda_{\mathrm{d}}^{\text{non-crit}}} = \lambda_{\mathrm{DU}} * \mathrm{MTBD} \quad \text{[Eq.7]}$$

where MTBD is the *mean time between demands.*

If we assume that the non-critical demands can reveal all failure modes and that the state of every component can be verified, we may conclude that functional testing gives no added value to the reliability unless they are performed very frequently (which in many cases is impractical).

5.2 Multiple components subsystem

For a subsystem with redundancy, an example is used to illustrate how the PFD can be calculated. Considering a subsystem of three pressure transmitters working as a 2oo3 configuration, it is assumed that the pressure transmitters are needed for a particular type of operation, for example for mud control during drilling operations, to raise an alarm upon a specified pressure setting. The same transmitters are also needed in the execution of a shutdown function. It is further assumed that the non-critical demands (i.e., when exceeding the pressure set point) on average occur 4 times per year ($\lambda_{\mathrm{d}}^{\text{non-crit}} = 4.57 \times 10^{-4}$ per hour).

The failure rate of each transmitter is set to 5×10^{-6} per hour, and we assume that no periodic functional testing is scheduled. This strategy mandates that the state of each transmitter is confirmed by inspection after each non-critical demand. We further assume that the non-critical demands can reveal all failure modes, that the inspection after a demand can verify all component states, $\theta = 1$, and that the time needed to perform inspection and maintenance is negligible. If a transmitter failure is revealed by diagnostics, repair is initiated immediately and the contribution from DD failures may be neglected. Common cause failures (CCFs) are not considered in this example, but may easily be added using e.g., the standard beta factor model.

According to Eqs. 3 and 5, the PFD for this 2oo3 configuration subsystem is [10]

$$\mathrm{PFD} = \int_0^{\infty} \lambda_{\mathrm{d}}^{\mathrm{non-crit}} e^{-\lambda_{\mathrm{d}}^{\mathrm{non-crit}} t} (1 - 3e^{-2\lambda_{\mathrm{DU}} t} + 2e^{-3\lambda_{\mathrm{DU}} t})\, \mathrm{dt} = 1 - \frac{3\lambda_{\mathrm{d}}^{\mathrm{non-crit}}}{\lambda_{\mathrm{d}}^{\mathrm{non-crit}} + 2\lambda_{\mathrm{DU}}} + \frac{2\lambda_{\mathrm{d}}^{\mathrm{non-crit}}}{\lambda_{\mathrm{d}} + 3\lambda_{\mathrm{DU}}}$$

[Eq.8]

By inserting the non-critical demand rate and failure rate in Eq.13, we obtain $\mathrm{PFD} = 6.8 \times 10^{-4}$.

6. Follow-up in the operational phase

IEC 61508 and IEC 61511 require the reliability of SRSs is maintained throughout the whole operational life. All factors that influence the reliability, such as the number of recorded failures and the time between periodic (and, as suggested in this paper, random) tests must therefore be monitored, analyzed, and used as basis for making decisions about reliability. For low demand systems, the approach suggested by Lundteigen and Hauge [11] may be used.

The approach advocated in this paper mandates monitoring of non-critical demands. New (experienced) demand rates and PFD need to be compared with the assumptions and calculations that were made in the design phase. A stepwise approach may be as follows:

1. Record the number of non-critical demands and the total operation time.
2. Estimate the actual demand rate as well as the actual PFD by the recorded operational information.
3. Compare the actual PFD and the predicted PFD.
4. If the actual PFD is greater than the predicted PFD, introduce measures (e.g., additional functional tests) to improve the SRS reliability and proceed to 5, otherwise stop.
5. Make a new PFD estimation with the actual demand rate and new measures.
6. Compare the new PFD estimation and the predicted PFD.
7. If the new PFD estimation is greater than the predicted PFD, introduce more measures, then proceed to 5 again, otherwise stop.

This procedure may be performed with a certain interval (e.g., once per year) to maintain the due SRS reliability.

In the daily follow-up of SRSs, we must avoid too long time between tests. With too long time since the last response to a non-critical demand, it may be necessary to perform a functional test to not compromise the reliability requirement. We may introduce the maximum allowed time between demands/tests, t_{max}. If this time is reached, we perform a functional test to make sure the time between demands/tests is not greater than the maximum allowed time between demand/tests, thus the PFD does not exceed a certain value. The t_{max} may be the mean time between demands, but it can also be argued that it is a value within a specified confidence interval.

A more detailed discussion of SRS follow-up is not made in this paper, but it is certainly a further research field.

7. Concluding remarks

In this paper, the possibility of using information from responses to demands in SRS reliability assessment is investigated. The conditions under which demands may be treated as functional tests and credited in PFD calculation are discussed. The study has been focused on SRSs operating in the so-called medium-demand mode, particularly when the SRS is subject to critical demand with a low-demand rate and non-critical demands with a medium-demand rate. By considering only the non-critical demands, PFD calculation formulas are established.

The paper has been written in response to a question raised by the SRS manufacturers: Can responses to demands and inspections (to some extent) replace the functional tests. Despite the effort made in this paper to give new insights into this topic, there are several questions that remain unsolved. Some examples include:

- More insights into the pros and cons of using responses to demands as tests are needed, and they should be founded on industry cases beyond those considered in this paper.
- With more reliance on demand rate to quantify the PFD, follows the need for a systematic recording of non-critical demands. Even if IEC 61508 and IEC 61511 point at the need to monitor the (critical) demand rate for the purpose of validating the assumptions under which the SIL requirements have been founded, the authors have seen little evidence that such a practice has been implemented.
- The paper points at the need for inspections after each response to a non-critical demand. In practice, this is time consuming and effort is needed to ensure that the confirmation about all component states is automatically collected by the SRS.
- The paper has not addressed the possibility of taking spurious activation into account in the PFD calculation, even if such responses may also replace (to some extent) the need to perform functional tests. The issue has been discussed by a few authors [12]. However, this is a premature field, and cautions should be taken to avoid a design that credit frequent spurious activations as a means to achieve high reliability.
- New quantification approaches give new requirements to the follow-up of SRSs in the operational phase. A brief approach has been indicated in this paper, but more research and case studies are needed before implemented by the industry.

The approach does not cover all requirements needed to comply with a SIL requirement: Quantifying PFD (or PFH) is just one step out of many that are needed to comply with a SIL requirement. Manufacturers also need to build systems that fulfill the requirements for architecture constraints and systematic safety integrity. End users must demonstrate that all the necessary conditions and assumptions that affect operation and maintenance are maintained throughout the SRS lifecycle.

It is the authors' opinion that more research is needed before an industry practice may be suggested for the topic of this paper and some of the other identified issues.

8. References

[1] IEC61508, "Functional safety of electrical/electronic/programmable electronic safety-related systems," International Electrotechnical Commission, Geneva, 2010.

[2] IEC61511, "Functional safety - safety instrumented systems for the process industry," International Electrotechnical Commission, Geneva, 2003.

[3] ANSI/ISA 84.00.01: 2004 (IEC61511 Mod), Application of safety instrumented systems for the process industries, ISA, Raleigh, NC, USA, 2004.

[4] O. Olamilehin, "Fault tree analysis applied to SIL," Master thesis, NTNU, Trondheim 2010.

[5] M. A. Lundteigen and M. Rausand, "Spurious activation of safety instrumented systems in oil and gas industry: Basic concepts and formulas," *Reliability engineering and system safety*, Volume 93, p. 1208-1217, 2008.

[6] OLF-070, "Application of IEC 61508 and IEC61511 in the Norwegian petroleum industry," The Norwegian oil industry association, Stavanger, Norway, 2004.

[7] H. Jin, M. A. Lundteigen and M. Rausand, "Reliability performance of safety instrumented systems: A common approach for both low- and high-demand mode of operation," *Reliab Eng Syst Safety* (2010), doi:10.1016/j.ress.2010.11.007

[8] PDS method handbook, "Reliability prediction method for safety instrumented systems," SINTEF, Trondheim, Norway, 2010.

[9] M. A. Lundteigen and M. Rausand, "Partial stroke testing of process shutdown valves: How to determine the test coverage," *Journal of Loss Prevention in the Process Industrie*s, Volume 21, 2008.

[10] M. Rausand and A. Høyland, *System reliability theory: Models, statistical methods and applications (2nd. ed.),* Wiley, New York, USA, 2004.

[11] M. A. Lundteigen and S. Hauge, "Management of safety integrity in the operational phase," *Inside functional safety*, Issue 1, 2010.

[12] E. Munkeby, "Effect of safe failures on the reliability of safety instrumented system," Master thesis, NTNU, Trondheim 2008.

Analysis of Independent Protection Layers Used in Three Typical Chemical Processes

Herbert W. H. Meyer, III
Lyondell Chemical Company, a LyondellBasell Company

8280 Sheldon Road
Channelview, TX 77530

Herbert.Meyer@LyondellBasell.com

Keywords: IPL, LOPA, administrative safeguards

Abstract

Independent protection layers (IPLs) used to mitigate the most severe consequences are analyzed for three chemical processes: Olefins, Low Density Polyethylene, and an intermediates chemical plant. The IPLs are categorized by type (i.e., interlocks, PSVs, independent alarms, etc.), function (preventative versus response), and design (engineered versus administrative safeguards). An analysis of management systems that are in place to ensure the integrity of the IPLs is also included. Categorizations for all three plants revealed similar results and suggest that the analysis extends across the chemical industry. The data shows that the chemical plants depend much more heavily on IPLs that prevent an incident rather than respond to an incident once hazardous material is released. The survey also found that there is a significant reliance on administrative safeguards to achieve an acceptable level of risk. It is generally understood that engineered safeguards carry higher credits because mature systems (i.e., testing and inspection, auditing, and management tracking through key metrics) have been developed to ensure their availability. Recommendations are made on how administrative safeguards can adapt these systems to improve their robustness.

1. Introduction

An analysis is conducted of the hazards of three typical chemical processes which had process hazard analyses (PHAs) conducted using the layer of protection analysis (LOPA) risk assessment method. These are an olefins unit, a low density polyethylene process, and an intermediate chemical process. The analysis looks at IPLs used to protect these processes from the highest-level consequences of all hazards (i.e., personnel safety, community impact, environmental damage, mechanical damage, and lost production). However, nearly all scenarios had a safety hazard given the nature of the materials being processed.

The IPLs are first categorized by type (i.e., safety instrumented functions, relief valves, alarms with operator response, etc.). The analysis further categorizes the IPLs by the degree of avoiding the hazard from propagating to the endpoint (i.e., prevention, mitigation, or emergency response). For our purposes, prevention is defined as an IPL averting a breach of the primary containment, and mitigation is defined as reducing the severity of the consequence following a breach. Mitigation occurs by preventing the scenario from propagating to the endpoint, thus permitting a less severe consequence. [1] Mitigation IPLs are typically devices such as blast walls, secondary containment or fire suppression systems. Thirdly, the IPLs are categorized as either engineered or administrative safeguards, with the former requiring no human decision or intervention.

Finally, the survey looks at the management systems that are in place to ensure the availability of these IPLs. Recommendations are made in accordance with how best practices can be adopted from well-managed IPLs to improve the availability of others.

2. Processes Evaluated

Three chemical processes are evaluated in this study. The first is an olefins unit which is a typical steam cracker that is fed a range of feedstocks. Multiple cracking furnaces are employed and IPLs are counted for each furnace. The tank farm and utilities are not analyzed in this study.

The second process is a low density polyethylene (LDPE) unit which utilizes a hypercompressor to bring the reaction to pressures of approximately 40,000 PSI. In addition to the reaction section, the study also considers equipment associated with the extruder and pelletizer. Product storage, product movement and utilities are not considered.

The last process is a unit that manufacturers chemical intermediates primarily for the urethane industry. The process employs several reaction and distillation steps, but only considers equipment inside the boundary limits. Therefore, feed and product storage as well as utilities are not considered.

3. Survey Methodology

All three chemical processes had current PHAs on file which utilized the LOPA risk assessment methodology. Only scenarios with the most serious consequences were considered in this study. This included scenarios resulting in one or more severe personnel injuries, minor injury to the community, serious offsite impact to the environment, or mechanical damage or lost production of greater than $5,000,000.

This study captures how often IPLs are identified as protective layers even if one device is considered for multiple scenarios. Therefore, there may be a higher IPL count than there is actual number of protective devices. In a few cases, the PHAs recommended the implementation of an IPL that did not currently exist. In those cases the recommended IPL was counted for this survey.

4. An Analysis of IPLs Used in the Three Processes

4.1 Engineered and Administrative Safeguards

Engineered safeguards are defined as those IPLs designed to function without human intervention. They can be either active (e.g., safety instrumented functions, basic process controllers, relief devices, etc.) or passive (e.g., blast walls, dikes, fireproofing, flame arrestors, etc.) in design.

Administrative safeguards are those that are designed to require human decision or action. While these are generally thought to always be active in nature requiring an operator to take action within a certain time, they can be passive. Passive administrative safeguards can be pre-established policies which prevent certain activities from occurring. Examples include rules prohibiting crane lifts over equipment bearing hazardous chemicals or policies excluding access to areas of the process during certain activities. Batch processes may employ these safeguards to exclude movement to the next step until certain conditions are met. Passive administrative safeguards should almost always be written and well understood. These types of safeguards can sometimes be very effective in preventing the hazardous event. Rather than counting passive safeguards as IPLs, some companies choose to consider these as part of the system and lower the initiating frequency to account for them.

Table 1. Engineered Safeguards

Independent Protection Layer (IPL)	Olefins	LDPE	Intermediates
Detonation/Flame Arrestors	0	0	0
Relief Devices to Flare	312	0	316
Relief Devices to Atmosphere	0	58	0
Safety Instrumented Functions	108	34	150
Secondary Containment	0	0	0
Fireproofing	0	0	9
Blast Wall/Bunkers	0	3	0
Basic Process Control - automatic	109	15	18
Car Seals/Mechanical Stops/Check valves	9	0	18
Fire & Gas Suppression - automatic	17	0	3
TOTALS	**555**	**110**	**514**

Table 1 shows the use of engineered safeguards for the three processes. The three processes are similar in that they rely predominately on relief devices and safety instrumented functions. For the three processes relief devices and safety instrumented functions represented 83% of the engineered safeguards. For the Olefins unit, automatic process controls represented a significantly used IPL.

The most frequently used IPL for administrative safeguards is an alarm requiring operator response, whether from fire and gas detection systems or process alarms. Table 2 shows that these represent 68% of all administrative safeguards. However, the three processes also relied heavily on operating procedure requirements; for the three processes this IPL was used 27% of the time for all administrative safeguards.

Table 2. Administrative Safeguards

Independent Protection Layer (IPL)	Olefins	LDPE	Intermediates
Fire and Gas Detection – operator response	5	0	19
Independent Process Alarms – operator response	197	95	211
Remote Isolation Valves - manual	0	0	0
Emergency Response	0	0	0
Operator Procedure Requiring Action	110	11	87
Operator Surveillance	2	28	1
Other	3	0	1
TOTALS	**317**	**134**	**319**

Comparing engineered and administrative safeguards, engineered safeguards were somewhat more relied upon accounting for 60% of all IPLs. Only the LDPE process depended more heavily on administrative safeguards.

4.2 Prevention and Mitigation IPLs

The same IPLs that were previously analyzed are reevaluated to determine which ones are preventative in nature (avoid progression of the scenario and maintain containment) and which ones are mitigation IPLs (prevent propagation to the endpoint of the consequence after containment is lost). As for the prevention IPLs shown in Table 3, no single type provided the most protection. IPLs for safety instrumented functions, alarms with operator response, basic process control and operating procedure requirements were fairly equally represented. However, for mitigation IPLs shown in Table 4, the IPLs predominately used are relief devices representing 88% of all mitigation IPLs.

Table 3. Prevention IPLs

Independent Protection Layer (IPL)	Olefins	LDPE	Intermediates
Safety Instrumented Functions	108	34	150
Independent Process Alarms – operator response	197	95	211
Operator Procedure Prevention Guidance	110	11	82
Basic Process Control - automatic	109	15	18
Car Seals/Mechanical Stops/Check valves	9	0	18
TOTALS	**533**	**155**	**479**

One point to keep in mind is that relief devices routed to a flare achieve a high destruction level of released material. As such, little hazardous material is released beyond the flare, thereby posing little risk to personnel or the environment. When considering the preventive IPLs and those IPLs which route material to the flare it reveals that 92% of IPLs contain or destroy the material before release.

Table 4. Mitigation IPLs

Independent Protection Layer (IPL)	Olefins	LDPE	Intermediates
Fire and Gas Detection – operator response	5	0	19
Fire & Gas Suppression - automatic	17	0	3
Detonation/Flame Arrestors	0	0	0
Relief Devices to Flare	312	0	316
Relief Devices to Atmosphere	0	58	0
Operator Surveillance	2	28	1
Operator Procedure Mitigation Guidance	0	0	5
Remote Isolation valves - manual	0	0	0
Emergency response	0	0	0
Secondary Containment	0	0	0
Blast Wall/Bunkers	0	3	0
Fireproofing	0	0	9
Other	3	0	1
TOTALS	**339**	**89**	**354**

4.3 An Analysis of Management Systems Used to Ensure Availability of IPLs

A review was also conducted of the management systems in place that are used to ensure the availability of IPLs. There were differences (see table 5) between those used to maintain engineered safeguards and those for administrative safeguards. IPLs for relief valves, safety instrumented functions, equipment subject to mechanical integrity inspections, and other engineered safeguards generally had good systems for testing and inspection. In addition, audits driven by corporate standards are in place to ensure that these tests and inspections are done appropriately and timely. Furthermore, many have documented key performance indicators (KPIs) to provide management with a simple indicator of the performance of these systems. Many of these KPIs tally overdue tests or inspections.

Table 5. Management Systems

Independent Protection Layer (IPL)	**Testing / Inspection**	**System in Place**	**Audit of System**	**Audit System in Place**
Detonation/Flame Arrestors	Preventative Maintenance	Yes	Corporate Audit	Yes
Relief Devices to Flare	Test and Inspect	Yes	Corporate Audit	Yes
Relief Devices to Atmosphere	Test and Inspect	Yes	Corporate Audit	Yes
Safety Instrumented Functions	Functional test	Yes	Corporate Audit	Yes
Secondary Containment	Mechanical integrity inspection	Yes	Corporate & Loss Prevention Audits	Yes
Fireproofing	Mechanical integrity inspection	Yes	Corporate & Loss Prevention Audits	Yes
Blast Wall/Bunkers	Mechanical integrity inspection	Yes	Corporate Audit	Yes
Basic Process Control - automatic	Preventative Maintenance	No		No
Car Seals/Mechanical Stops/Check valves	Periodic Check	Yes	Corporate Audit	Yes
Fire & Gas Suppression - automatic	Preventive Maintenance	Yes	Corporate & Loss Prevention Audit	Yes
Fire and Gas Detection – operator response	Preventative Maintenance	Yes, but not operator response	Corporate & Loss Prevention Audit	Yes, but not operator response
Independent process alarms – operator response	Preventative Maintenance	No		No
Remote Isolation valves - manual	Preventative Maintenance	Yes	Corporate & Loss Prevention Audit	Yes
Emergency response	Periodic drills	Yes	Corporate Audit	Yes
Operator Procedure requiring action	Operator training and certification	No		No
Operator Surveillance	Supervisor review	Yes	Corporate Audit	Yes

One exception to the rigorous management of engineered safeguards is BPCS functions which are not managed with the rigor of safety instrumented functions. One reason for this may be that safety instrumented functions are subject to unrevealed (covert) failures and thus must be tested to identify faults. In general, most BPCS controls are witnessed by operating technicians to be operating continuously and faults are readily identifiable. However, within the framework of BPCS controls exists overrides, permissives, and interlocks which are subject to unrevealed failures. These could benefit from more formal management systems which can ensure their availability. LyondellBasell now requires that BPCS controls be verified to ensure that they are normally operated in automatic and that their function be documented in operating procedures.

On the other hand, IPLs of many administrative safeguards lacked these tests, audits and KPI measures. Only a few alarms are functionally tested routinely and operator response is not routinely embedded in training and certification documentation. Procedural requirements also suffer from the same problem. Furthermore, management systems to audit the performance of these IPLs are not as mature as those used for engineered safeguards. One of the requirements for an IPL is that it must be auditable [1], but this also implies that it should be audited. Without testing and audits, there is no system for generating KPIs for management review. A systematic method for evaluating administrative safeguards credited in a LOPA can be found in reference [2]. Within LyondellBasell, standards now require that alarms used for IPLs have the alarm response documented in procedures, operator certification that includes an understanding of alarms, and audits conducted periodically to confirm this understanding.

One possible reason for the differential way in which many administrative safeguard IPLs have been managed is that there have been relatively long histories on how to manage engineered safeguards such as relief valves and safety instrumented functions. Most companies have developed corporate standards not only for their design but also for their testing and maintenance. The general understanding of IPLs is more recent and companies have not in all cases adopted new standards regarding how these should be managed collectively. LyondellBasell is one company that has written corporate standards to systematically and rigorously manage all IPLs.

5. Conclusions

Three different chemical processes are surveyed for their use of IPLs to reduce the risk from the most severe consequences. The similarity of the data suggests that the results extend across the chemical industry.

Conclusions from this work suggest:

- Preventive IPLs and those that route material to flares are used much more often that other mitigation IPLs and do much to minimize the impact of hazards.

- Engineered safeguards are generally well tested and audited. One exception to this is basic process control functions which may have unrevealed (covert) failures. These should be functionally tested on a periodic basis.

- PHAs need to verify administrative safeguards are understood, effective and audited before assigning them credit.

- Administrative safeguards for the most severe consequences should be embedded into the training and certification process for technicians.

- Alarms used for IPLs should be functionally testing on a periodic basis in the same manner that safety instrumented functions are.

- Additionally, scenario drills should be developed around the most severe consequences which use administrative safeguards to ensure that the knowledge is retained. PHAs are an excellent source for identifying these potential scenarios.

- Checklists should be considered for complex administrative safeguards. These checklists should be initialed and time stamped by the person performing the work.

- Differential means should be used to highlight administrative safeguard IPLs in operating procedures.

- Audits should be conducted to verify retention of knowledge related to administrative safeguards of IPLs.

- KPIs should be developed for IPLs related to administrative safeguards to provide management insight into their performance.

- Corporate standards for managing all IPLs should be considered to ensure consistency in how they are managed.

6. References

[1] American Institute of Chemical Engineers, Center for Chemical Process Safety; Layer of Protection Analysis: Simplified Process Risk Assessment. New York, 2001.

[2] Freeman, Raymond, "Procedural-Based Controls in Layer of Protection Analysis," *Process Safety Progress*, Volume 27, NO. 4, December 2008.

LOPA and Human Reliability – Human Errors and Human IPLs (Updated)

William Bridges
Process Improvement Institute, Inc. (PII)
1938 Riversound Drive, Knoxville, TN 37922, USA
wbridges@piii.com

Dr. Tony Clark
Process Improvement Institute, Inc. (PII)

Keywords: human factors, Layer of Protection Analysis, LOPA, process safety, PSM, human error, independent protection layer, IPL, SIS, SIL, safety instrumented system

Abstract

Layer of Protection Analysis (LOPA) is a simplified risk assessment method that provides an order of magnitude estimate of the risk of a potential accident scenario.[1] Humans can be the cause on an accident scenario (the Initiating Event [IE]) or humans can serve or participate as an independent protection layer (IPL). In either case, estimating the likelihood of the human error and measuring the human error rate at a site are troublesome tasks within the LOPA framework, which is one of the reasons some companies do not give any credit for a human IPL.

Identifying and sustaining independent protection layers (IPLs) is the heart of LOPA. Each IPL must be:

- independent of the initiating event (IE) and the other IPLs
- capable (big enough, fast enough, strong enough, etc.)
- maintained or kept in practice/service
- validated/proven to provide the probability of failure (PFD) chosen

and all of the above must be documented and audited periodically to ensure compliance with these definitions.

There are many types of IPLs, and some are more trustworthy than others, hence the difference in the PFD of IPLs. As just mentioned, one possible type of IPL is a Human IPL. These include preventative steps that may stop a scenario from progressing once it is initiated, but more typically the human IPLs are responses to alerts or alarms or troubling readings and sample results.

This paper (based on a similar paper from 2010[2]) discusses the data needed for adequately counting the human in a LOPA (and other risk assessments), and includes discussion of the theory of human factors. One key focus of the paper is on practical means for collecting raw data in a plant setting for substantiating the error rates for the site, and especially for crediting a human IPL. The method for data collection covers the training requirements that should be met, proof drills for response to alarms, simulations and tests, and frequency of proofs, and of course the effect of human factors on human error rates. **Actual plant data and tests are included in the paper to provide the reader with some examples of how a simple data collection and validation method can be set up within their companies.** This paper also provides an overview of an alternative method for estimating the PFD of a Human IPL, based on plant and scenario specific factors (such as stress factors, complexity, and communication factors).

1. Human Error Fundamentals

Human errors are sometimes mistakenly called procedural errors. This is not true anymore than saying all equipment failures are due to design errors. Over the past 5 decades of research and observation in the workplace on human error, we have come to know that human error probability depends on many factors. These factors are described in more detail elsewhere[3] but they include:

- Procedure accuracy and procedure clarity (the number one most sited root cause of accidents):
  - A procedure typically needs to be 95% or more accurate to help reduce human error; humans tend to compensate for the remaining 5% of errors in a written procedure.
  - A procedure must clearly convey the information and the procedure must be convenient to use.
  - Checklist features – These should be used and enforced either in the procedure or in a supplemental document.
  - For human response IPLs, a checklist is not always practical; in fact a trouble-shooting guide (rather than a prescriptive, linear procedure) is usually required since we do not always know what condition a process will be in before a serious problem begins that requires humans to respond (as a human-based IPL)

- Training, knowledge, and skills
  - Employees must be selected with the necessary skills before being hired or assigned to a department.

  - Initial Training – There must be effective, demonstration based training on each proactive task and each reactive (e.g., response to alarm) task.
  - Ongoing validation of human action is required and usually must be repeated. For human IPLs, the action must be demonstrated to be "fast enough" as well. For the Site-Specific Data approach, validation can be done in actual performance (measuring errors to responses to actual alarms) or in drills/practice. Or the human performance can be validated by Expert Judgment, Generic data, or by Prediction. Appendix B and C provide examples of two methods for validating human IPLs (and estimating human IEFs).
  - Documentation – the actual performance of the humans or alternative validation approaches must be documented and retained to demonstrate the error rates chosen are valid.

- Fitness for Duty – Includes control of many sub-factors such as fatigue, stress, illness and medications, and substance abuse.

- Workload management – Too little workload and the mind becomes bored and looks for distraction; too many tasks per hour can increase human error as well.

- Communication – Miscommunication of an instruction or set of instructions or of the status of a process is one of the most common causes of human error in the workplace. There are proven management systems for controlling communication errors.

- Work environment – Factors to optimize include lighting, noise, temperature, humidity, ventilation, and distractions.

- Human System Interface – Factors to control include layout of equipment, displays, controls and their integration to displays, alarm nature and control of alarm overload, labeling, color-coding, fool-proofing measures, etc.

- Task complexity – Complexity of a task or job is proportional to the (1) number of choices available for making a wrong selection of similar items (such as number of similar switches, number of similar valves, number of similar size and shaped cans), (2) number of parallel tasks that may distract the worker from the task at hand (leading to either an initiating event or failure of a protection layer), (3) number of individuals involved in the task, and (4) judgment or calculation/interpolation, if required. For most chemical process environments, the complexity of the task is relatively low (one action per step), but for response actions (human IPLs) there are almost always other tasks underway when the out-of-bounds reading occurs or the alarm is activated, and as mentioned earlier, we cannot predict what state the rest of the process will be in when then alarm comes on.

In addition to these human factors, other considerations for use of a human as an IPL include (1) time available to perform the action and (2) physical capability to perform the action.

When considering human IEs and IPLs, the site must ensure that the human factors just listed are consistently controlled over the long-term and that they are controlled to an adequate degree during the mode of operation for the particularly LOPA scenario. For instance, if the workers are fatigued following many extra hours of work in a two-week period leading up to restart of a process, then the human error rates can increase by a factor of 10 or 20 times during startup and so a LOPA scenario for startup would need to take these into consideration.

Revealed versus Unrevealed Errors for Human. As with equipment failures, human errors can lead to a revealed fault in the system (e.g., the flow does not start) or to an unrevealed fault, such as the block valve downstream of a control valve is left closed, but the failure is not revealed until the control valve is used. If the error is revealed, then the error can be corrected or compensated for. If the restoration/correction time is sufficiently short, then the probability of being in the failed state is much less for a revealed failure than for an unrevealed failure, which is only discovered upon testing or inspection.

2. General Relationship of Human Factors to LOPA

Every risk assessment must consider the likelihood and effect of human factors. For LOPA, poor human factors can lead to higher human error rates that increase IE frequencies and that increase the PFD of a human IPL (and of other IPLs indirectly). Table 1 on the next page summarizes the human factor issues that relate directly to LOPA. The table contrasts the impact of good and poor human factors on initiating event frequency (IEF) and on the PFD of human-based independent protection layers (IPLs). If all of the factors were optimized, one would expect about 0.01 per task as an IEF and about 0.05 PFD (which would normally be rounded up to 0.1) for response to alarms or call for actions (human IPLs), but such an analysis must be made on a case-by-case basis, especially for the PFD of IPLs.

3. Types of Human Errors of Importance in LOPA

Human error comes in many forms, but the primary ones of interest in LOPA are:

Human errors that directly relate to human IEs and human IPLs

- Errors in following proactive procedures, such as startup of a unit, that results in an initiating event of a LOPA accident scenario
- Errors in responding to a call for action; if performed correctly such actions would interrupt the scenario and prevent the consequence

Table 1. Considerations for Getting Low Human Error Rates

Issue	Requirement for claiming a low error rate for humans causing an accident (usually skipping a step or doing one wrong)	Requirement for using a low probability of failure on demand for a human as a protection layer against an accident (usually a response to alarm)	Comments
Time to perform action	NA	The total time to: detect deviation, diagnose the problem, decide on proper action, and take action must be less than the time required to reach the consequence of interest or MART. The term "maximum allowable response time" (MART) is used throughout this guideline.	The reliability of the indication (sample & analysis in lab, field reading, etc.) or annunciation (alarm of the deviation) may limit the value for the human action IPL. Operator must be reliably available to quickly respond.
Capable	NA	It is physically possible to perform the control action required to forestall the consequence of interest, and the capability has been demonstrated and is not the limiting factor in the human action (e.g., the operator is strong enough to do the required task)	If the final control requires closing a manual block valve, the action and strength required must be verified in the field (some valves take several operators more than 1 hour to close, even with a valve wrench or cheater). This usually requires preventive maintenance or equipment use as part of a 'critical action.'
Procedure	Step must be in procedure. Not completing the step as stated and in that sequence (most aspects of errors of omission and commission) is the IE. The step is a critical step and is indicated as critical in the procedure or with a warning or caution. The procedure must follow "best practices" for control of human error; best practices relate to content accuracy, page format, and step presentation. Follow the current rules for procedure development and writing.[3]	Step should be in a trouble-shooting guide or similar contingency procedure (including emergency shutdown procedure), that is in paper form or reliably available on demand electronically (including via linkage to trouble-shooting guides within a distributed control system [DCS] for an alarm point) that describes how to respond to a process deviation. The response procedure follows best practices as well; follow the current rules for procedure development and writing.[3]	Note that we assume there is a 100% chance of skipping the required step or doing it substantially wrong if the step is not in the written procedure or if the step written is wrong or in the wrong sequence.
Checklist	Checklist in place and its use is enforced.	NA	Without a checklist, the error rate goes up by a factor of 3 to 5 times.

Issue	Requirement for claiming a low error rate for humans causing an accident (usually skipping a step or doing one wrong)	Requirement for using a low probability of failure on demand for a human as a protection layer against an accident (usually a response to alarm)	Comments
Initial Training	There is initial training focused on this step as a critical step	There are initial training and drills on the response to a deviation; and the deviation is annunciated.	Note that we assume there is a 100% chance of skipping the required step or doing it substantially wrong if the step is not emphasized in initial training.
Ongoing validation (or practice) of human action	Steps must be practiced, but for seldom-used procedures, this practice (normally in the form of a talk-through for shutdown or startup of a continuous unit) can be just prior to use. Practice rate increases reliability per action (due to increase in skill), but Actual Use Rate also increases the number of opportunities for failure. These factors can offset each other completely. Per CCPS, *Guidelines for IPLs and IEs (2011)*[4], there are limitations on error rates that account for "practice" versus "error rate per task demand" that are incorporated within the Human IEs.	Steps must be practiced routinely enough to assure the reliability of the action under the increased stress caused by the alarm/indication. The time it takes to complete the action must be validated, to ensure response time limits are not exceeded. For the Site-Specific Data approach to validation, each person who is expected to implement the action "on demand" must perform the practice/drill per the assigned frequency of about once per year. Per the Predicted Data approach to validation (see the brief example at this end of this paper for an example), the PFD for the human IPL is estimated by calculations using factors based on the quality of the underlying management systems, procedures, and training; and based on the level of stress involved with the response task. Other methods of validation may also be used, as described in the upcoming textbook from CCPS, *Guidelines for IPLs and IEs.*[4] It is normally possible to group similar IPLs along the classification of similar types of actions and similar available response times. When grouping is possible, then validating one action will in effect be validating all similar actions. This paper provides an example of grouping, if validating using Site-Specific drills of response actions.	Note that we assume there is a 100% chance of skipping the required step or doing it substantially wrong if the step is not practiced often enough (to be determined by the site).

Issue	Requirement for claiming a low error rate for humans causing an accident (usually skipping a step or doing one wrong)	Requirement for using a low probability of failure on demand for a human as a protection layer against an accident (usually a response to alarm)	Comments
Documentation of validation of human action	Documentation methods depend on the validation method chosen. Validation based on Expert Judgment or Predicted Data approaches would require the site to maintain sufficient documentation, including calculations and description of methods, for the individuals IEs or groupings of IEs. The Site-Specific Data approach to validation requires more effort since the site would need to document each practice (i.e., each startup and each shutdown) and document failures rates for 'critical steps'. For this same validation method, the site would keep the file of the validation associated with the task and also with each worker. Though more time consuming on documentation, this method may produce more side benefits that out weight this cost.	Documentation methods depend on the validation method chosen. Validation based on Expert Judgment or Predicted Data approaches would require the site to maintain sufficient documentation, including calculations and description of methods, for the individuals alarm response actions. For the Site-Specific Data approach to validation, the site would need to document each practice (i.e., each response) and document failures rates for 'critical response action.' The site would also keep the file of the validation associated with the 'action' and also with each worker. If a sample plan method is used, the same data as above is still required, and further, the data should be checked each quarter to ensure against inadvertently using the same IPL and same operator too often.	Note that we assume there is a 100% chance of skipping the required step or doing it substantially wrong if the step is not practiced as often as necessary (to be determined by the site) or if there is no documentation of the completed refresher training or just-in-time training for an IE, or of drills for an IPL (if drills are part of the validation method).

Issue	Requirement for claiming a low error rate for humans causing an accident (usually skipping a step or doing one wrong)	Requirement for using a low probability of failure on demand for a human as a protection layer against an accident (usually a response to alarm)	Comments
Control of related physiological and psychological stressors (see SPAR-H, NRC, 2007)	• Fatigue – Company policy and enforcement to limit fatigue, including limit of hours worked per day and per week, and including restrictions on hours worked outside of work. Can affect error rates by factor up to 20X. • Other fitness of duty issues – Company policy and enforcement to limit effect of alcohol & drug abuse, illness, prescription drug effects, and personal life stress effects. Can affect error rates by factor up to 50X. • Workload – Workload needs to be managed to optimize worker performance. Enough stress (time/task load-based), but not overstress. (Workload is also related to the complexity of the work. Complexity increases mental workload, even if the number of tasks per hour is not affected.) Workload evaluation may be necessary. Can affect error rates by a factor up to 10X. • Communication –The second most common cause of human error, the site must guard against miscommunication by having a policy and rules and follow-through for controlling verbal and visual communication. Rules include repeat-back on instructions received and use of common jargon. Poor control of communication rules can increase human error rates by 10X. • Work environment – Temperature, lighting, noise, distractions, ventilation, etc., has been optimized to improvement worker performance. Can affect error rates by factor up to 5X. • Human-System Interface – Includes control-display-information integration, labeling, error-proofing designs, color-coding, alarm management, etc. Poor human-system interface control can increase human error rates by 10X or higher (especially if "norms" are violated, where a negative effect of 20-50X is assumed). • Complexity – Complexity of a task or job is proportional to the (1) number of choices available for making a wrong selection of similar items (such as number of similar switches, number of similar valves, number of similar size and shaped cans), (2) number of parallel tasks that may distract the worker from the task at hand (leading to either an initiating event or failure of a protection layer, (3) number of individuals involved, and (4) judgment or calculation/interpolation, if required. For most chemical process environments the complexity of the task is relatively low (one action per step), but for response actions there is almost always other tasks underway when the out-of-bounds reading or the alarm is activated. Complexity is difficult to predict (since it is not known when a human action will be needed), but higher complexity can increase error rates by 2X to 10X.		All of the listed human factors (beyond the procedure and training and skills listed above) have a large, independent influence on human error rates. A large negative swing in any one of the factors can increase human error rate by a factor of 3 to 50 times.

Issue	Requirement for claiming a low error rate for humans causing an accident (usually skipping a step or doing one wrong)	Requirement for using a low probability of failure on demand for a human as a protection layer against an accident (usually a response to alarm)	Comments
Qualitative hazard evaluation	There has been a hazard evaluation of the scenario for skipping the step and doing it wrong and a hazard evaluation team has evaluated the consequences and safeguards.	The human action in this case is a safeguard that is called into play once a sequence of events is initiated. The human action (following the trouble-shooting guide or the emergency shutdown procedure) is to make a change of state in the system to prevent propagation of the event to the stated consequence. The hazard evaluation has evaluated the human action (and any annunciation device unavailability) in judging that this "critical responsive action" is capable, reliable, and audited.	It is essential that a company finds accident scenarios that can arise during all modes of operation, including startup, shutdown, emergency shutdown, and online maintenance.
Use of multiple successful human actions within a scenario	Making multiple mistakes (including shortcutting multiple steps) is the same probability as making a single mistake, if the "safeguard" steps amount to ensuring the same state in the system. The human who makes the errors sees this as performing the task by method B instead of method A. Method B could have different steps or less steps that method A.	If one person in a workgroup is assumed to make the mistake that is the initiating event, then it is assumed that no one in the workgroup can be part of the protection layers. Further, supervisors are also part of the workgroup, due to the trust/relationships that build quickly. Seemingly unrelated workgroups, such as maintenance and operations, usually build enough trust to rule out counting as separate groups. For a group/individual to be counted as a safeguard against the errors by another human error, the detection and response group must not share staff, must have an incentive to find mistakes and report/respond, and near miss reporting and auditing both indicate that independent measurement and action is occurring. An exception that is normally allowed is to re-use a workgroup or individual if the second action/response is separated by distance (location) and/or separated by enough time and/or there is a reminder alarm for the same alarm (the alarm that was missed earlier) that recurs often enough in the allowed time to still avoid the consequence.	It is difficult to use a human response safeguard, if a cause of the predicted accident sequence is a human error. A human reliability analysis (HRA) is typically done to verify the human protection layer probability in this situation.

Human errors that indirectly relate to LOPA component-based IEs and IPLs

- Errors when following management systems that lead to higher failure rates of components

The examples below illustrate both types of human error that directly affect LOPA. (Human errors that indirectly effect component type of IEs and IPLs are discussed near the end of this paper.)

3.1 IE Caused by Human Error

In NUREG/CR-1278[5]– The Human Reliability Handbook (Dr. Alan Swain was the primary author), error rates were tabulated from various sources, including some empirical data. For errors that could be considered initiating events have been extracted and included in slightly amended form below as Table 2 below.

Table 2. Errors of Omission versus Error Probability[5]

Omission of Item	Human Error Probability
When procedures with check-off provisions are correctly used:	
(1) Short list, less than or equal to 10 items	.001
(2) Long list, greater than 10 items	.003
When procedures without check-off provisions are used, or when available check-off provisions are incorrectly used:	
(3) Short list, less than or equal to 10 items	.003
(4) Long list, greater than 10 items	.01
(5) When written procedures are available and should be used but are not used:	.05

This table shows the estimated probabilities of errors of omission (per item of instruction) when use of a written procedure for a task is required. The probabilities in the tables above assumes that human factors are optimized as well, which includes providing an accurate and easy to follow procedure, good control of fitness for duty factors (including controls against fatigue), good controls against miscommunication, training/practice (or actual performance) at least once per year, etc., as described earlier.

Since startup (for a continuous process) is when most accidents occur (50% of major accidents in the chemical-related industry occurs during startup mode of operation)[6, 7, 8], and since the causes are typically due to human error, it is appropriate to use that as an example. If all human factors are optimized, the probability of human error taken could be 0.003 per step, which we would round up for Basic LOPA to 0.01 errors per step. It may appear then that 0.01 is a conservative value for human error, but consider that there are few (if any) chemical plant sites that have optimized most of the human factors. And poor control of certain human factors can have a multiplying effect on the human error of 3 to 50 times, so the real error rate for use of a startup procedure after a maintenance turnaround could be as high as 0.2 (as was approximately the case during the BP Texas City Accident in March, 2005). ***It is dangerous to use the value of 0.01 errors per step***

per opportunity as representative of the error rates at a given site without site-specific data. Some analyst believe that in the absence of site data, these values can be applied, but perhaps it is best to scale the values upward (use higher base error rates) if the site determines that the human factors are ***not*** optimized. This concept is described at the end of the paper.

3.2 Human Error Related to Human-Based IPLs

Again, from The Human Reliability Handbook [5], estimates were provided for the probability of the human to respond correctly and in time following an alarm (annunciation). Figure 1 below (extracted from the handbook) is for response by control room personnel to an abnormal event. This represents the time it takes to diagnose the problem given there are no other alarms and given that all human factors are optimized, including practicing the response once per year or more often. The error probabilities given are the joint probability of failing to notice the alarm and failure to diagnose the situation correctly. Note that this study was done for nuclear plant control rooms that always have several operators available to potentially respond. Many chemical plants do not have this level of control room staffing.

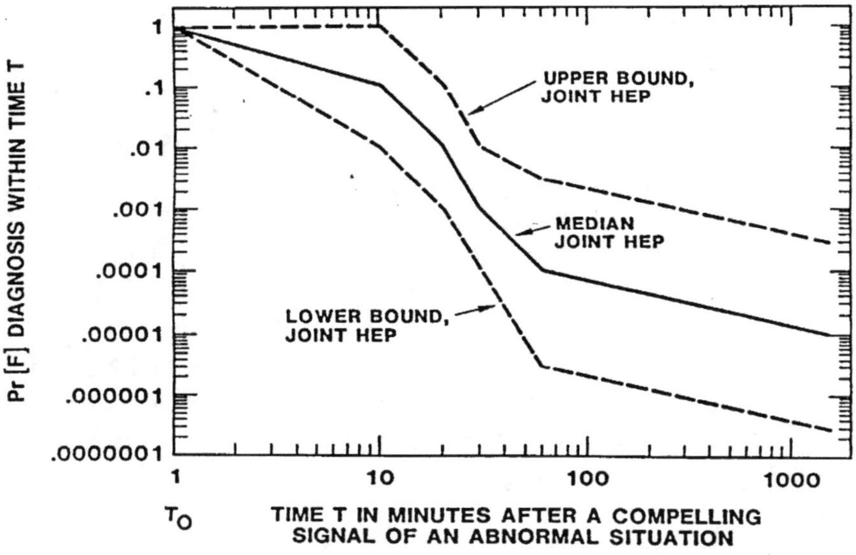

Figure 1. Errors Probability versus Time Available to Respond[5]

NOTE: HEP refers to Human Error Probability. Pr refers to the probability of making a diagnosis error.

Note from Figure 1 that the more time to detect and diagnose the problem, the lower the chance of error. But Figure 1 only tells part of the story. After diagnosis, the operator, perhaps with help of others, must then respond appropriately to prevent the scenario from propagating to the next protection layer or to the ultimate consequence. The sum of the time to detect, annunciate, diagnose, and complete the proper response must happen within the time available, is best termed "maximum allowable response time (MART)," which is used in this paper (and in the upcoming CCPS text, *Guidelines for IPLs and IEs (2011)*[4].

MART is the maximum duration for a given IPL to detect an out of limit condition, fully execute its safety function, and bring the process to a safe state. It is a property (constraint) of many IPLs which the LOPA team imposes based on knowledge of the scenario timeline and feeds directly into setpoint selection for ISF (and other active IPLs) and checking frequency for proactive human actions such as routine rounds. It is useful where time is a critical element of IPL design (i.e., applies to active but not passive IPLs). The MART must be known for a site to judge if a specific human response can qualify as an IPL (given other criteria of an IPL is met; see LOPA 2001[1] for guidelines on qualifying an IPL).

When an operator on rounds observes a failure that he/she is specifically looking for (e.g., small leak in a pump seal or flange) he/she must assure the leak is repaired. The LOPA team specifies this MART based on the anticipated progression of the failure. MART would be significant for an analytical test which detects that the process is leading to (but has not yet achieved) an unsafe state (e.g., wrong composition feeding a post reaction tank based on low conversion where there is still time before the tank overheats). For other analytical tests such as identity check before offloading a railcar the procedure step order is critical and time is not (thus MART doesn't apply).

It is generally assumed that if the problem is detected (such as by hearing an audible alarm) and if the diagnosis is correct, then the action is likely to be "more" accurate than the detection/diagnosis, as long as there is time to fully complete the response action. The norm therefore is to use a PFD of 0.1 if the response can be performed within the MART and if the MART is equal to or greater than 20 minutes (so the complete detect-diagnose-response steps are accomplished in 20 minutes. Note that some organizations use a lower limit of 10 minutes or even 5 minutes for MART for Human IPLs if the detect-diagnose-response steps are all accomplished in a control room. A value of better than this (lower error rate) is normally not allowed in Basic LOPA. From an advanced topic perspective, however, Human Reliability Analysis (HRA) will often use a PFD of 0.01 for MARTs greater than 40 minutes but care must be taken to assure that controls used for annunciation or response are appropriately improved to SIL integrity. Performance of HRA and documenting the basis for lower probability values requires a level of effort and a degree of education/experience far beyond the required learning to perform Basic LOPA[4].

4. Human Errors as Related to IEs

Human errors that result in IEs of major process incidents are usually occur during startup; one definitive study (Rasmussen, 1989)[7] and a survey of the major accidents in the past 20 years[8] show that roughly 65% of **major** process accidents in chemical-related process plants occur during startup and on-line maintenance. Further, errors during normal shutdown or emergency shutdown together represent about 10% of the IEs caused by human errors. (Other studies show that most **minor releases**, such as packing leaks, occur during normal operation.) Human errors during normal operation result in less than 25% of the and include such issues as performing a step wrong during operation in manual mode, such as when a process controller is bypassed or by external impact (one major accident during normal operation occurred when a fork-truck hit and broke off a drain valve).

The two primary ways to find accident scenarios caused by human error (before the accident occurs) are:

- Prediction of the accident scenario in PHA of all modes of operation, especially startup and shut down procedures. Most PHA/HAZOPs world-wide are weak on finding accident scenarios during non-routine modes of operation. Overcoming this weakness is critical and related methods are explained in Chapter 9.1 of the *Guidelines for Hazard Evaluation Procedures, 3rd Edition* [9] and in several papers [10, 8].
- Getting a lot of near misses reported and investigated; usually having at least 15 times the number of near misses reported as accidents occurring is needed to have a 90% chance or higher of preventing the major accidents [11, 12]. The ratio of near misses report to accidents occurring is relatively low at most companies world-wide, so this is low hanging fruit for many companies.

For LOPA, the ways to estimate and validate human IEFs (and validate PFDs of IPLs) include expert opinion, using industry values, calculation from human reliability factors, or direct measurement at the site. This paper provides examples of how to validate an IPL by Predicted Data (a brief example at the end of the paper) from human reliability factors and by using direct measurement at the Site-Specific Data, respectively.

To help minimize human errors that could be IEs, many companies will review a normal shutdown or normal startup just before its next use. This has proven one effective way to reduce human error in chemical-related industries. Nuclear power plants also perform such "just-in-time (JIT)" refresher training on normal shutdowns and normal startup procedures. If a site records the errors that are made during such tasks, then the company can validate the average IEF for that task and so the site will have a better estimate of the error probability. Further, if the measured rate is too high, they can take proactive steps to lower the error rate. Errors can be measured either in actual practice in the field or via simulators. They can also be roughly estimated by analysis of near misses reported (if the number of near misses reported is at least 15 times higher than the number of loss events or accident [11, 12].

Below is an example of data from one site that was collected to estimate the IEF for mistakes made during normal startup:

> ***EXAMPLE - Initiating event frequency data collection and use:*** In one company, the operating group was able to provide the data on the following page for errors made during startup of various units. Based on this data, the average probability of human error was 0.0071 and the number of errors per year that could lead to a known accident sequence was 0.17 (but note that if all errors are counted, the average error rate would be 0.29/yr). For basic LOPA,

this value is rounded up to an IEF of 1/yr, for initiating events arising from these and similar procedures at the site. ***See Table 3 on the next page for the raw data.***

Chemical-related process plants should collect such data to understand and validate the initiating event frequency caused by human error. This data would then be used to validate the values for IEF used in LOPA and other risk assessments.

5. Validation of Human IEs and IPLs by Site-Specific Data

A 0.1 PFD value for a human response indicates that the correct response occurs at least 9 out of 10 times (or no more than 1 wrong response for every 10 attempts). Most organizations will have identified many human responses involving a number of personnel, as part of their LOPA studies. Some organizations believe that if they have a procedure and a training program in place, that they can claim the PFD value of 0.1 for a Human IPL. This is no truer for a human IPL than it is for an active component-based IPL.

As required for all IPLs, a human IPL must be validated. The Preferred approach to validation is direct measurement or testing of the human response (under controlled conditions or drills); but other methods of validation can include Expert Judgment, using data from other comparable settings (Generic Data method), and estimation of the PFD of human IPLs by mathematical modeling (Predicted Data method, see an example of this method later in this paper).

On the next few pages are the options for validating human IPLs with **site-specific data**. These include:

- 100% testing for each human responder and for each action to be taken
- Sample plan testing of random combinations of human responders and actions.

One key focus of this paper is discussion of practical means for collecting raw data in a plant setting for substantiating the error rates for the site, and especially for crediting a human IPL. The method for data collection covers the training requirements that should be met, proof drills for response to alarms, simulations and tests, and frequency of proofs, and of course the effect of human factors on human error rates. *Actual plant data and tests are included in this paper to provide the reader with some examples of how a simple data collection and validation method can be set up within their companies.*

Table 3. Site data on Human Errors related to Initiating Event Frequency (IEF)

Task for which a Potential Initiating Event exists	*Number of steps in the procedure*	*Number of Uses of the procedure*	*Number of Human Errors Noted*	*Number of Years of Data*	*Number of Errors That Initiated a Potential Accident Sequence*	*Human Error Probability per Step*	*Error Rate (probability of error per year)*
1. Startup of Process A	46	4	2	6	1	0.0109	0.17
2. Startup of Process B	71	5	3	6	1	0.0085	0.17
3. Startup of Process C	55	4	2	6	2	0.0091	0.33
4. Startup of Process D	56	4	0	6	0	0.0000	0.00
					Average:	0.0071	0.17
						IEF for LOPA for site:	**1/yr**

As mentioned earlier, if a site has a very good system for reporting and investigating near misses, then this can be used to find site specific data for human errors (including both IEs and failure of human IPLs). Another way to collect site-specific data on error rates is to **measure them with tests or drills of the action**; and for human IPLs discussed later, the results may need to be adjusted to account for actual stress levels of a response. This is not commonplace in the chemical industry, but the US Nuclear Regulatory Agency (NRC) in 10 CFR 55.45 and 55.59[13] requires that all power plant operators be tested once per year on abnormal procedures. This testing is mostly related to humans involved as IPLs. 10 CFR 55.45 and 55.59 also allude to the possibility of using a test of a representative sample of human responses, but we address this option later in this section of the paper.

This response can be demonstrated by walkthroughs in the field or using simulators of the process unit. The nuclear power industry uses this approach for validating response by control room operators, but many in the chemical industry think this will take "too much time." As an example of the effort required, one nuclear power plant allocates about 200 hours per year per operator for refresher training activities, including 60 hours per year per operator for demonstration of skill in responding to critical alarms (from Tennessee Valley Authority [TVA] internal requirements to meet the performance-based requirements of NRC regulations 10 CFR 55.45 and 55.59). This equals about 3% of the work-year for an operator. (The example below and also the example of the sample-plan approach discussed later shows how this investment of time to measure response effectiveness can be reduced by a factor of 10 or more.)

Consider the following as an example of the application of this method of testing responses for 100% of the triggers by 100% of the operators at a chemical plant:

> ***Background:*** The operating area being evaluated has 20 operators (spread across 4, rotating shifts of work) and 130 identified (from a hazards analysis and LOPA) human response IPLs.
>
> ***Validation Test***: The "tests" are documented on a set of index cards that call out various alarm conditions; these are the events (triggers) for a scenario identified in the hazards analysis (and LOPA) to be protected by the given human response IPL.
>
> ***Demonstrated Result:*** A correct answer (success of the IPL) is the desired response to the alarm scenario. A wrong answer (failure of the IPL) is any other response than the one desired or a response that takes too long (longer than the MART, defined earlier).
>
> ***Estimation of Resources Requirements for 100% Testing Scheme:*** Below is an estimate of how much test effort is needed to ensure that training and retraining programs are sufficient to validate a 0.1 value for the PFD of all of these identified human response IPLs:

1. **Determine the number of test to be performed.** This would be the number of human response IPLs multiplied by the number of people who are expected to respond (perform as the human IPL) at some point in the future during their own shift. This example would yield 2600 (20 x 130 = 2600) discrete tests for one test of each combination of trigger and human responder (which makes up the 2600 IPLs).

2. **Determine the test frequency**. It is difficult to get consensus on this value. One documented example is from the nuclear industry. The US NRC regulation for control room operators (10 CFR 55.45 and 55.59) requires annual demonstration of proficiency in response to critical alarms and signals. On the surface, testing once per year per alarm may seem unlikely to give a 90% chance of proper response to every alarm, except for the fact that response to one alarm is normally similar to response to other alarms, so the operator is in essence getting more **practice on similar alarms** as they perform each demonstration. Operators under this regimen of recertification have shown a 90% chance or better of proper response (detection, diagnosis, and action) within 10 minutes of an annunciation of an event of interest (from TVA tests, but also inferred from, nuclear power plant data[5], since the basic control room regimen of testing each operator with each action each year was essentially the same in the 1970s as they are now). For this example, a frequency of 1 validation per year is chosen.

3. **Determine the time required to perform each test.** Assuming 10 minutes of allowed response time for success (per alarm/trigger), then the test time for the organization would be 26,000 minutes or about 430 staff-hours (22 hours per operator per period; most likely the period would be once per year). With a frequency (test period) of once per year per alarm, this equates to about 1% of the normal staff-hours for a worker in the USA. (Note: An equal amount of time would also be required to record/document the training record for the test [once the tracking system for such demonstrations is set up], but this is likely not a load the operator will have to bear. Also note that testing only a sample of these responses, discussed later, would reduce the load considerably.)

5.1 Actual Example of Using Tests/Drills (site-specific data collection) to Validate Human Response IPLs

Up until now, the actual effort to collect such data has not been well documented in the literature, though many chemical companies, refineries, and nuclear power plants do validate human response using this method. Recent research by three chemical companies (one in Malaysia and one in Texas, USA and one in Canada) documented the effort required to validate human response IPLs using Site-Specific Data. The following is an excerpt of the research results.

5.5.1 Validation Setup:

A simple test was use in measuring the response to an alarm condition. The test was not meant to measure the probability of detection of the alarm, but rather was meant to measure the time and success in determining and accomplishing the proper response to critical alarms as part of human IPLs. Three chemical plants belonging to large organizations performed the test.

The test involved having multiple operators in one unit of one plant/site (one for each company) perform responses to critical process alarms. These alarms were related to human IPLs. The actual response and time of response was measured, but essentially the tests were setup as a "pass or fail" – in other words, the test were to determine if the operators were able to respond as desired/expected, within the allotted MART. To run each test, the plants printed a data card (the size of an index card) and handed it to an operator chosen at random. Figure 2 is an example of such an index card.

Human IPL Validation Test/Drill			
Response Task:	Max. Allowable Resp. Time (MART)		Response Time:
LAH for Tank 105	*15 minutes*		5:20 minutes
Date of Test:	Time/Shift:		Employee Number:
1/23/10	07:35/A		23122
	Pass/Fail:	Pass	

Figure 2: Example of card used to administer validation of a single human IPL

Note that the card contains an estimate of the MART – as defined earlier in this appendix and elsewhere in this guide; this is the time an operator has to perform the task once the alarm is received until it is too late to take any further action. The time it took to print and hand out the index card was minimal.

5.1.2 Validating/Testing:

A human response IPL "failed" if the operator could not perform the required action to prevent the hypothetical outcome within the MART (defined earlier). The person administering the test timed the operator response and recorded the results. (Again note that these tests did not validate the probability of an operator failing to detect an alarm.) Each test took 10-15 minutes to administer and less than 1 minute to record the data. The validation was performed by multiple operators on multiple shifts for randomly selected alarms. The tests were administered by a shift supervisor, a shift engineer, or in some cases, a process safety coordinator. It is likely another operator could administer most proof tests (validations) and then the site management could audit some percentage of the tests to help ensure against bias. If the operators test each other, then the time to administer a test is likely not significant enough to measure since they have to be "there" regardless for their other duties. The total time for the test varied, but the sites that performed the test considered the time to administer the test to be minimal; the largest effort was simply for someone other than the operator to be there to "independently" measure the operator's response time.

For the most part, the tests came with little warning and occurred on all shifts. (It is anticipated that unless a sample plan is used, each operator will perform roughly one response related to each human IPL each year.) The time to respond was recorded on the index card and later transferred to a summary spreadsheet

Based on such raw data, the site was able to (1) evaluate the degree to which they were controlling human factors for the units, (2) identify which human responses qualify as IPLs, and (3) validate that the response is accurate enough and quick enough to qualify for the PFD used as the credit for the IPL (which for basic LOPA, the PFD is limited to a value of 0.1). Table 4 provides a sample of the site-specific data for several similar Human IPLs from the three sites.

For the Malaysia and Canada sites, the data was from the same operating area consisting of multiple operators across 4 rotating shifts of 8 hours per shift and for the Texas site, the shift was 12 hours.

All of the IPLs passed (all operators performed each action correctly within the allotted time) during these tests/drills. The labor to perform the test took less than 15 minutes per test (including documentation time). After the initial resistance at each site to performing the test, the subsequent tests were not resisted and in fact the operations staff embraced the testing/drills since they saw many side benefits from the test/drills, including the re-enforcement of "what to do" with all parties involved (the person doing the test, the person recording the test, and the other operators who noticed the test in progress). Lead operators and supervisors administered the test; very little training or coaching was necessary to have the drills done properly.

All three sites believe it is possible to use a sampling of human error for similarly qualified humans doing similar response or proactive tasks. (This is because the responses for all IPLs were the same when the same operator acts on different alarms or when different operators act on the same alarms.) A sample plan of perhaps only 5% to 10% of the number of human-task pairs may be necessary to have a valid statistic for human error for a "type of action." Sampling is valid for human actions because of how the mind processes information and how human take the necessary actions for similar situations. Obviously, sampling can greatly reduce the measurement and documentation load for validation of human error rates. If sampling is used, the sites suggested that:

- The site should first screen which response can be grouped together into general types of response IPLs. Then a lesser percentage will need individual validation drills.

Table 4. Site-Specific Validation of Human Response IPLs[a]

IPL No.	Response Task	Number of Test Performed	Average Response Time (minutes)	Maximum Average Response Time (minutes)	Number Failures	PFD (average)	LOPA PFD
Company A (USA), Site 1							
IPL 1	ABP: High Temp in Generator	6	2.3	10	0	0	0.1
IPL 2	ABP: Loss of Acid Flow to Generator	12	2.2	10	0	0	0.1
Company B (Canada), Site 1							
IPL 1	Low Seal Gas Pressure to Turbo-Exchanger Bearings	5	5.7	15	0	0	0.1
IPL 2	High Lube Oil Temperature – XX Compressor	5	6.3	30	0	0	0.1
IPL 3	Low Level Emergency Alarm -- Steam Drum	5	4.9	15	0	0	0.1
IPL 4	High-High Lube Oil Temperature – XX Compressor	5	6.1	30	0	0	0.1
Company C (Malaysia), Site 1							
IPL 1	High Level on Ammonia Absorber Column	10	5.1	15	0	0	0.1
IPL 2	Low Temperature Alarm on Ammonia Compressor Discharge	10	4.9	15	0	0	0.1
IPL 3	Low Level in CO_2 Compressor Interstage KO drum	10	7.0	15	0	0	0.1
IPL 4	High Level on High Pressure Carbamate Heat Exchanger	10	6.1	15	0	0	0.1
IPL 5	High Pressure High Pressure Carbamate Condenser	10	5.0	15	0	0	0.1
IPL 6	High Pressure on Steam Controller to Rectifying Column Recirculation Heater	10	5.4	15	0	0	0.1

[a]Data provided by 3 companies (USA and Malaysia)

- Perhaps do just 1 or 2 drills per shift per group per year for the simpler ones on some periodic basis; that gives a chance to test feasibility (make sure valves are not locked, make sure a valve wrench or other tools are available, etc.).

Human performance sampling is discussed in more detail later in this paper.

5.1.3 Adjustment of Results for Stress

As mentioned, this data (as with all drills) is collected during a simulation of a call for action. In a real event, ***the stress to perform the task correctly would increase the average error rate***. NRC estimates[14] the stress for this type of pre-emergency response action (versus emergency response and evacuation) will likely not be "extreme" but it will be "high," in which case a conservative estimate is that error rates would double from the test/drill case. It is likely not possible to get a drill that accurately mimics the stress of a real alarm event, so there will likely always be a need to adjust data for increases in errors due to stress. It is likely more appropriate to double the observed error rates (observed PFDs) rather than doubling the observed response time. But in either case, the IPL data collected above must still "pass" when adjusted for stress for an IPL to be validated using Site-Specific Data.

5.1.4 Comments from Site Personnel on Collection of Site-Specific Data

Below are responses from the sites to questions related to the measurement:

- ***How much labor did each performance test take (on average)?***
  - The majority of the labor was with the actual operator going out to the field to 'take action' on the 'alarm'. Prep time was 1 hour for the initial recording of 10 data points for each of 2 to 6 IPLs, the prep time for each test was minimal as the cards were prepared beforehand and the recording takes less than 10 seconds.
  - The test time was about 15 minutes per operator per test.
  - Recording time was negligible.

- ***How much work do you think this would be longer-term (after getting in the habit and working any bugs out)?***
  - Not much difference in time as the testing process was picked up after the 4th to 5th data point in the unit, so it did not take long to get to "minimal effort." One benefit of the test was the re-enforcement of "what to do" with all parties involved (the person doing the test, the person recording the test, and the other operators who noticed the test in progress).
  - The hard part is getting that up-front commitment; after doing a few tests, doing more was no issue.

- ***Do you feel such data is necessary for each human response IPL? (In other words, do you think a 5-10% sample of related type of responses would suffice?)***

- The response for all the IPLs were the same when the same operator acts on different alarms or when different operators act on the same alarms. If this is the case, we probably can use a sample to test a "general" response to a type of alarm.
- The site should first screen which response can be grouped together into general types of response IPLs. Then a lesser percentage will need individual validation drills.
- Perhaps do just 1 or 2 drills per shift per group per year for the simpler ones on some periodic basis; that gives you a chance to test feasibility (make sure valves not locked, make sure valve wrench or other tools are available, etc).

- ***What improvements should be made to the tests/validation scheme?***
  - If we make the simulation a little bit sophisticated, we may be able to induce stress and surprise (maybe create some disorientation or carry out the tests in not-so-smooth situations – such as when Gene Hackman commenced a missile drill at the end of a minor crisis in the movie *Crimson Tide*) so that we can get closer to the actual critical situations. This would obviously need to be under very controlled conditions.
  - Perhaps with simulators or other methods, we could also include "detection and diagnosis" of the alarm in the same test.

- ***Who administered the test?***
  - This varied across the three sites, but were either plant production engineer, shift supervisor (mostly), PSM Engineer, or Shift Executive (Engineer) although these latter two were more in an observer capacity.

- ***Who do you think should admin the tests in future?***
  - Same or a Lead Operator, which is basically an expert from the operator ranks.
  - Shift Supervisor level. In the Malaysian company, this is the highest level for non-degree holders in operations (or it can also be a newly graduated engineer).

- ***What unit was this in?***
  - This was a small batch plant in Texas, Ammonia Plant and Urea Plant in Malaysia, and a specialty continuous chemical plant in Alberta Canada.

- ***Overall feeling of the exercise?***
  - The staff said it was easy to do and didn't have much difficulty in setting it up.
  - Initially, probably due to not fully seeing the whole picture, it was seen as a burden to do. Convincing was required to explain the potential benefits. But the plant staff rose to the occasion and the actual implementation was not so bad judging by what was able to be achieved even with the unplanned shutdowns that happen to occur in the month when the validation tests were run.

5.2 Approach to Using a Statistical Sample Plan for Validation of Human IPLs.

It is important to ensure that a 0.1 value is indeed valid before using a human IPL. Rather than testing 100% of the responses by 100% of the operators, it is normally valid to use a sampling plan. This is especially true for groups of responses that are similar in action and response time. US NRC alluded to a "sampling" of human response in 10 CFR 55.59, indicating that this may be acceptable for validating human response to triggers (i.e., for validating Human IPLs). Statistical techniques developed decades ago are used to establish a basis of acceptance of all kinds of products, raw materials, components, etc. (See Walpole, 2006, and other standard textbooks for typical approach on statistics and sampling.) These methods can also be used to validate Human IPLs.

The sample plan approach must group similar type of actions and similar response time requirements. For a sample plan approach, choice of responder and trigger must be chosen at random. The lot size is the product of the number of responders multiplied by the number of similar response actions in the group.

As a rough rule of thumb, the sample should be about 10% to 5% of your total population of data, but not smaller than 30 samples and not greater than 350 to 500 samples. The sample size and number of failures before the PFD is invalid is related to the confidence level and margin of error that is acceptable to the organization. A confidence level of 95% and an error margin of 5% indicate that the result of your testing (validation of a Human IPL) will be within +/-5% of the true PFD 95% of the time the validation testing is performed.

The correct sample size is a function of those three elements – your universe (how many people multiplied by the number of actions; each pairing makes up a Human IPL), the desired error margin, and the preferred confidence level. For IPL validation purposes, it is likely reasonable to use a 5% error margin at 95% confidence. Below are typical sample sizes (the first at a 10% error margin, the second at 5%):

50 in the population, sample 33 or 44
100 in the population, sample 49 or 80
200 in the population, sample 65 or 132
500 in the population, sample 81 or 217
1000 in the population, sample 88 or 278

The trend above approaches a limit that hardly moves above a 350 sample size no matter how large the population, for a 10% error margin (and approaches a limit of 500 for a 5% error margin). The sample size can also be approximated using Equation 1 below:

Equation 1: SS (for infinite population) = $$\frac{Z^2 * (p) * (1-p)}{c^2}$$

Where:

- Z = Z value (e.g. 1.96 for 97.5% confidence level for single-sided, normal distribution; note that though human action is pass/fail and so is typically described using binomial distributions, for large populations/groups, a normal distribution approximates a binomial distribution)
- p = percentage picking a choice, expressed as decimal (0.9 used for Human IPL sample size determination)
- c = confidence interval, expressed as decimal (e.g., .05= ±5%)

This calculation then must be corrected to account for a finite population (validation SS):

Equation 2: Validation SS (finite population) = $$\frac{SS}{1 + \frac{SS - 1}{\text{population}}}$$

Meeting acceptance criteria for a human response means that the procedures, training, retraining, communication control, and all other human factors are achieving the desired result of no more than 1 wrong response in 10 demands. Rejection means that the PFD of 0.1 is not valid (or that the control of human error needs improvement for the PFD to remain valid). When applying this sampling plan to a validation of administrative IPLs, the "acceptance" criteria (the desired response) is the response that prevents the consequence that is being considered in the LOPA, e.g., the operator response to the alarm prevents the overflow.

Sampling plan schemes rely on developing a valid definition of a "group." For Human IPLs, the group is a combination of similar operators (likely all "qualified, independent" operators are treated equally for this purpose) and similar responses to similar triggers. Creating this grouping takes careful consideration by a multiple-disciplinary team (with heavy emphasis on the operators on the team composition), to ensure the grouping of IPLs makes sense (i.e., use expert judgment). Although it is likely that all operators can be lumped into the same statistical group (if they are selected at random for validation drills of IPLs), the triggers and responses will need to be grouped to ensure that validation of one trigger/response is essentially the same as validating other triggers/response within the same group.

Next, the sample size and pass/fail targets must be estimated. This is based on the size of the groupings, the confidence level desired in the result, and the expected error margin (distribution) of the results.

- One method for determining sample size is to use Equation 1 and 2 discussed earlier. After the sample size if determined, then the pass/fail target (to prove if the hypothesis of a PFD of 0.1 is valid or invalid) can be estimated (refer to statistical textbooks for this derivation).

- Another recognized standard for determining the sample size and the pass/fail targets is to use ANSI Z1.4; this standard is now incorporated by reference within the US MIL standards for lot sampling and pass/fail determination. Examples calculations and the resulting savings achievable by use of sample plans are also provided in ANSI Z1.4.

Example Calculations: Using the same theoretical example as before with 20 operators and 130 alarms that require response, the total population of combinations of responses and operators is 2,600.

Case A: For one case, assume that all actions and all operators are equivalent and so the population makes up one group. In this case, at a confidence level of 97.5% and a confidence interval of 5%, the sample size adjusted for the finite population would be 131. So, only 131 combinations of operators and alarms/actions would have to be validated (tested) each period (and a typical period is each year). This is about 5% of the total population of human IPLs and so the validation would likely only take .05% of a staff year per operator, or 1 hour per year per operator (a very small investment in time for validation of IPLs). For this sample of 131 validation test, if 8 or more fail to accomplish the action in the required time, then All human IPLs have failed their validations. In addition, this is without counting for the stress of an actual response. If the error rate is doubled to account for stress, then the number of acceptable failures is cut in half, so for the sample size of 131, if 4 or more fail, then All human IPLs have failed their validations. If the validation failed, then the site would likely enlarge the sample size and re-test and/or find out where the problems are occurring and work to improve the response success. Regardless, the workload would be less with sampling and the company would obtain valuable insights into where to focus improvement efforts for Human IPLs.

Case B: For this case, assume that not all alarms/actions are equivalent, but still assume as in Case that all operators are equivalent. Further assume that the alarms and actions can be separated into 10 groups of 13 similar alarms/actions each. In this case, the population of any one group is 20 operators multiplied by 13 alarms or actions for a total of 260. In this case, at a confidence level of 97.5% and a confidence interval of 5%, the sample size adjusted for the finite population would be 91. So, 91 combinations of operators and alarms/actions from each of the 10 groups would have to be validated (tested) each period (and a typical period is each year). This is about 35% of the total population of human IPLs and so the validation would likely only take 0.35% of a staff year per operator, or 6 hour per year per operator (a small investment in time for validation of IPLs). For this sample of 91 validation test, if 6 or more fail to accomplish the action in the required time, then all human IPLs in this grouping of 13 alarms or actions have failed their validations. In addition, this is without counting for the stress of an actual response. If the error rate is doubled to account for stress, then the number of acceptable failures is cut in half, so for the sample size of 91, if 3 or more fail, then all human IPLs in this grouping have failed their validations. If the validation of a group of alarms failed, then the site would likely enlarge the sample size for that group and re-test and/or find out where the problems are occurring and find ways to improve the response success. Regardless, the workload would be less with sampling and the company would obtain valuable insights into where to focus improvement efforts for Human IPLs.

Table 5 provides various Sample Size and the Maximum Number of Failures (pass/fail target) using standard statistical methods (such as Equation 1 and 2); this table includes the data for Case A and B above.

6. Estimation of PFD Using Generic Human Factor Data and Calculation Approach

An alternative to collecting your own data to validate the PFD for a human IPL is to calculate the PFD by combining expert judgment and generic and predicted data (see CCPS text, Guidelines for Independent Protection Layers and Initiating Events, Appendix B[4] and the paper by Stack [15]). The calculated value is then used to validate the PFD value to be used in LOPA.

The PFD is calculated by combining a baseline unreliability value with Performance Shaping Factors (PSFs), also called human factors. The **baseline unreliability** value depends on the type of activity and generally is **set at 0.01 for a response activity** when used for basic LOPA. *(Note: the baseline error rate can be as low as 0.001 for an initiating event activity, where time pressure is not a factor, such as following a procedure to startup a unit.)* The calculation steps use the following PSFs that are specific to the response task:

- Number of steps in the procedure
- Number of activities or alarms per year
- External stress factors
- Communication factor
- Feedback factor

In addition, this value must be adjusted for a general score on global human factors issues at the site, such a control of fitness for duty (most notably control or not of fatigue), work environment, and human system interface. This global human factor issues adjustment is described at the end of this section.

6.1 Example of Calculation Approach, before Adjustment for Global Human Factors Issues

The example below illustrates this approach.

Number of steps in the procedure: If the number of steps in a procedure is less than 10, then the PSF multiplier = Number of steps/10, otherwise the factor is =1.0.

Table 5: Sample Size and Acceptance Criteria Estimation for Human IPL Validation (single-sided Z test; for normal distribution)

Employee	Alarms	pop = actual size of population	SS = sample size based on infinite population	Z value	Confidence Level	p = anticipated PFD	c = confidence interval (related to expected range of results; +/-)	Validation SS = sample size adjusted for actual population size	% of actual population sampled	Expected average number of failures per test period using the sample size	Acceptable failures to have Z confidence (95% if Z = 1.645) that the entire population has less than 1-p error rate (accounting for one side of the confidence interval [+%] of the test plan)	**Therefore, the PFD of the human response IPL is not 0.1 or better if the number of failures are equal to or greater than:**
20	130	2600	59	1.28	90%	0.9	0.05	58	2	5.8	3	4
20	13	260	59	1.28	90%	0.9	0.05	48	19	4.8	2	3
20	130	2600	97	1.645	95%	0.9	0.05	94	4	9.4	5	6
20	13	260	97	1.645	95%	0.9	0.05	71	27	7.1	4	5
20	130	2600	138	1.96	97.5%	0.9	0.05	131	5	13.1	7	8
20	13	260	138	1.96	97.5%	0.9	0.05	91	35	9.1	5	6

Number of activities or alarms per year: An activity frequency factor used as the PSF multiplier for the number of activities or alarms per year is obtained from Equation 3:

Equation 3: Activity Frequency factor = $(12/\text{frequency per year})^{0.25}$

External stress factors: The External Stressor Factors are provided for stress, conflict with normal expectations, and distractions with values ranging from 2 when an operator has significant distractions up to 15 in which some tasks in the procedure need to be executed in close proximity to the portion of the unit that is in a more dangerous mode for this accident scenario.

Communication factor: The effectiveness of communications has been shown to be a critical factor where a key operator is responding to an alarm. Values of 1 are assigned in cases where an alarm is received and acted upon by a single person up to a maximum 5 in situations where an alarm is received and acted upon by multiple people in different locations and the communication between the people is verbal by phone or radio.

Feedback factor: Feedback allows the operator to take additional action if the intended results do not occur, if there is sufficient time. If there is no feedback, then a PSF figure of 1 is used, with the lowest value being 0.01 if the feedback is immediate and compelling.

Example: Consider the case of response to a high temperature situation in a monomer tank, which could lead to a runaway reaction and possible tank failure. The selection of values and the resulting calculation of the human response IPL for this scenario are presented in Table 5.

The calculated PFD for the human IPL in this case (interim result) is:

0.01 x 1 x 1.57 x 6 x 5 x 0.1 = **0.047**

Therefore a PFD of 0.10 would be valid in LOPA.

However, if the tank farm fire water monitor were activated from the control room, by the control room operator, then the alarm would be received and acted upon by a single person (so the Communication PSF Value becomes 1) and is checked by a video monitor in the control room (so the Feedback PSF Value becomes 0.05), then the calculated PFD for the human IPL in this case (still an interim result) would now be:

0.01 x 1 x 1.57 x 6 x 1 x 0.05 = **0.005**

Therefore a value of 0.01 should be valid for use of this human IPL in a LOPA scenario.

Table 5: Estimation of PFD of Human IPL Using Calculations

Step #	Validation Step Description	Value	PSF (multiplier)	Probability	Perception, Diagnosis, and Response Activities
1	Activities Involved (type of action required) - this sets the **baseline probability**			0.01	Board operator acknowledges alarm & contacts field operator to put fire monitor spray onto the monomer tank to control high temperature detected/alarmed in the tank
2	Number of steps	10	1		10 steps in the actual procedure, including immediate actions, diagnosis, and feedback checks
3	Number of activities or alarms per year	2	1.57		Estimate of 2, since operator is trained or tested annually on what to do and there is also about one event per year
4	External Stressor Factors		6		Field operator does not have to go to monomer tank (hazardous zone) to verify the temperature; there is remote display from multiple indicators in the control room; independent of alarm. Field operator does not have to go near the monomer tank. Control operator still under threat of taking wrong actions.
5	Communication Factor		5		Board operator and field operator have two way push to talk cell phones to work through the trouble-shooting guide
6	Feedback Factor		0.1		Board operator would observe tank temperature after water spray application by field operator, and field operator still required to observe water spray from safe distance, so feedback is good
			Interim Result	0.047	<< Calculated PFD for this Human IPL
			FINAL RESULT	**0.10**	**<< PFD to use for this Human IPL (must be equal to or less than 0.1 to allow this Human IPL)**

6.1 Example Adjustment for Global Human Factors Issues for the Site and Mode of Operation

The value above did not pre-adjust the baseline human error rate for the mode of operation and for the specific site. Global Human Factors Issues that are different for each (1) mode of operation and (2) site include:

- Fitness for Duty
- Work environment
- Human System Interface

Many approaches have been proposed for qualitative scoring (excellent, good, fair, poor) and for quantitative estimation of global human factors issues, such as SPAR-H[14]. These approaches share a common strategy of combining the multiplying effect that poor human factors can have on baseline error rates. For an IPL, the baseline rate is normally set at 0.01 because of the success rate reduction imposed by a limited time to take detect, decide, take action, and verify the response. As stated in Table 1, each human factor deficiency can have a multiplying effect on the number of human errors:

Fitness for Duty Control

- Fatigue – Can affect error rates by a factor of up to 20 times.
- Other fitness of duty issues – Can affect error rates by a factor of up to 20 to 50 times

Work Environment Control

- Can affect error rates by a factor of up to 5 times

Human-System Interface Weakness

- Can increase human error rates by 10 times or higher (especially if "norms" are violated, where a negative effect of 20 to 50 times is assumed, but we are assuming that a LOPA scenario evaluation or PHA/HAZOP analysis prior to LOPA would have already eliminated any violation of human-system interface norms.)

It has been noticed that even in accidents involving a combination of human system weaknesses, the error rates for humans appears to reach a plateau of 20% maximum (or 1 mistake in 5 steps). Therefore, 0.2 is the upper bound for a PFD for a human response IPL and 0.01 is the lower bound. One approach is to use the estimates of the "local" factors for fitness for duty control, work environment control, and human-system interface weaknesses to adjust the interim human error rate calculated in section 6.1, but limit to the bounds just stated. Table 6 lists typical multipliers we use for Global Human Factors Issues.

SPAR-H typically has larger multiplying effects, but we adjusted these to match the 0.01 to 0.2 range for PFD for a human response IPL.

Table 6: Adjustment Factors for Estimation of PFD of a Human Response IPL Global Human Factors Issues (based on SPAR-H[14])

Human Factor Category	Human Factor Issue/Level	Multiplier for Cognitive & Diagnosis Errors
Fitness for Duty	Unfit (high fatigue level, illness, strong medication, not physically capable of job today) Degraded Fitness Nominal	10 - 20 5 1
Work Environment	Extreme (high noise, heat, vibration; too cold) Good	5 1
Human-Machine Interface (includes tools)	Missing/Misleading (violates what is normally expected) Poor Nominal	10 - 20 5 1

Adjustment Approach:

1. Rank the Human Factors Issue/Level using the qualitative terms in the middle column of Table 6.
2. Select the matching multiplier in the right-hand column.
3. Multiply the Interim PFD (as calculated in Table 5) by the multipliers, but do not exceed a PFD of 0.2
4. If using Basic LOPA, do not use a human response IPL that has a PFD greater than 0.1.

One obvious benefit of this adjustment approach is that it raises attention to the gains possible on improving human factors to drop these multipliers closer to 1.

7. Human Errors as Related to Impact on Non-human IEs and IPLs (such as SIS and Relief Systems)

In addition to having a direct impact as an IE or as failure of a human IPL, human error can of course impact all other IEs and IPLs as well. This is because ***all*** safeguards are ultimately controlled and maintained by humans. Typically, a baseline human error rate of 1 error in 50 steps is reasonable, or for excellent control of human factors, a baseline error rate of 1 error in 100 steps may be achieved (as described earlier). This baseline error rate then can increase if the control of human factors slip, as described earlier in this paper. Increasing human error rates in inspection, calibration, maintenance, quality control of part, commissioning (or re-commissioning activities) will impact the failure rate of engineered IPLs or cause more frequent initiating events.

Below are some examples of how human factors control the reliability of all IPLs and IEs:

Example related to IE – wrong materials of construction received: If the humans fail to control or detect errors during selection and use of materials of construction, then perhaps the wrong grade of steel or other material will be used in a process environment that is highly sensitive to materials of construction, such as highly corrosive service or cryogenic service. So, the IEF will be many times higher than expected.

Industry data in the 1990s indicated that the materials delivered were different than what we specified about 3 to 7% of the time. Positive material identification (PMI) was employed to detect and correct such failures, but plant data suggests that using 100% PMI at receiving lowered the composite error rate to 1% and that another 100% PMI of "as-welded in the field" lowered the combined error rate to about 0.3%. This still left a large number of components that may be of the wrong materials. In the past 10 years, the use of PMI throughout the entire supply chain has dramatically improved and third-party inspectors are now expert at PMI checks as well. This about a 10 fold reduction in the number of parts received that are the wrong materials of construction; the error rate is now below what is measureable by plant staff doing 100% or 200% PMI checks.

Example related to IPL – high integrity SIS: Safety Instrumented Systems (SIS) can be designed to provide risk reductions of 1, 2, or 3 orders of magnitude. However, even the best designed SIS can be negated if (1) the root valves for the sensors are left closed or (2) a bypass valve around an emergency isolation valve is left open. Other errors are also possible; for instance the humans may be capable of revaluing trip points, etc., in the safety-rated PLC, given enough time to learn these shortcuts and given enough spurious trips to make them want to use such a shortcut. The current versions of SIS standards do not explicitly require validation of the control of human factors and the resulting human errors that can leave the valves in the wrong position or otherwise increase the PFD of the SIS. The current versions of the standards also do not explicitly require validation of the likelihood of other systemic errors, such as plugging of sensor ports by contaminants through the system, which would be a common cause failure mechanism for all ports. Although a SIL 4 is allowed by IEC 61508, with a PFD of $<10^{-4}$ to $\geq 10^{-5}$, it is very unlikely to control human errors low enough to install or maintain a SIL 4. Similarly, due to the complexity of the analysis of human errors associated with in situ testing of SIS, basic LOPA only allows use of SIL 1, unless the systemic errors of leaving SIS bypassed is accounted for rigorously in the SIL verification step.

Example related to IPL – pressure relief systems: Many relief system IPLs can have a PFD of 0.01 or better, assuming the rest of the process does not interfere or restrict the flow. However, if you have a block valve upstream or downstream of a relief device, then leaving it closed will negate the entire value of the relief system. A human reliability analysis of existing plant practices is needed to establish that the PFD is better than 0.1, and excellent sustained control of human factors is necessary to maintain a PFD of .01 or better for a relief system, because of the relatively high probability that a human will leave a block valve closed upstream or downstream of a relief valve. Common industry data for human error rates indicates that the error of leaving such a block valve closed will likely be the dominating failure mode for the relief system, so including block valve upstream or downstream of a relief device (though allowed by ASME with administrative controls) will need to be evaluated by a site before assigning a PFD for the combination system (comprised of the relief system and block valves). The conservative

PFD of 0.1 is therefore used as the default value for relief devices that have blocks valves upstream or downstream of relief devices, unless a human reliability for the plant proves otherwise. A preferred design is to install dual relief valves with a single valve arrangement (3-way valve) that helps ensure against blocking flow to the on-stream relief valve.

8. Conclusion

Control of human factors is key to achieving high reliability of processes. Poor human factors lead to higher initiating event rates and less or no value for IPLs. Proper control of human factors can be achieved by following industry best practices (not minimal regulations and codes/standards).

Failing to control human factors will result in the time invested in LOPA to be wasted, and more importantly will lead to more accidents. Simply installing or claiming IPLs will not change the critical dependency all plants face on controlling risk – risk control always depends on human factor control.

9. References

1. "Layer of Protection Analysis (LOPA) Guideline," CCPS/AIChE, 2001.
2. Bridges, W., "LOPA and Human Reliability," *6th Global Congress on Process Safety*, AIChE, 2010.
3. Tew, R. and Bridges, W., "Human Factors Missing from PSM," *6th Global Congress on Process Safety*, AIChE, 2010.
4. "Guidelines for Independent Protection Layers and Initiating Events," AIChE/CCPS, 2011 [pending].
5. Swain, A. D., Guttmann, H. E., "Handbook of Human Reliability Analysis with Emphasis on Nuclear Power Plant Applications, Final Report," NUREG/CR-1278, 1983, US Nuclear Regulatory Commission.
6. Rasmussen, J., and Rouse, W., "Human Detection and Diagnosis of System Failures," New York, Plenum Press, 1981.
7. Rasmussen, J., "Chemical Process Hazards Identification," *Reliability Engineering and Safety Systems 24*, Elsevier Science Publishers Ltd., UK, 1989.
8. Bridges, W. and Clark, T., "How to Efficiently Perform the Hazard Evaluation (PHA) Required for Non-Routine Modes of Operation (Startup, Shutdown, Online Maintenance)," *7th Global Congress on Process Safety*, AIChE, March 2011.
9. "Guidelines for Hazard Evaluation Procedures, 3rd Edition", CCPS/AIChE, 2008.
10. Bridges, W., Kirkman, J. and Lorenzo, D., "Strategies for Integrating Human Reliability Analysis into Process Hazard Evaluations," *International Conference*

on Hazard Identification and Risk Analysis, Human Factors and Human Reliability in Process Safety, *CCPS/AIChE,* January 1992.

11. Bridges, W., "Gains in Getting Near Misses Reported," *8th Conference, ASSE-MEC*, Bahrain, 2008.
12. "Guidelines for Investigating Chemical Process Incidents, 2nd Edition," CCPS/AICHE, 2003.
13. "Training Requirements for Nuclear Power Plant Operators," 10 CFR 55.45 and 55.59, US Nuclear Regulatory Commission,
14. Gertman, D.; Blackman, H.; Marble, J.; Byers, J. and Smith, C., "The SPAR-H Human Reliability Analysis Method," NUREG/CR-6883, U.S. Nuclear Regulatory Commission, Office of Nuclear Regulatory Research, Washington, DC, August 2005.
15. Stack, R. and Delanoy, P., "Evaluating Human Response to An Alarm for LOPA or Safety Studies," *6th Global Congress on Process Safety*, AIChE, 2010.

A Surgical Risk Reduction Approach to Reduce Global Process Safety Risk: The "PSM Gamma-Knife"

Bruce K. Vaughen
Cabot Corporation
700 E US Highway 36
Tuscola, IL 61953
(217) 253-5775
bruce_vaughen@cabot-corp.com

Pieter Kruger
Cabot Corporation
700 E US Highway 36
Tuscola, IL 61953

Tim Hauser
Cabot Corporation
700 E US Highway 36
Tuscola, IL 61953

Keywords: Risk, Reduce, Process Safety Management, Monte Carlo, Operational Discipline, PHA, LOPA, SIL, PSM

Abstract

Businesses with more than one site strive to minimize operational variability across their entire organization. However, uniformly implementing company-wide standards is difficult because each site maintains its own personality – its own culture. When a business strives to reduce its global process safety risk, a significant incident that exposes a particular site's weakness may result in a simple corporate decision forcing all sites to implement the same process safety improvement. This paper shares the basic concept and the novel use of a risk minimization tool, commonly used to reduce financial risks in an investment portfolio, to help decision-makers minimize their multisite, global process safety risk without adding undue stress on the rest of the organization. This approach helps quantify each individual site's process safety risks into a global business process safety risk model, uses a Monte Carlo simulation to determine which site-specific factors contribute the most to the overall business risk, and then helps decision-makers prioritize and choose site-specific risk reduction efforts that minimizes the global risk. The global process safety risk reduction model used to show how to use this approach is based on a simplified risk equation containing site-specific terms for frequency, consequence and Operational Discipline. Often the most contentious term in risk estimating is the frequency, especially with significant scenarios that have never happened before and, hence, have a very low likelihood. The beauty of the Monte Carlo simulation is that the user can account for the uncertainty in the frequency by choosing a frequency probability function based on a range of possible best- to worst-case consequence scenarios. From the surgical perspective, a person (the whole organization) with cancer is often adversely affected when treated with chemotherapy.

The results from this approach, using the image of a "PSM Gamma Knife," show how selective site-specific risk reduction treatments specifically address each site's process safety risk to help reduce the overall risk without adversely affecting the health of the whole organization.

1. Introduction

This paper is designed to share a novel approach to help quantify a business decision-making process to reduce its overall process safety risk. This risk modeling approach drives business consensus on events with low likelihood, uncertain frequencies, helping reduce upper-level management disputes by quantifying the decision-making process on site-specific risk reduction efforts. The quantified Monte Carlo simulation results provide easy-to-interpret decision-making graphs based on the agreed-upon inputs (i.e., Tornado charts), such that case studies are easily compared using the different risk distribution profiles. The authors hope that the example provided below provides a framework for decision-makers to model their risk reduction efforts and help reduce their business' overall process safety risk on a site-by-site basis that does not add undue stress to the entire organization.

2. A Multi-Site Risk Model

To help quantify and describe the surgical risk reduction approach to reduce global risk, the following simplified risk equation is used with the risk represented as a function of three variables: the event's consequence, the event's estimated frequency and the site's safety culture and subsequent adherence to the Operational Discipline characteristics [1]:

$$\text{Risk} = \frac{\text{Frequency} \times \text{Consequence}}{\text{Operational Discipline}} \qquad \text{[Eq. 1]}$$

A goal of a multi-site global process safety risk reduction effort is to reduce the overall business risk. Assuming that each site i has a unique, independent risk term and its own site-specific safety culture, the "sum" of these individual site-related risks could be used as a measure of the overall or global business risk, and could be represented by the following equation where the business used in this example has n separate sites, i:

$$\text{Risk}_{\text{Business}} = \sum_{i=1}^{n} \frac{F_i C_i}{OD_i} \qquad \text{[Eq. 2]}$$

The authors recognize that the general terms in this simple risk equation are not independent - for example, poor OD on planned maintenance could increase the frequency of failure of an engineering control. However, by choosing this simple equation, the general surgical risk reduction approach can be clearly described with the simulation results presented below. More complicated models can be used once the basic concepts are understood. The purpose of this paper is to explore the novel use of a risk minimization tool on the risk model shown in Equation 2 to help decision-makers quantify and minimize their global risk.

3. The Surgical Risk Reduction Approach

This section describes a risk minimization tool that helps multi-site businesses prioritize their risk-reduction strategy with a quantified decision making process. This analysis drives consensus on the model's inputs, and, based on the model's results (its outputs), reduces the likelihood of disputes that may occur when deciding on the next business process safety risk reduction effort.

The tool provides a global risk reduction view that defines and identifies the individual variables impacting each site's risk. Once the variables with the greatest impact are identified, decision-makers can prioritize which effort to work on first. The solution may involve significant engineering efforts (and capital expense), or even a significant re-evaluation using Inherently Safer Processes (ISP) and design [3, 4] when consequence-related reduction efforts need to be first. If the focus is placed on reducing the frequency, efforts could proceed towards a Layer of Protection Analysis (LOPA) with its corresponding Safety Integrity Level (SIL) analysis to more accurately define frequency [5]. The results from this approach benefit the overall business since these efforts will be site-specific, not global, and will provide the business with it best return (a reduced global risk) on its site-specific investments.

Assuming that the risk of single site's hazardous event is described by Equation 1, the event's risk is directly related to the event's frequency (F), directly related to event's consequence (C), and is inversely related to site's "Operational Discipline." Engineering and administrative controls, such as safety systems and routine interlock verification tests, can be used to ensure that the Frequency is as low as possible. Inherently safer design principles or comprehensive emergency response programs can be used to reduce the event's Consequence. The site's Operational Discipline (OD) provides a measure of its "safety culture," with poor OD reflected in a number less than 100%, and hence, an increased risk. Poor OD, such as letting critical system interlock tests lapse, is represented with a number less than one in the denominator. As was noted earlier, poor OD would increase the likelihood of the event, as well, which would raise the event's risk even more.

As is shown in Figure 1, the standard risk matrix visually represents an event's frequency against its consequence. This type of matrix is often used during Process Hazards Analysis (PHA) efforts to guide the PHA Team through its hazards analyses to help prevent undesirable events. The goal for the final low risk location on this matrix is the best combination of the lowest F and the lowest C that can be practically obtained. We note here that PHA Teams usually focus on improving administrative and engineering controls, which directly affect the Frequency and Consequence terms. Implementation and adherence to maintaining the PHA Team's recommendations, if any, is then addressed separately by ensuring that the site has effective Operational Discipline in its PSM program [2].

3.1 The Problem: The Traditional "Global" Treatment

The surgical risk reduction approach is obtained if one imagines the business, the organization, as a body with a head, torso, arms and legs, as is shown in Figure 2. Each individual site is represented as a part of the body, and each site has its own safety "cancer" that must be removed to help improve the overall business. However, using this surgical perspective, a person with a localized cancer is adversely affected when treated with a global chemotherapy solution (in particular, hair loss, stomach upsets, etc.). The tool described below can be thought of as a "PSM Gamma Knife," similar to a surgical Gamma Knife [6], to show how selective site-specific risk reduction treatments address each site's process safety risk without adversely affecting the health of and overall risk to the whole business.

3.2 A Solution: Perform Local Risk Reduction Surgery

This novel risk reduction approach quantifies the different site-specific issues as inputs on the global risk-reduction model (shown in Equation 2) through a Monte Carlo simulation that estimates the best site-specific risk reduction approaches to minimize (reduce) the overall business risk. This risk minimization approach is used for many other types of risk reduction efforts, including those for financial asset portfolio risk management [7]. Although the Monte Carlo simulations and this surgical risk reduction approach are not new, increased desktop computing power, available user-friendly risk-minimization software and quantitative process-safety risk models make the simulation easier to apply in other areas such as process safety.

3.3 The "Surgical" Risk Reduction Approach

The global process safety risk model, shown in Equation 2, contains site-specific frequency and consequence terms, and includes an estimate of the site's Operational Discipline (OD). Although OD is difficult to quantify, a relative measure can be developed, agreed upon by the business, and used to help understand the site-specific issues and address the overall business need. Examples of OD-related metrics are presented elsewhere [8]. Note that the frequency and consequence issues are usually determined through the efforts of Process Hazards Analysis (PHA) Teams [9].

3.3.1 Calculating Global Risk

For those familiar to the trials and tribulations when performing detailed PHAs, the most heated arguments often occur around estimates of the worst case scenario's event frequency. The beauty of the Monte Carlo simulation is that the user can account for the uncertainty in the frequency by choosing a frequency probability function based on the range of possible best- to worst-case consequence scenarios. This aspect of the simulation provides a much better likelihood "range" beyond what is probably the most problematic aspect when determining the location of the risk on the Risk Matrix (Figure 1).

The results from a Monte Carlo simulation using a standard risk minimization software package can be shown using a risk distribution chart. For the example presented below, the original overall business risk profile, the Base Case, is shown in Figure 3 and presented in Table 1. The more uncertainty, the more the overall risk will exceed its tolerable limit. The goal of a global risk reduction effort is to reduce the amount of risk exceeding this limit (to reduce the area of the curve beyond the limit). As is shown in Figure 3, the business has quantified its "tolerable" risk, estimated its site-specific issues, and established a risk profile of "acceptable" (in this case, below 50) and "unacceptable" risk (above 50). The goal of the risk reduction decision-maker is to select and invest in site-specific programs that move the overall profile further to the left such that more risk is below 50 – where less of the risk's uncertainty is below the limit.

3.3.2 The Tornado Chart

Once the site-specific issues are entered into the model, the Monte Carlo simulation uses pre-chosen uncertainty ranges for the frequency and iterates the calculations to generate the risk profile shown in Figure 3. Among many of the statistical charts that can be generated, the one called a "Tornado Chart" provides a clear visual graph that easily shows which site-specific efforts should be addressed first to help reduce the overall business risk. As is shown in Figure 4, if you imagine the bars rotating around the central axis, the graph would look like a tornado, hence its name: a "Tornado Chart." The goal is to reduce the "length" of the bars, which represent the relative impact of the particular site-specific term on the overall global risk profile. By focusing on the site-specific terms that have the "most impact" (i.e. the terms noted at the top of the tornado with the longest bars), managers can focus their resources on these efforts and make the greatest impact on reducing the global risk. The goal is zero (another familiar mantra). The results from the Base Case, and hence, recommendations for the decision-makers, are summarized and shown in Table 3 and are discussed in greater detail below [10]. Please do not conjure thoughts on the actual nature of tornados, which are rather unpredictable and very destructive, especially since this modeling approach is performed under a highly predictable and (hopefully) a very constructive environment.

3.3.3 Comparing Global Risk Reduction Case Studies

Two case studies are evaluated below, presenting a simple example of how focused site-specific risk reduction efforts produce a reduction in the risk profile's uncertainty (an overall risk distribution curve shift). The results are shown schematically in Figure 5, with the Base Case (red, the original risk profile, Figure 3, mean risk of 45 ± 6.3), Case 1 (blue, a subjective risk-reduction effort, Figure 7, mean risk of 40 ± 6.0), and Case 2 (green, the surgical risk-reduction effort, Figure 8, mean risk 35 ± 5.2). This simple example shows the concept for reducing global risk using this quantitative approach; more complicated interactions and models can be used to remove an inherent qualitative (and sometimes emotional) bias when comparing risk reduction options. The risk distribution profile shift from red to green quantitatively shows where the business risk-reduction priorities must be, hopefully helping eliminate disputes that naturally arise when managers compete over limited resources.

4. Applying the Surgical Risk Reduction Approach

This section provides a simple example of the surgical risk reduction approach, starting with a quantified risk matrix designating the business' chosen tolerable risk region. Two case studies, based on an original risk evaluation, show the difference between decisions based on a subjective risk reduction approach and decisions based on the surgical, quantified approach. The results from both cases are compared to show the overall effect on the global risk. Depending on the situations being compared, the software provides graphs to help decision-makers prioritize their efforts. As another saying goes: if you wish to explain your choices and decision to upper management, draw a picture. Hence, sites with limited resources can use this approach to justify that their resources are being used effectively to help minimize the overall business risk.

The business in this example, "Inpain," contains five different manufacturing sites making the same product. These sites, numbered 1 through 5, are located in the US and across the world, representing different global locations with different local, regional, and governmental expectations and cultures. These sites determined their risk using identical business-based risk measurements and plotted their risk evaluation location (their F and C) on the risk matrix, as is shown in Figure 7. Although each site uses similar raw materials, similar processing equipment and technologies to make similar final products, it is apparent that each site has a different a site-specific risk profile using the common business-guided risk criteria. This is a clue that each site has different risks, due in part to different local cultural impact and to different local resource constraints. As was noted before, when globally-mandated standards are created to address the ills of one site and to ensure consistent and uniform risk-reduction efforts across the entire business, the standard's implementation may add undue stress to the other sites (recall the image of chemotherapy described for Figure 2 above). Our goal: "Think Global. Act Local"

4.1 A Quantified Risk Matrix and Tolerable Risk Ranges

Inpain quantified its risk matrix by assigning each Frequency (F) and Consequence (C) level a weighted number from 1 (low) to 10 (high), with 2, 4 and 7 representing the levels in between. Entering a different numerical range, such as a linear 1 (lowest level), 2, 3, 4, and 5 (highest level), changes the numerical values in the simulation, but does not change the subsequent results when comparing case studies. Inpain's leaders chose its tolerable risk criterion at 50 (a numerical cut-off), and designated an "Orange Zone" for its "undesirable" risk range of 20 to 49. In other words, all risks determined at 50 or above are unacceptable and must be reduced. The entire risk ranking ranges, from acceptable (low) risk to unacceptable (high) risk, is shown in Figure 6.

4.2 Methods for Reducing Global Risk

As is shown by the arrows in Figure 7, Case 1 represents a subjective decision to reduce both Site 1's Frequency term from "Frequent" to "Probable" and Site 2's Consequence term from "Catastrophic" to "Critical" (the business certainly doesn't like the idea of having a "Catastrophic" event!). Hence, based on these decisions, the Site 1's site-specific efforts could be to reduce the frequency by determining additional administrative and engineering controls and for Site 2 to reduce the consequence by proposing Process Safety Information (PSI) design changes based on ISP principles [4].

Case 2 represents the risk reduction decisions based on the quantified, surgical approach using the simulation software results based on the original case [10]. As is shown in, Figure 8, Case 2 shows the choices recommended by the surgical approach using the inputs from the base case listed in Table 1 and shown in Table 3: reduce Site 1's Consequence term from "Serious" to "Marginal," and reduce Site 2's Frequency term from "Occasional" to "Remote." These quantitative result recommendations are designed to minimize the global risk, and are different (even appearing contrary!) to the subjective choices shown in Figure 7 for Case 1. Although both of these Cases are consistent with the goal to reduce the "Orange Zone" risks across the business, Case 2's solution is different.

Note that resourcing the site-specific Frequency or Consequence reduction efforts for either Case 1 or Case 2 will depend on the site-specific resources. A global business case for Case 2 (green) could be made using the visual chart shown in Figure 5, where the uncertain risk area is reduced more - the "best" global risk reduction effort - moving the overall business risk profile further to the left. As is shown in Figure 5, the overall risk profile has moved from to the left with Case 1, and further to the left with Case 2: there is less overall risk uncertainty in the "Unacceptable" range - greater than 50. In addition, because of the simple nature of this example, the business could be satisfied with its Case 1 results and not be interested in further risk reduction, or the subjective case could have been Case 2 in the beginning. However, with more complicated businesses, the quantitative technique shown in this paper could help decision-makers better separate and choose between many competing site-related efforts for the best overall business impact.

5. Conclusion

This paper is designed to share a novel approach to help drive consensus on risk modeling inputs using uncertain frequency estimates. The model and simulation includes complex input variables to help reduce upper-level management disputes by quantifying the decision-making process on site-specific risk reduction efforts. The Monte Carlo simulations provide excellent quantitative graphs based on the agreed-upon inputs, such as Tornado charts and case study comparisons and showing the different Risk Distributions. It is hoped that the example provided in this paper provides a framework for others to model their risk reduction efforts and reduce their overall process safety risk.

6. References

[1] Klein, James A., and Bruce K. Vaughen, "A Revised Program for Operational Discipline," AIChE, Process Safety Progress, Vol. 27, No. 1, March 2008, 58-65.

[2] Vaughen, Bruce K., and James A. Klein, and, "Improving Operational Discipline to Prevent Loss of Containment Incidents," AIChE, Process Safety Progress, accepted for publication October 2010, in press.

[3] Hendershot, Dennis, "What Does Inherently Safer Mean?," AIChE, Chemical Engineering Progress, August 2010, pp. 23-25.

[4] Center for Chemical Process Safety, "Inherently Safer Chemical Processes, A Life Cycle Approach," 2nd edition, AIChE, New York, NY, and John Wiley & Sons, Hoboken, NJ, 2009.

[5] Center for Chemical Process Safety, "Layer of Protection Analysis; Simplified Process Risk Assessment, AIChE, New York, NY, 2001.

[6] Klingemann. P., "3D Medial Axis Transform for Radiation Therapy," Report BSL2004-002, Bioimaging Systems Laboratory, Department of Electrical and Computer Engineering Virginia Polytechnic Institute and State University, Blacksburg, VA (2004).

[7] Ragsdale, Cliff T., Spreadsheet Modeling and Decision Analysis; A Practical Introduction to Management Science, Third Edition, South-Western College Publishing, Thomson Learning™, United States, 2001.

[8] Vaughen, Bruce K., Kelly Lock, Tim Floyd, "Improving Operating Discipline Through the Successful Implementation of a Mandated Behavior-Based Safety Program," Process Safety Progress, 29(3), pp. 192-200 (2010)

[9] Center for Chemical Process Safety, "Guidelines for Hazard Evaluation Procedures," 3rd edition, AIChE, New York, NY, 2009.

[10] @RISK, Palisade Software, Palisade Corporation, Ithaca, NY, http://www.palisade.com/risk/, accessed 09-Dec-2010.

Figure 1

The Risk Matrix

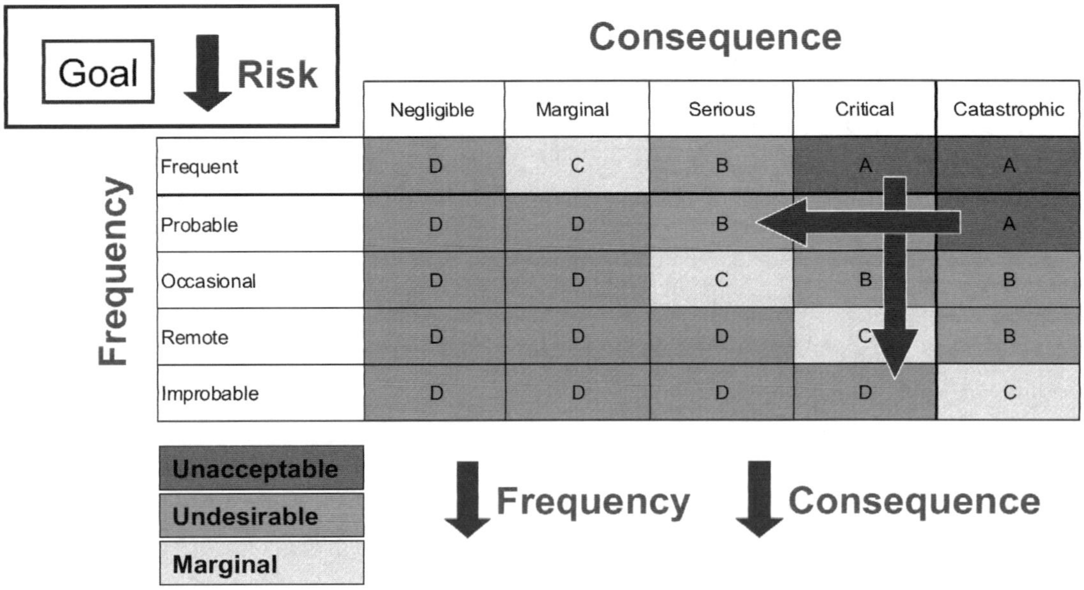

Figure 2

Image of the Surgical PSM "Gamma Knife" (Think Global. Act Local.)

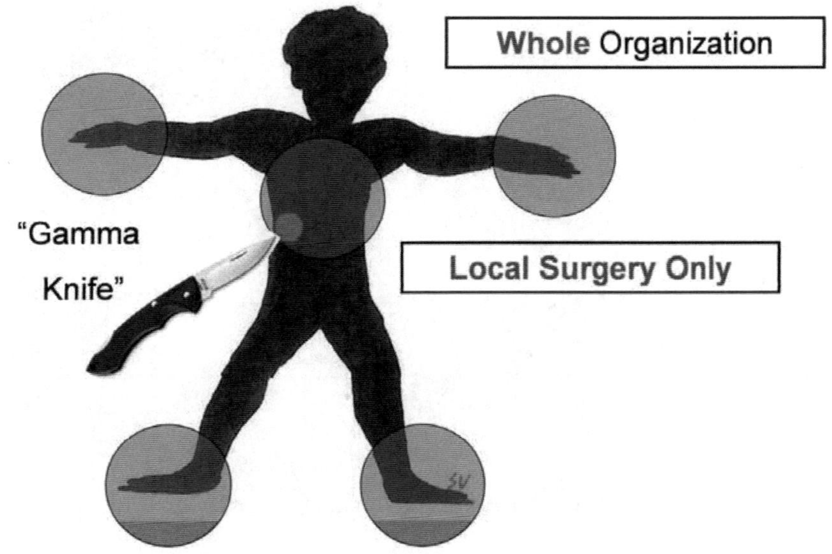

Figure 3

Histogram of the Base Case Risk Uncertainty Distribution

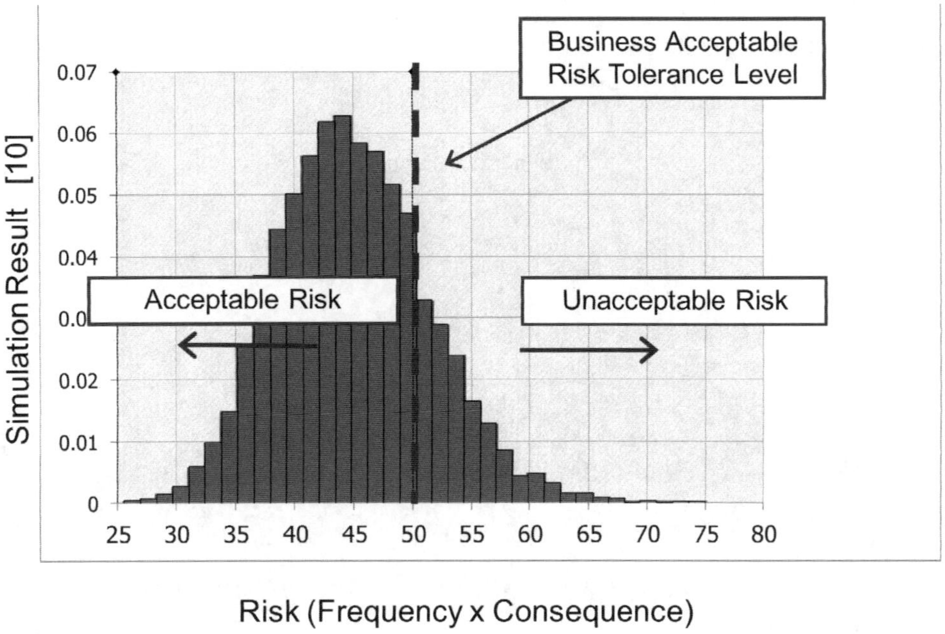

Figure 4

The Base Case "Tornado Chart" Chart

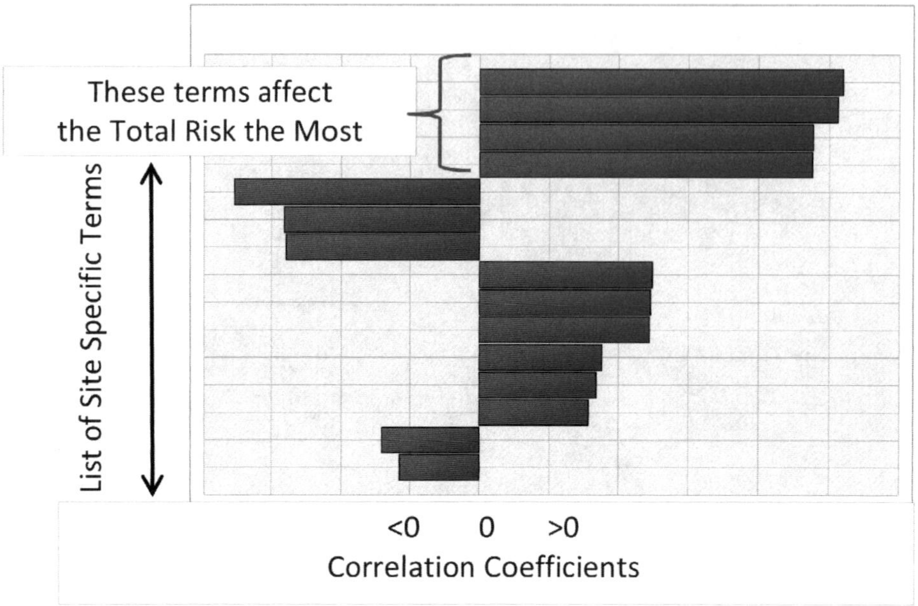

Figure 5

Histograms from Each Case Study's Uncertainty Distribution

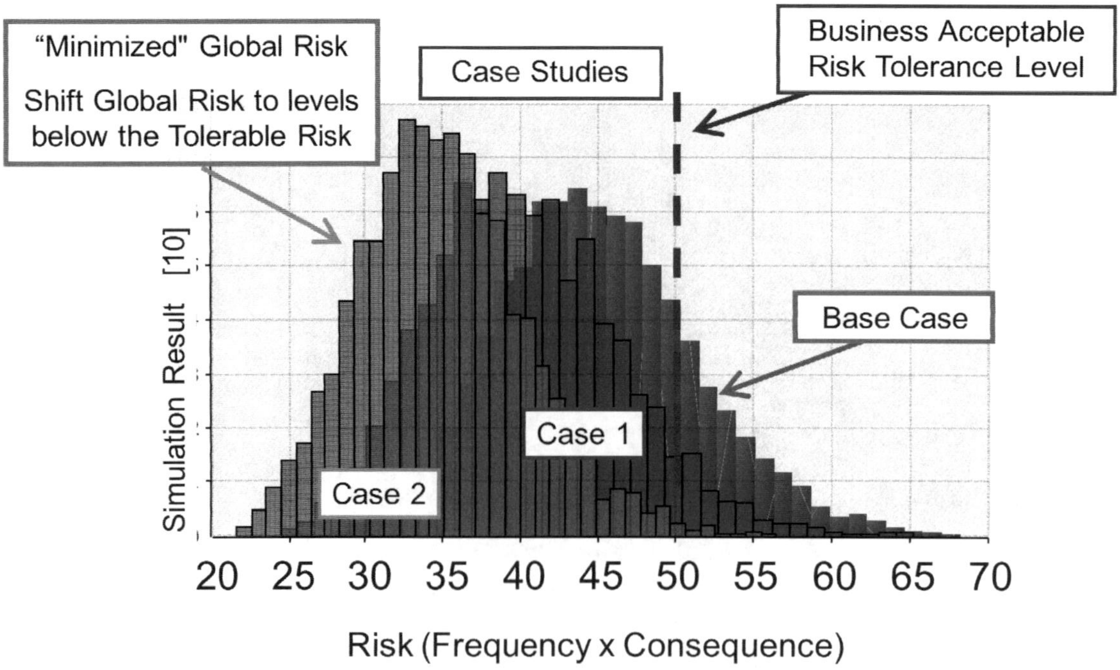

Figure 6

An Example of Risk Ranges

Risk Range	
Unacceptable	50 to 100
Not Desirable	20 to 49
Acceptable with Controls	10 to 19
Acceptable	< 10

Tolerable Risk
< 50

Figure 7

Case 1: A Subjective Risk Reduction Approach

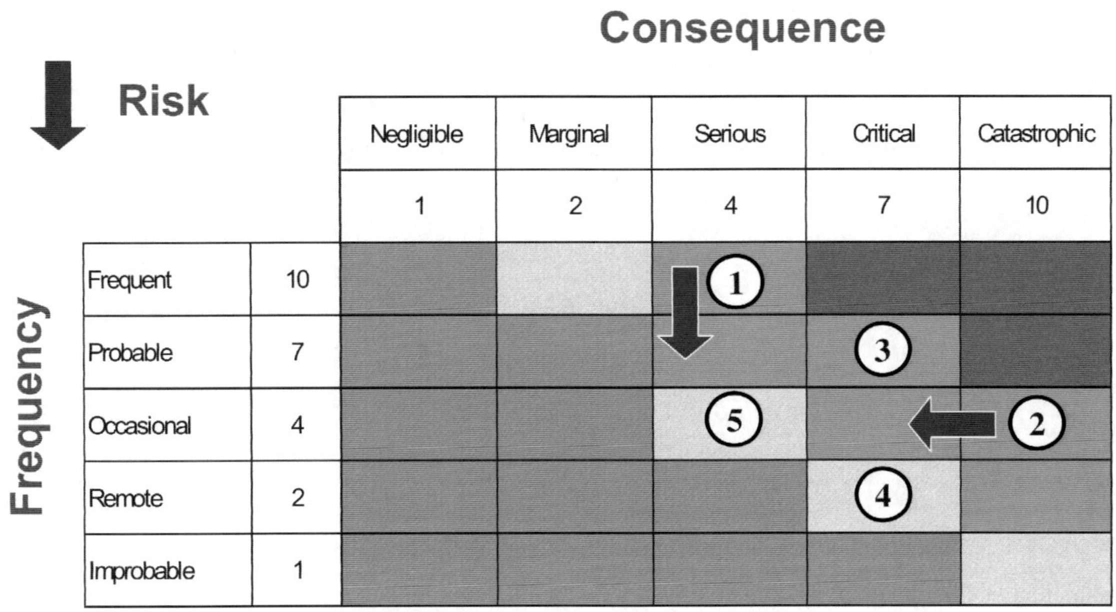

Figure 8

Case 2: The Surgical Risk Reduction Approach

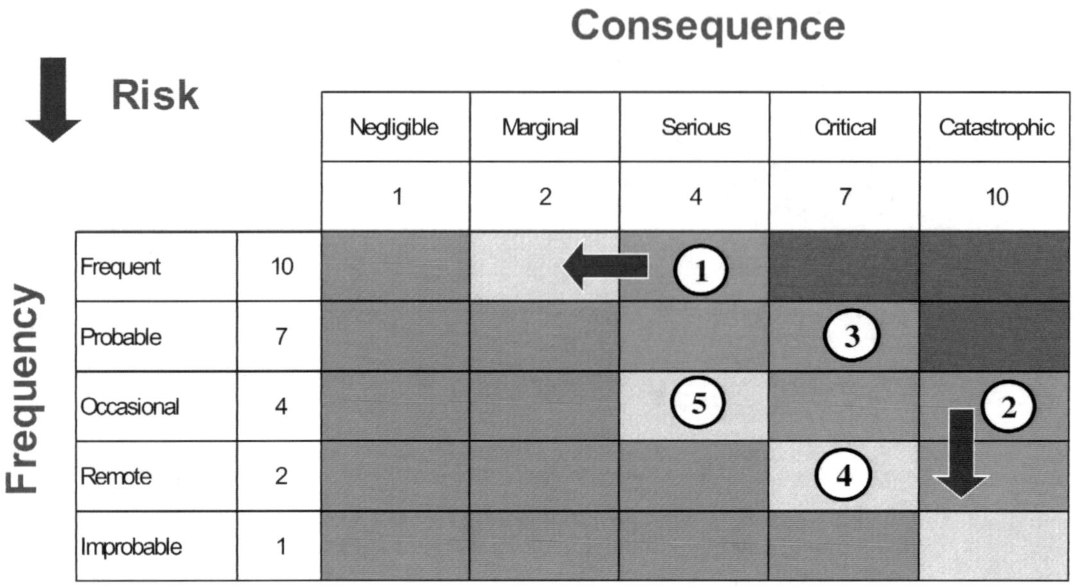

Table 1

Base Case Simulation Input Parameters [10]

Please refer to Adobe acrobat .pdf file: "Surgical Risk - Table 1"

Table 2

Base Case Simulation Formulas [10]

Please refer to Adobe acrobat .pdf file: "Surgical Risk - Table 2"

Table 3

Base Case Simulation Results [10]

Please refer to Adobe acrobat .pdf file: "Surgical Risk - Table 3"

Table 1

Base Case Simulation Input Parameters [10]

Site	Consequence	Minimum	Most Likely	Maximum	Frequency	Minimum	Most Likely	Maximum	OD	Minimum	Most Likely	Maximum
1	4	2	4	7	10	7	10	10	0.75	0.5	0.75	0.95
2	10	7	10	10	4	2	4	7	0.75	0.5	0.75	0.95
3	7	4	7	10	7	4	7	10	0.75	0.5	0.75	0.95
4	7	4	7	10	2	1	2	4	0.75	0.5	0.75	0.95
5	4	2	4	7	4	2	4	7	0.75	0.5	0.75	0.95

	Risk
1	53
2	53
3	65
4	19
5	21

42	Average Risk

Table 2

Base Case Simulation Formulas [10]

	A	B	C	D	E	F	G	H
1								
2		Site		Consequence	Minimum	Most Likely	Maximum	
3		1		=RiskTriang(E3,F3,G3,RiskStatic(F3))	2	4	7	
4		2		=RiskTriang(E4,F4,G4,RiskStatic(F4))	7	10	10	
5		3		=RiskTriang(E5,F5,G5,RiskStatic(F5))	4	7	10	
6		4		=RiskTriang(E6,F6,G6,RiskStatic(F6))	4	7	10	
7		5		=RiskTriang(E7,F7,G7,RiskStatic(F7))	2	4	7	
8								
9				Risk				
10		1		=RiskOutput(,D9,1)+D3*I3/N3				
11		2		=RiskOutput(,D10,1)+D4*I4/N4				
12		3		=RiskOutput(,D11,1)+D5*I5/N5				
13		4		=RiskOutput(,D12,1)+D6*I6/N6				
14		5		=RiskOutput(,D13,1)+D7*I7/N7				
15								
16				=RiskOutput("Total Risk")+SUM(D10:D14)/5	Average Risk			
17								

Table 2

Base Case Simulation Formulas [10]

	A	B	C	I	J	K	L	M	N	O	P	Q
1												
2		Site		Frequency	Minimum	Most Likely	Maximum		OD	Minimum	Most Likely	Maximum
3		1		=RiskTriang(J3,K3,L3,RiskStatic(K3))	7	10	10		=RiskTriang(O3,P3,Q3,RiskStatic(P3))	0.5	0.75	0.95
4		2		=RiskTriang(J4,K4,L4,RiskStatic(K4))	2	4	7		=RiskTriang(O4,P4,Q4,RiskStatic(P4))	0.5	0.75	0.95
5		3		=RiskTriang(J5,K5,L5,RiskStatic(K5))	4	7	10		=RiskTriang(O5,P5,Q5,RiskStatic(P5))	0.5	0.75	0.95
6		4		=RiskTriang(J6,K6,L6,RiskStatic(K6))	1	2	4		=RiskTriang(O6,P6,Q6,RiskStatic(P6))	0.5	0.75	0.95
7		5		=RiskTriang(J7,K7,L7,RiskStatic(K7))	2	4	7		=RiskTriang(O7,P7,Q7,RiskStatic(P7))	0.5	0.75	0.95
8												
9												
10		1										
11		2										
12		3										
13		4										
14		5										
15												
16												
17												

Table 3

Base Case Simulation Results [10]

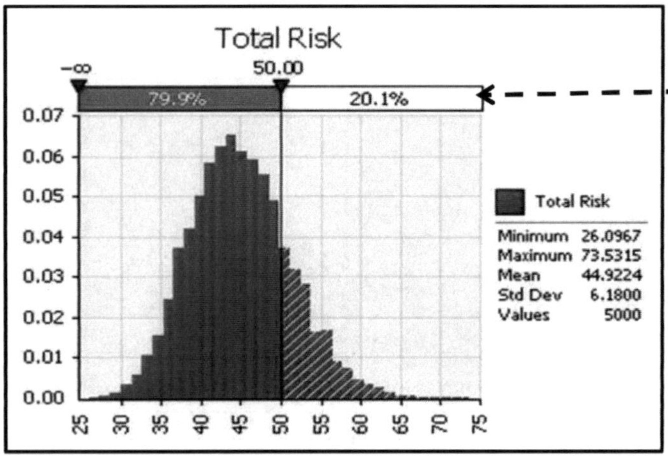

Simulation Summary Information	
Workbook Name	Inpain Proof.xlsx
Number of Simulations	1
Number of Iterations	5000
Number of Inputs	15
Number of Outputs	6
Sampling Type	Latin Hypercube
Simulation Start Time	1/23/11 6:39:09
Simulation Duration	00:00:05
Random # Generator	Mersenne Twister
Random Seed	134793784

20% of Risk is "Uncertain"

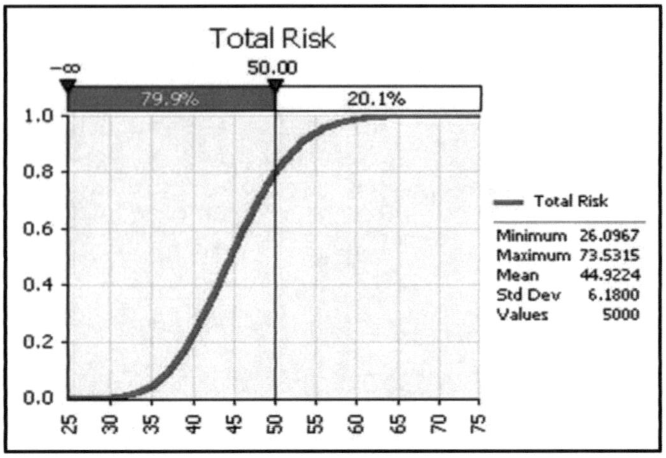

Summary Statistics for Total Risk			
Statistics		**Percentile**	
Minimum	26.09669731	**5%**	35.43664508
Maximum	73.53153619	**10%**	37.15669851
Mean	44.92235858	**15%**	38.45928204
Std Dev	6.180020823	**20%**	39.58897539
Variance	38.19265737	**25%**	40.5806913
Skewness	0.27768607	**30%**	41.3906723
Kurtosis	2.989006075	**35%**	42.29541377
Median	44.61408495	**40%**	43.07632563
Mode	43.46098677	**45%**	43.83133489
Left X	35.43664508	**50%**	44.61408495
Left P	5%	**55%**	45.41028804
Right X	55.53176517	**60%**	46.21898752
Right P	95%	**65%**	47.12322802
Diff X	20.09512009	**70%**	47.97712011
Diff P	90%	**75%**	48.95205999
#Errors	0	**80%**	50.00776205
Filter Min	Off	**85%**	51.4664099
Filter Max	Off	**90%**	53.08181234
#Filtered	0	**95%**	55.53176517

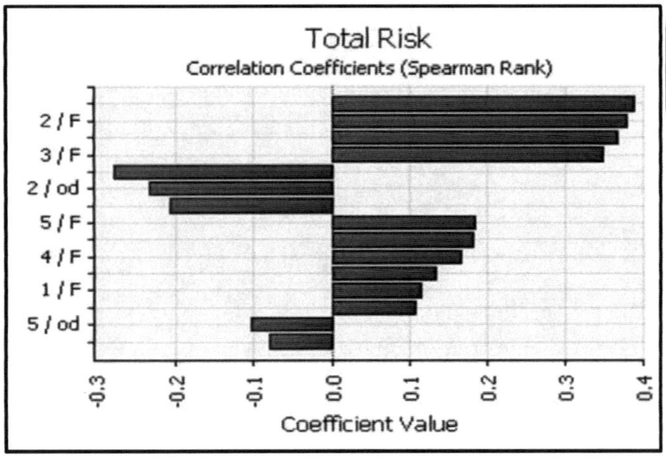

Regression and Rank Information for Total Risk			
Rank	Name	Regr	Corr
1	2 / F	0.415	0.379
2	1 / C	0.413	0.387
3	3 / F	0.387	0.349
4	3 / C	0.385	0.368
5	3 / od	-0.285	-0.278
6	2 / od	-0.229	-0.231
7	1 / od	-0.228	-0.207
8	5 / C	0.205	0.181
9	5 / F	0.199	0.184
10	4 / F	0.195	0.166
11	2 / C	0.140	0.107
12	1 / F	0.134	0.114
13	4 / C	0.130	0.133
14	5 / od	-0.110	-0.103

Tornado Results

Site 2 - Reduce F
Site 1 - Reduce C

(Case 2 Simulation)

Portfolio Risk Management for Process Safety

Georges A. Melhem, Ph.D.
ioMosaic Corporation
93 Stiles Road, Salem, NH 03079 USA
sales@iomosaic.com

R.P. Stickles, P.E.
H.R. Forgione
ioMosaic Corporation

Abstract

A majority of PSM tools and risk methodologies are focused on risks at individual sites or process units. For example, Process Hazard Analysis (PHA) risk-ranking matrices are frequently cast in terms of consequences that occur at a plant level. However, corporations with many operating facilities need to understand and manage process risk across the entire enterprise.

Effective management of process plant risks is likely to involve a mix of several strategies including risk avoidance, risk reduction, risk transfer (loss insurance) and risk retention (intentional and unintentional). In practice the mix will result from activities at a number of levels within the firm, ranging from risk managers who deal with insurance aspects to engineers responsible for the safe operation of individual plants.

This paper introduces a framework for process facility risk analysis that enables the use of a combination of these strategies at a variety of organizational levels.

1. Introduction

Effective management of risk in organizations operating chemical process plants is likely to include a mix of several strategies:

Risk Avoidance – for example, eliminating a risk by choosing a different process route to avoid use of potentially hazardous intermediates or process conditions.

Risk Transfer – insuring any loss of either property or profits. An element of the risk is retained depending on the deductible and the coverage limits. Some risk may also be retained by choice or by insurance vehicle (for example, by use of an insurance captive).

Risk Retention – the deliberate choice to finance retained risks either from internal funds or contingency borrowing. Risks can also be retained unintentionally through lack of thorough systematic analysis.

Risk Reduction – the use of a variety of programs including training, improved organization and engineering modifications to eliminate certain elements of risk.

In practice, the mix will result from activity at a number of levels in the firm, ranging from risk managers who will be dealing with insurance aspects (probably on an aggregated corporate basis) to engineers responsible for the safe operation of sections of individual plants. In the following sections we present a framework for the analysis of process plant risk which enables the mix of risk strategies to be analyzed and determined on a coherent basis at a variety of organizational levels.

2. Evaluation of Risk Reduction Measures

The selection of the appropriate risk reduction measures involves choices between expenditure to improve plant safety and expenditure to improve the safety organization. It also involves issues of timing. These measures are introduced with the intention that they be implemented on a corporate level rather than on the plant or location level. Attempting an overly aggressive risk reduction program may overwhelm available human and financial resources, making it necessary to schedule work according to its safety priority.

Resources can either be devoted to management systems, such as training and other safety-related programs or to equipment modifications including both personnel protection and changes to facilities. These demands for resources must be judged alongside funding requests for production-oriented expenditure to improve operational reliability or increase throughput and yield. An important part of the approach introduced in this paper is applying a common basis for the evaluation of both safety-related and production-related spending.

3. Process

Any program to implement risk reduction entails:

- selection of the order in which work is to be done
- effective project management of implementation

While it is preferable to have plans in place to deal with all safety-related inadequacies, available resources may require prioritization of these issues. Modifications will also need to be scheduled within constraints set by plant maintenance, regular shutdowns and operations-related improvements. The key to effective use of available resources is good project management.

To determine the order of work to be completed, it is necessary to qualify the potential losses from incidents which could arise if the work were not done. The first steps in project management are to define the scope of work and derive a cost and feasibility schedule for implementation. The combination of quantified loss potential and cost estimate permits a cost/benefit analysis to be performed, which serves as the basis for prioritization of the work. Figure 1 illustrates the activities leading to the cost/benefit analysis.

This approach for plant modifications can also be applied to training and organizational recommendations. For example, the cost of developing and running a specific training course can be estimated. If the course is designed to address particular objectives, then specific performance improvements can be targeted and measured by testing or feedback programs after training. The improvements in performance can be related to a lower frequency of accidents or reduced consequences if they do occur.

4. Steps in Implementing Risk Reduction

Risk reduction is typically achieved by a three step process:

- identify the sources of risk and potential mitigation
- quantify the level of risk and required level of reduction
- select the most cost-effective projects for implementation

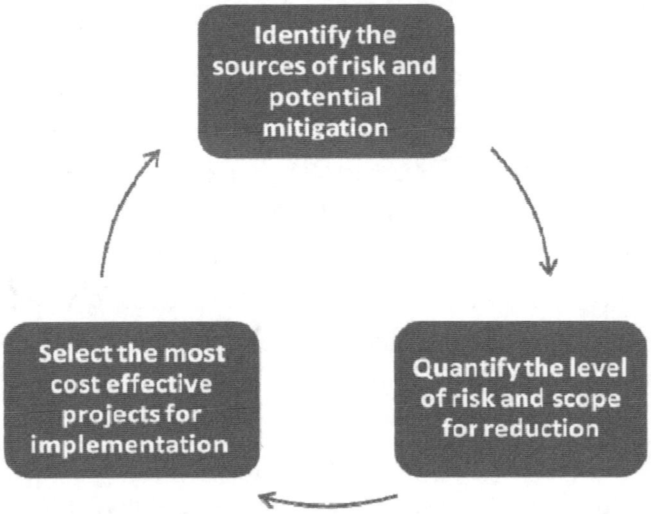

Risk identification and quantification in the process industries can be tackled by several well-established methods [Ref 1, 2, 3, and 5]. Many protocols have been developed for major risk surveys which entail experienced engineers on a site visit reviewing both the plant and the human aspects of operational safety. Thorough major risk surveys are an excellent basis for identifying the principal sources of risk as well as recommending practical steps to make improvements.

A more systematic approach to risk identification is the use of PHA methodologies in which the plant is examined equipment-item by equipment-item and line by line through the use of plant piping and instrument diagrams (P&IDs). The method identifies the design intent of each item of equipment and credible deviations from the design intent that could occur due to equipment failure or operator error. Widely used throughout the process industries, a Hazard and Operability (HAZOP) study is an example of this approach.

5. Risk Assessment Criteria

5.1 Facility Level Analysis Procedures

Risk qualification of both surveys and PHA studies can be achieved in some measure using risk ranking criteria to prioritize findings [Ref 4]. Figure 2 illustrates a typical matrix used for this purpose. It comprises a matrix of consequences described by the terms *negligible*, *marginal*, *substantial*, *critical* and *catastrophic*. Each term is defined by an estimated loss range that includes both loss of property and loss of profits. The ranges can also be defined in terms of numbers of accidents or fatalities for safety ranking. The two scales are treated separately (for example, it is not necessarily implied that an event causing one fatality will result in losses of between $1 and $10 millions.) In this table, frequency is covered by a five point range; *frequent*, *occasional*, *infrequent*, unlikely and rare are similarly specified by ranges of annual frequency. The ranking table can be adapted to suit the needs of the client and the site being surveyed. In many cases, a 4x5 matrix is preferred to provide a tolerable risk based on low frequency for the highest consequence ranking.

This particular matrix is used to assign an action status to each recommendation based on relative risk. Therefore, the priority *high* suggests initiating work at once, medium implies some urgency, and *low* indicates less urgency (for example, perhaps an issue which can be dealt with during two or three year turnaround intervals). Items with a *very low* priority are usually considered optional for improving process safety. This process of status determination emphasizes that a safety recommendation is not something which can be made and then forgotten. Positive action is necessary, with some actions being taken sooner than others.

These plant level exercises usually generate many hazard cause/consequence scenarios which result in several risk reduction recommendations with associated costs. Figure 3 illustrates results of an overall ranking of a large HAZOP study covering a number of plant sections. As well as the overall figure, rankings for individual sections can be used to identify the areas of a facility which are of special concern. Figure 4 shows such a representation.

If greater precision is required, a Quantitative Risk Analysis (QRA) can be performed in which the frequency of potential events and their consequences are systematically enumerated. Frequencies are typically analyzed using a fault tree approach quantified with failure data from industry sources, modified to the individual circumstances of the plant in question. Consequences are calculated using dispersion models and probit functions for toxic releases or flammability limits and overpressure radii for explosive materials. Again, these methods are well-established and widely described in the literature [Ref 3].

The selection of options and the implementation of a risk reduction program require decision making on the basis of the steps already outlined. Whatever the degree of sophistication used for the quantification, it must be recognized that there is an element of judgment and assumptions involved and that all results should be regarded as an approximation with increasing resolution of the approach serving at best to reduce the margin of error.

5.2 Enterprise Wide Analysis

While many tools have been developed for identification and quantification of risk at the plant level, the techniques for managing process safety risk at the enterprise level have received less attention. Large domestic or multinational companies with many facilities and diverse operations must face the issue of how to prioritize expenditures for safety across their expansive operations. Important decisions at this level related to operational risk could include:

- Funding for large projects to reduce process risk
- Intentional retention of risk (self insuring)
- Avoidance of exposure from catastrophic safety or environmental events with major cash flow and/or stock value or company reputation impact

When thinking about risk tolerance, the management level of the organization can make a difference in terms of the severity criteria. What might be considered a catastrophic impact at the plant level could be absorbed on the corporate level by a corporation with substantial resources, albeit, not without significant ramifications. Nonetheless, the tolerance for risk when addressing corporate portfolio risk management has some merit. The 5x5 risk matrix previously shown in Figure 2 is constructed in an attempt to incorporate criteria that might be considered tolerable at the portfolio risk management level. Some frequency/consequence pairings shown as tolerable (very low) might not be considered acceptable safety goals at the plant level.

The frequency criteria scale in Figure 2 is expressed in terms of likelihood of events with the potential to cause the indicated impact (injury, fatality, etc.). These event criteria do not take into account other conditional events (probability of ignition or personnel in affected area) or affected population density. Impact scenario criteria can also be utilized for risk ranking, which attempt to take into account other conditional probabilities in a generalized way. For setting the scenario criteria frequency scale, API annual injury rates provide some guidance (as indicated in Table 2). This data suggests that scenario criteria frequencies should be at least an order of magnitude lower than event frequencies. The use of scenario criteria can be thought of as a proxy for estimating individual risk as it considers proximity, but not population density. Scenario criteria are becoming more favored at the plant level as more companies embrace the incorporation of Layers of Protection Analysis (LOPA) into their PHA methodology.

Use of event criteria are similar to assessing societal risk where population and the cumulative effect of multiple scenario impacts (as expressed in F-N curve criteria for societal risk) can shift the risk tolerability metric, especially in the low impact range. The examples in this presentation use scenario criteria for ranking the risk as the focus is on higher level risk management.

Table 1: Historical Injury Data

Consequences	API Annual Injury Rate (1994 – 1998 average)	Annual Impact Frequency Range
Recordable injury	$2x10^{-2}$	10^{-1} to 10^{-2}
Lost workday injury	$5x10^{-3}$	10^{-2} to 10^{-3}
Disabling injury	Not reported	10^{-3} to 10^{-4}
Fatality	$5x10^{-5}$	$< 10^{-4}$

Companies with many operating locations face the issue of what risk tolerability criteria to use when the risks from all facilities are aggregated. Risks are additive at the corporate level. The approach described here is designed so that an annual loss potential can be determined without resorting to risk-tolerance criteria that are orders of magnitude higher than at the plant level.

6. Safety Expenditures

In applying this approach, much can be done without going to the extent of a detailed planning and cost-estimating exercise on each significant safety mitigation expenditure. It is useful to classify the potential project work into items which can be managed with plant resources (small projects) and those requiring external support (larger projects).

Small projects are characterized as using existing plant stores and purchasing for materials and maintenance crew for the implementation. Work needs to be fitted into the existing scheduled and unscheduled maintenance work load. The funding is likely to be by means of an existing revenue budget for maintenance and operations with the appropriate level of local approval and control.

Larger projects are likely to involve extensive external procurement from vendors of specialized items requiring specification and quality control from the owner. Contracted services will be used for all or part of the work. This necessitates preparation of bidding documents, tender review and contractor supervision if the work is to be completed successfully. Funding is likely to be from annual capital investment funds which will need to be approved by upper management and controlled by a more complex procedure than is normal for small works.

As a first step, work can be separated into items which can be done:

- as part of regularly scheduled maintenance
- when particular items are taken out of service while the plant continues operations
- as a separate project with tie-ins installed at shutdown
- entirely during the regular plant shutdown

Similarly, expenditure ranges can be estimated using the order of magnitude criteria presented in Table 2. Projects or options with costs above $1 million are then separated from those less than $1 million for each asset in the portfolio. Such items for all operational assets are then aggregated for separate loss avoidance analysis.

Table 2: Order of Magnitude Cost Classification

	Safety Related Expenditures
Size	**Cost Range**
Small	<$100K
Medium	$100K to $1MM
Large	$1M to $10M
Major	>$10MM

Although this may seem crude when considering an individual project or loss, the range estimating approach can give useful results if a large number of improvement opportunities are involved, grouped for example by schedule requirements.

Figure 5 illustrates the cost ranking approach on the data from Figure 3. High and medium rank recommendations have been assigned to four cost ranges.

7. Loss Potential

Once risks have been identified by survey or PHA and ranked as illustrated in Figures 3 and 4, they provide the basic means of estimating the potential losses from the facility. Note that the Figure 2 matrix uses two ranking criteria; one for financial losses which include:

- business interruption
- asset losses
- accidents or fatalities

In this approach, we have avoided attempting to evaluate human life in financial terms for several reasons:

- it implies an acceptance of human fatality
- it leads to placing a value on human life with potential ethical issues arising if different values are assigned at different locations
- it suggests that fatalities are regarded as part of the natural cost of doing business

Nevertheless, it is clear that projects/expenditures aimed toward process modifications that reduce the chance of injury or fatality will also significantly lessen the potential for financial loss. This cost/benefit methodology has been developed to identify such projects to emphasize the proposition that safe operation is good business.

8. Cost/Benefit Methodology

A stepwise procedure can be utilized to study portfolio risk and the cost/benefit of safety expenditure. The steps are:

1. Segregating the proposals into discrete order-of magnitude financial loss avoidance ranges (using ranges from risk ranking matrix).
2. Compiling the scenario frequencies and estimated losses of each of the projects/proposals (frequency/consequence pairs) grouped by loss estimate range.
3. Developing a loss/frequency distribute (histogram) by subjecting the population of scenario-frequency/loss pairings to Monte Carlo simulation.
4. Converting the loss frequency distribution to a probability loss distribution and plotting it as a loss profile (cumulative probability versus loss).
5. Compare the cost and averted loss.

Steps 1 and 2 were described earlier.

Generation of the loss profile (step 3) shows the relationship between probability and the magnitude of the loss. The loss profile is obtained using a Monte Carlo simulation based on the ranges of frequency and consequence assigned to medium, large and major expenditures using the risk-ranking matrix. Use of the Monte Carlo modeling approach provides a random walk through the discrete frequency/consequent pairs enabling the development of a cost or injury category versus frequency distribution (e.g., a histogram). The histogram of a frequency distribution can be easily converted to a probability distribution by dividing the cumulative frequency for each loss category by the total frequency of all loss categories. The loss profile is basically a cumulative probability versus magnitude of loss (cost or injury) plot. The useful information that can be extracted from the loss profile includes:

- average annual loss (distribution mean) and probability
- maximum annual loss
- contribution by plant sector or facility location if scenarios are so grouped

Figures 6 and 7 illustrate the results obtained from a major risk survey study with results ranked in Figures 3 and 4. The loss profile is calculated separately for the rankings of financial loss and fatality. The PHA scenarios for each plant section have been taken separately to show their contribution to the overall loss.

The two charts show that sections of the complex may be important sources of financial loss but have low contribution to the safety risk (fatality) posed by the plant.

The same process can be applied to an aggregation of all proposed large safety expenditures across many units within a complex or many plants within a business entity. Figure 8 and 9 illustrate the loss profiles for multiple units within a large petrochemical complex. Hazard scenarios for each facility have been taken separately to show their contribution to overall loss. This perspective can be used to set priorities for using allocation resources to achieve the most benefit in managing overall risk across the company.

In step 4, the mean of the Loss Profile is computed, which gives the average loss. If the assigned frequency ranges are annual frequencies, this is the average loss per year. The probability corresponding to the mean value can be used to indicate the chance of this loss occurring. The loss history available for the particular plant can be used to verify that there are sufficient data points in relation to the probability of the mean.

The averted loss analysis (step 5) compares the average annual loss to the expected cost of project work. This may be done for the risk mitigation recommendations for the plant as a whole or for recommendations grouped by priority.

Due to Monte Carlo data set limitations, the calculations can be done satisfactorily only for groups of projects. An item-by-item comparison of potential loss and mitigation cost is possible only if detailed calculations of cost, consequence and frequency are available. The calculation must also take into account any residual risk remaining after the mitigation work is completed. In this way, the net loss averted by the expenditure is compared against the cost of the modifications.

Figure 10 shows the results of such a comparison based on the loss profile from Figure 6. The results show that including all proposed projects leads to costs which are likely to outweigh the loss averted. However, the small projects and the medium sized high priority projects do show a positive balance between loss aversion and cost. This is significant, since over 130 recommendations are shown to have a positive benefit-to-cost ratio.

As an alternative to simple payback, loss averted can be treated as revenue and discounted in the same way as if investment were made to increase positive cash flow. Figure 11 illustrates the discounting calculations, which are done over the remaining life of the plant, and may be much shorter than the period normally used in discounted cash flow (DCF) for new facilities. This approach may be useful for large and major projects grouped by plant location. In this case, if the timeframe to reach positive cash flow does not meet threshold target ranges, other options such a risk retention or risk transfer can be considered.

At this point it may be desirable to disaggregate specific expenditure proposals and utilize a cost/benefit analysis to sort recommendations so that choices can be made between alternatives. To this extent, it is necessary to define the potential losses and costs (as previously classified) just sufficient to distinguish between them. Figure 12 illustrates a ranking of 4 recommendations on this basis. The bars here show the range of loss aversion and cost as bands one standard deviation high centered about the mean value and assuming a normal distribution (1□ = ±34.1%). Recommendations 2 and 3, for example, have similar ranges for risk aversion but can be easily discriminated on the basis of cost. More detailed work is only justified to narrow the range of uncertainty when the loss and cost bands overlap.

As an alternative to a graphical presentation, the cost/benefit ratio can be tabulated as shown in Figure 13. Rather than immediately eliminating those recommendations with poor cost/benefit ratios, it may be worthwhile to reevaluate them. While they may appear financially less attractive than other improvements, they may nevertheless be important investments in terms of safety. These recommendations could be retained for future consideration depending upon their degree of impact on loss avoidance.

In such cases, the potential accident rate can be compared against cost as shown in Figure 14, which takes its input from the loss profile in Figure 6. This serves to rank the projects in terms of the effectiveness of investment for reducing accidents. The analysis may be completed by determining at what point in the ranking to divide the work into items which will be implemented and those which will be delayed. There are several methods which can be applied:

- Identify the relationship between accident aversion and cost for a part of the design required by regulation or code. Implement all work with a cost/benefit ratio greater than that corresponding to the regulatory requirement.

- Set a particular fatal accident rate (FAR) which will apply to the facility (corresponds to the mean of the fatality loss profile), and identify the minimum cost combination of work needed to reach the desired FAR.

Conclusion

An approach for analyzing portfolio risk that is semi-quantitative without requiring detailed averted loss and risk remediation cost estimates has been described. Consequence and frequency assignments are approximate and the treatment of population impact is subjective. The method is useful for the rapid assessment of large complex process facilities or locations with multiple plants. It provides management with a comprehensive summary of the economic implications of safety surveys or PHA study recommendations, yet requires only a small incremental effort.

This tool is most useful for companies with several plant locations as it can provide guidance in setting priorities for allocation of resources to achieve the most direct risk improvement firm wide. In addition, it identifies at an early stage, work which can be approved without the expense of a detailed risk analysis study.

With this straightforward approach, cost/benefit analysis can be added to the list of factors used to make safety related decisions. The results can be considered in the context of the overall corporate risk policy consistent with risk acceptability criteria and the preferred risk management strategy. The analysis provides a confirmation of the high risk areas of a plant, a means of establishing rational safety project priorities and defines the economic incentive for implementing the safety program.

References

[1] *Guidelines for Hazard Evaluation Procedure, 3rd Edition*, Center for Chemical Process Safety 1992

[2] Loss Prevention in the Process Industries, Lees F. P. 1980

[3] *Guidelines for Chemical Process Quantitative Risk Analysis*, Center for Chemical Process Safety 2000

[4] H. Ozog, *Risk Ranking For PHA, LOPA, and Facility Siting – Part II*, ioMosaic Corporation White Paper

[5] *Layer of Protection Analysis – Simplified Process Risk Assessment*, Center for Chemical Process Safety 2001

Figure 1: Evaluation Process

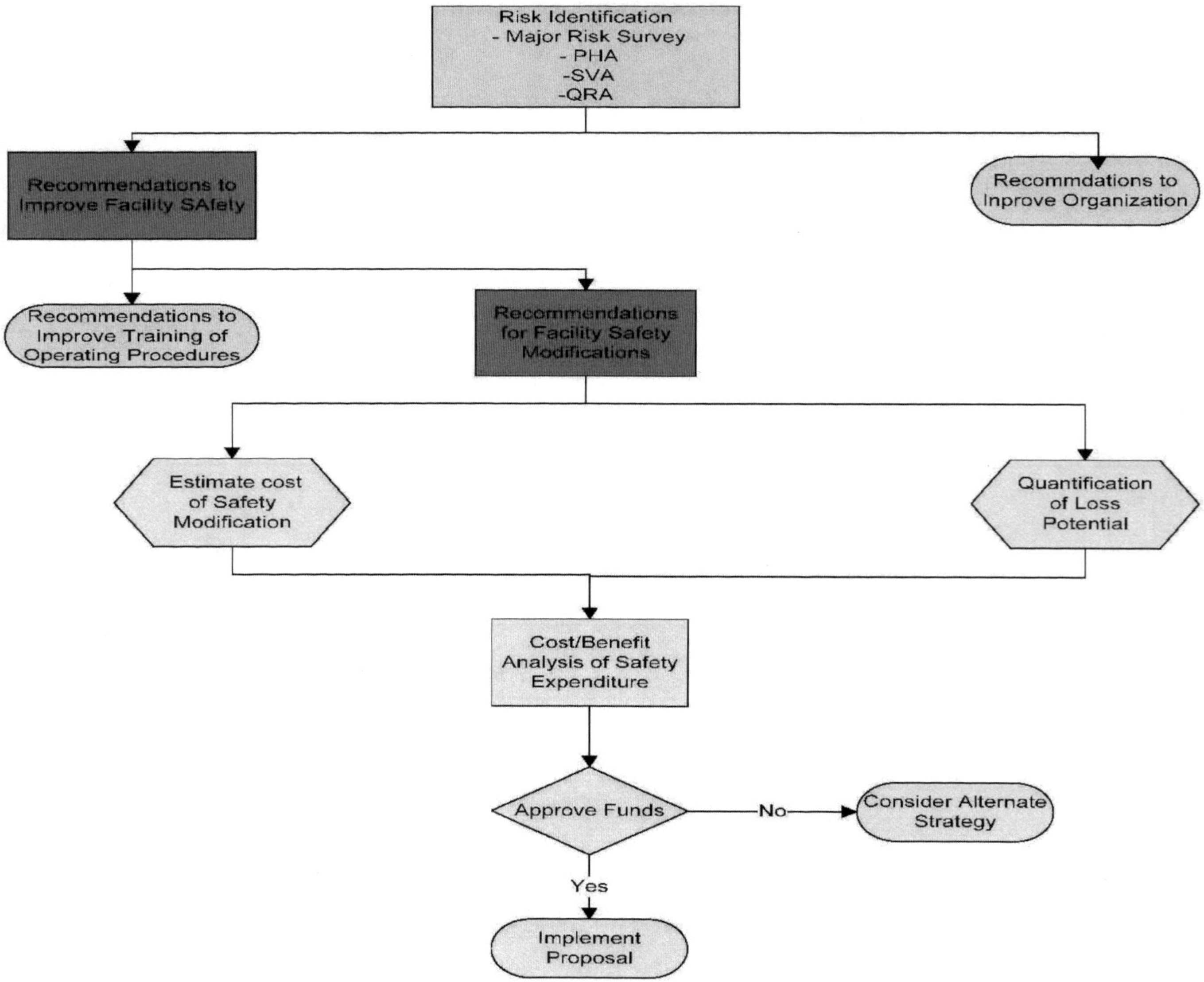

Figure 2: Matrix for Classifying Events

Five priority ranking used

		Significant	Serious	Substantial	Critical	Catastrophic
Impacts	Accident/Fatality Damage/Impact Cost ($1MM)	Reportable injury <0.1	Permanent Disability 0.1 - 1	Fatality 1 -10	1 to 3 Fatalities 10-100	Multiple Fatalities >100
Event Frequency	Rare $<10^{-4}$/yr.	Very Low	Very Low	Very Low	Very Low	Low
	Unlikely $10^{-4} – 10^{-3}$/yr	Very Low	Very Low	Very Low	Low	Medium
	Infrequent $10^{-3} – 10^{-2}$/yr	Very Low	Very Low	Low	Medium	High
	Occasional $10^{2} – 10^{-1}$/yr	Very Low	Low	Medium	High	High
	Frequent $>10^{-1}$/yr	Low	Medium	High	High	High

Figure 3: Risk Potential Classification

The PHA recommendations were classified as follows:

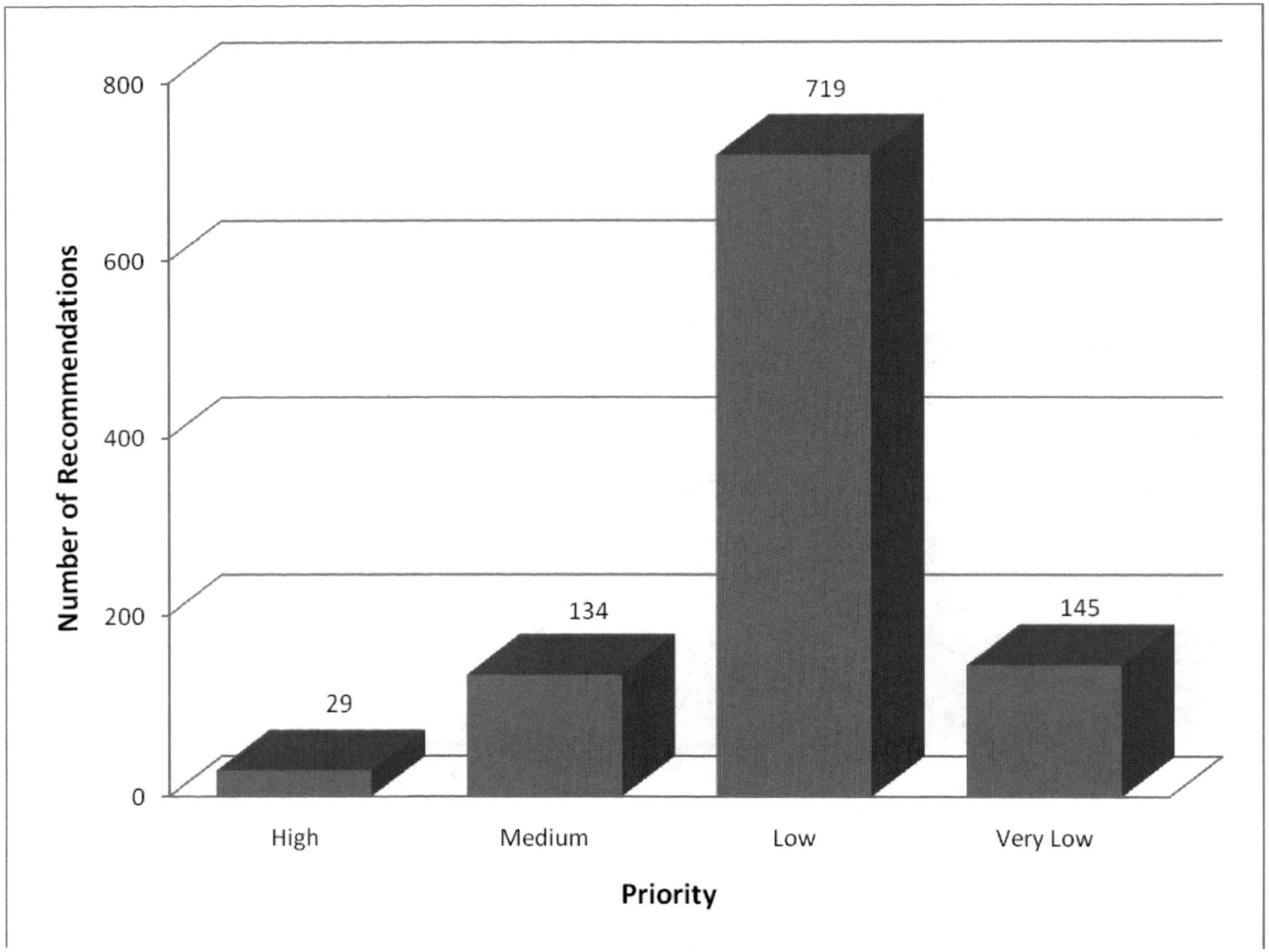

Figure 4: Risks by Section of Plant

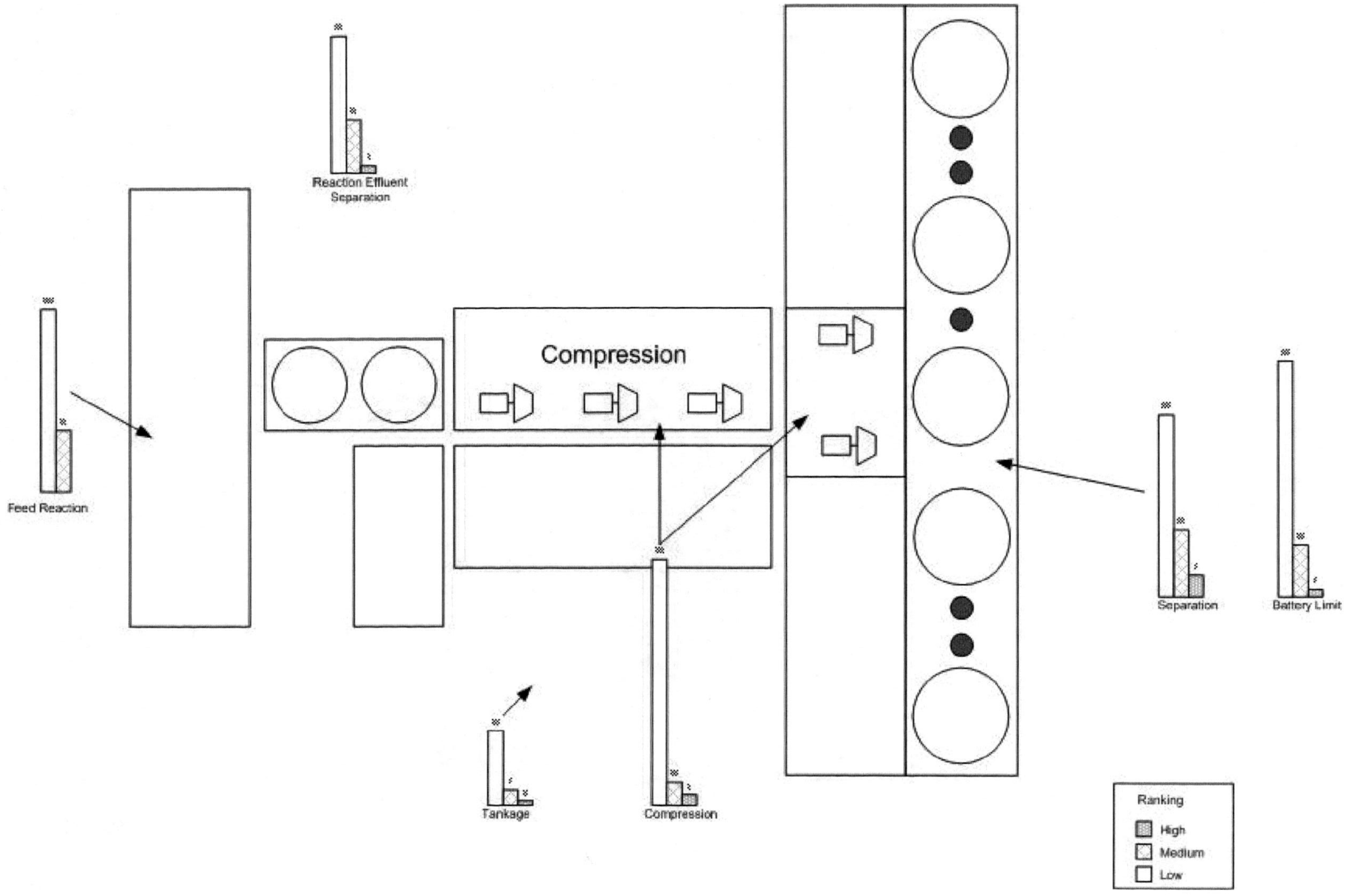

Figure 5: Risk Reduction Proposed Expenditure Classification

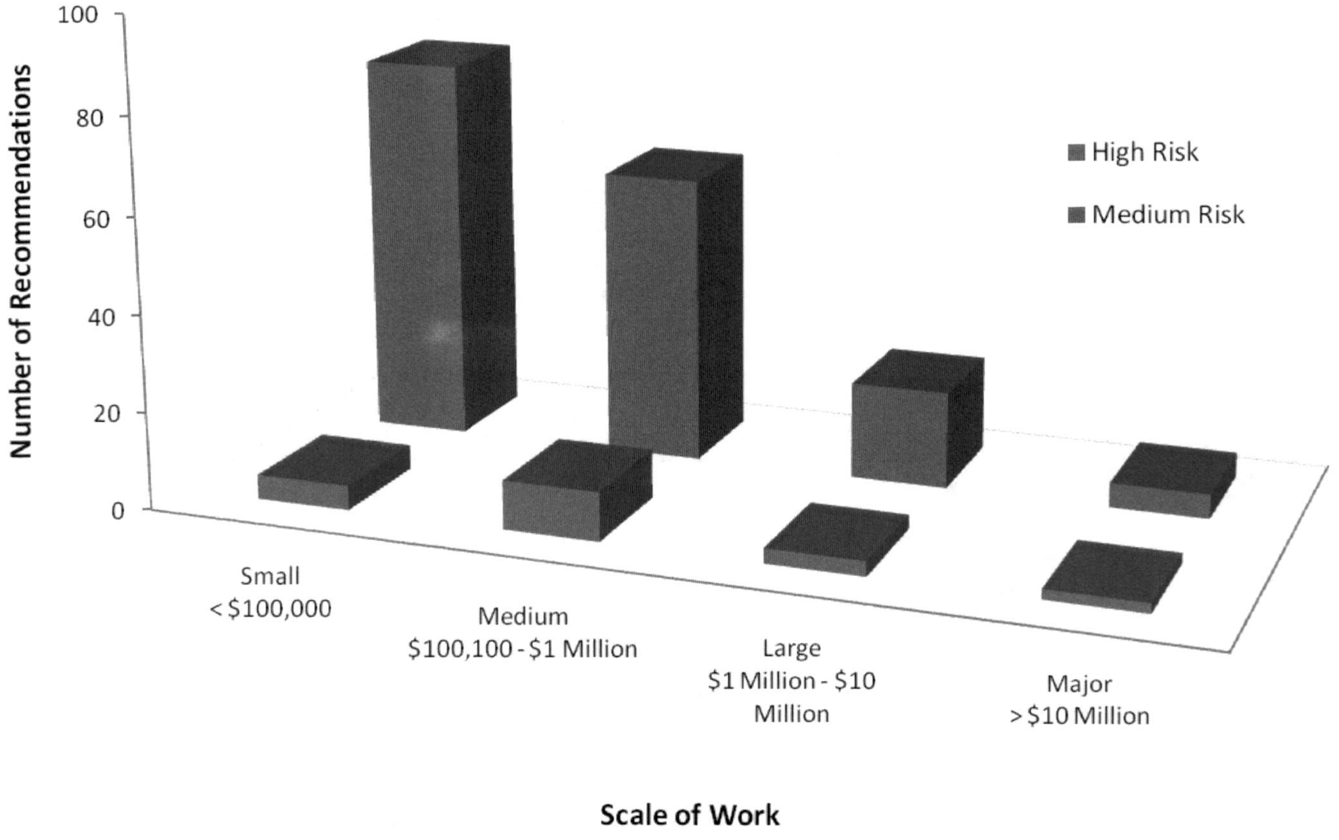

Figure 6: Facility Level Loss Profile

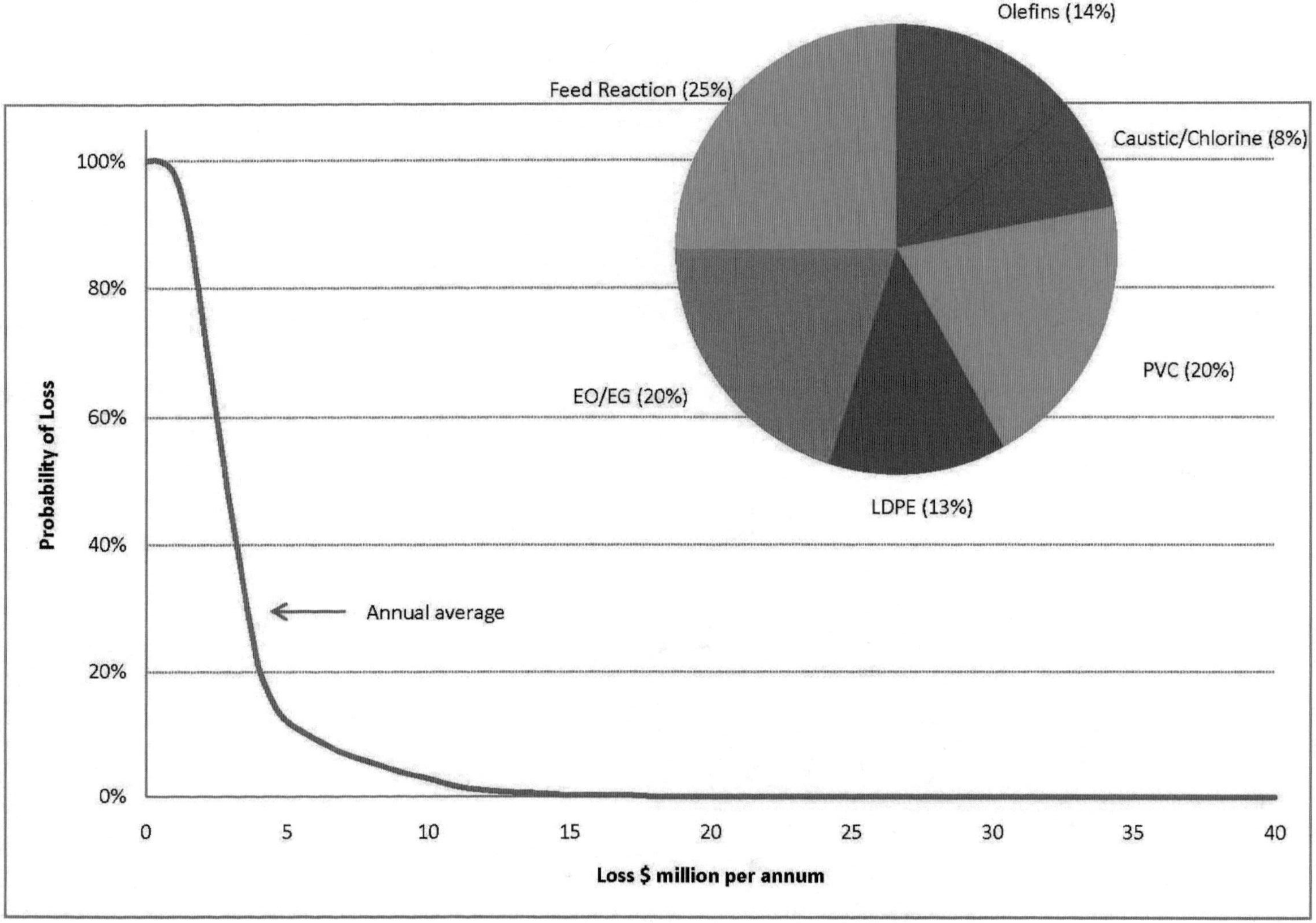

Figure 7: Facility Level Accident Profile

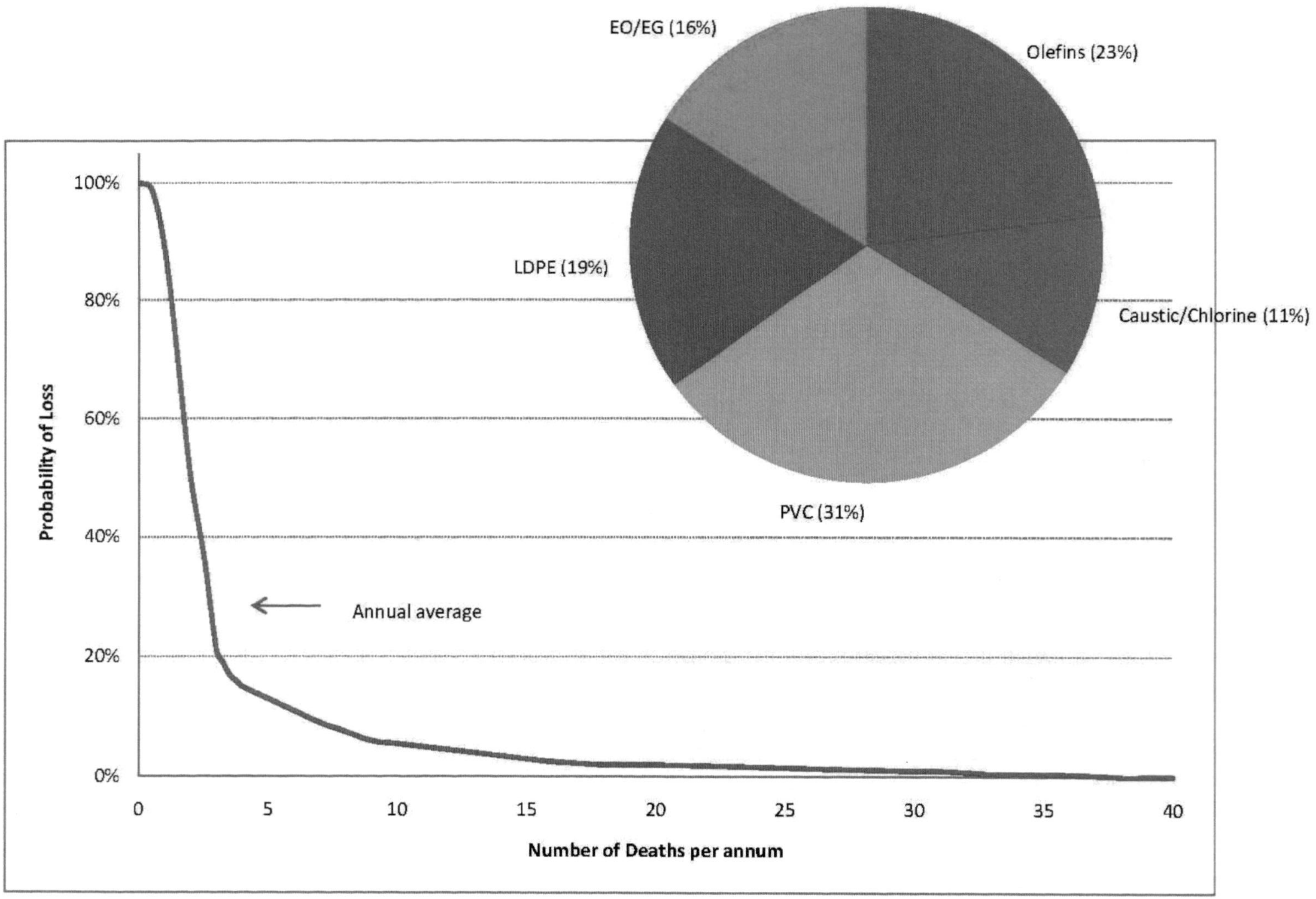

Figure 8: Corporated Level Loss

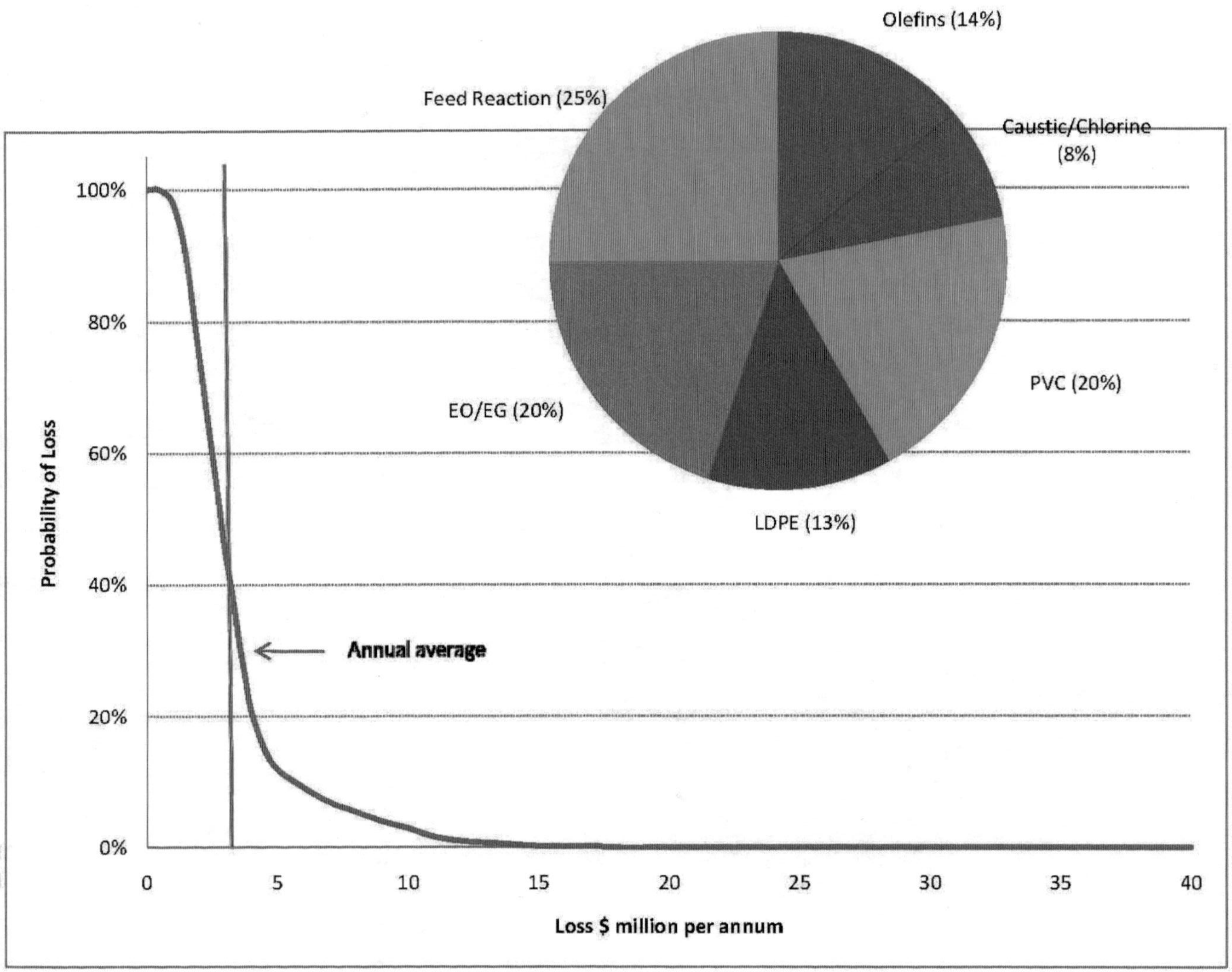

Figure 9: Corporate Level Accident Profile

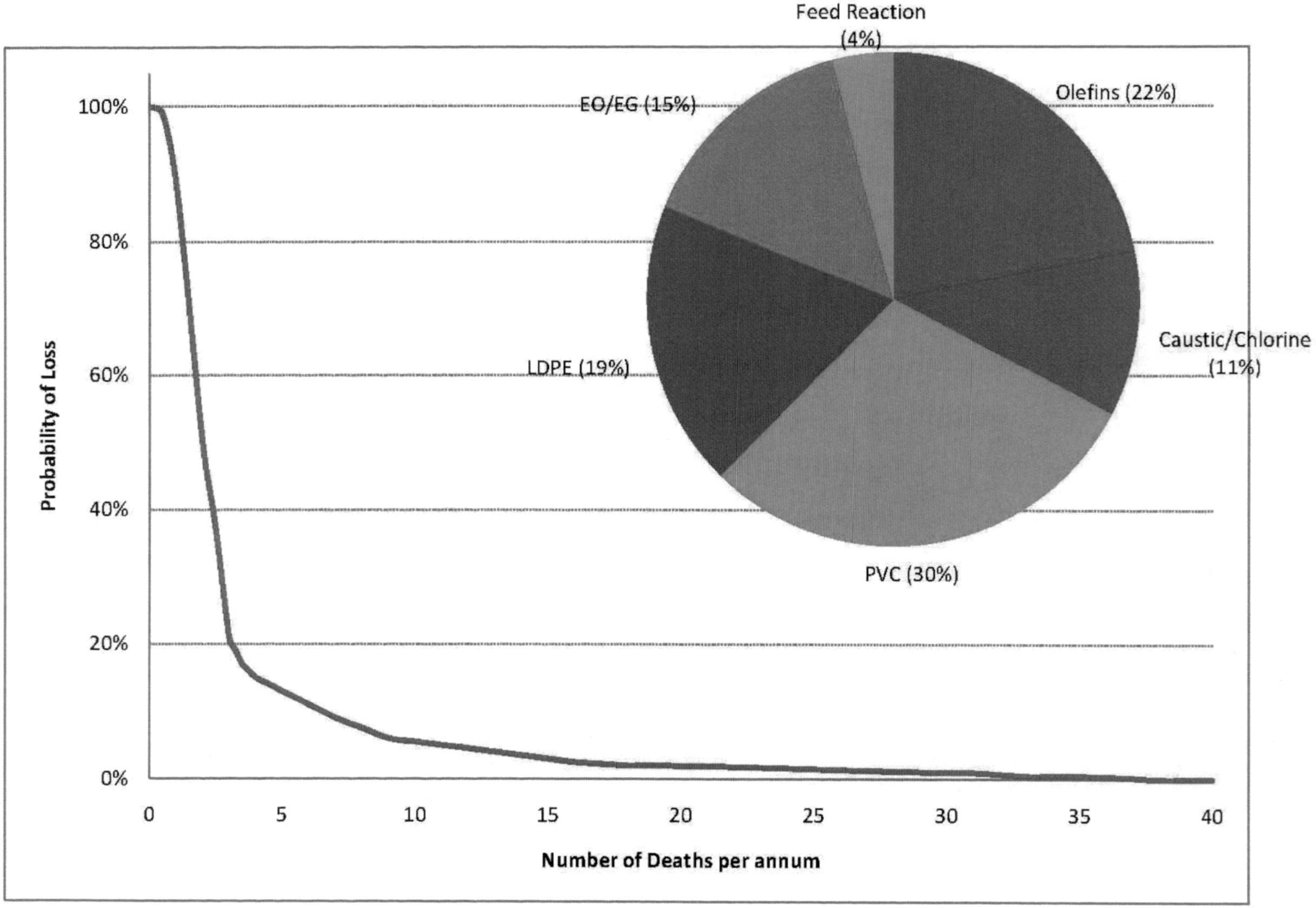

Figure 10: Analysis of Recommendations: Cost Benefit

Loss averted has been estimated from the recommendation ranking. Ranges of cost have been assigned depending on scope of work involved. Small projects are cost effective as are medium sized projects with high ranking.

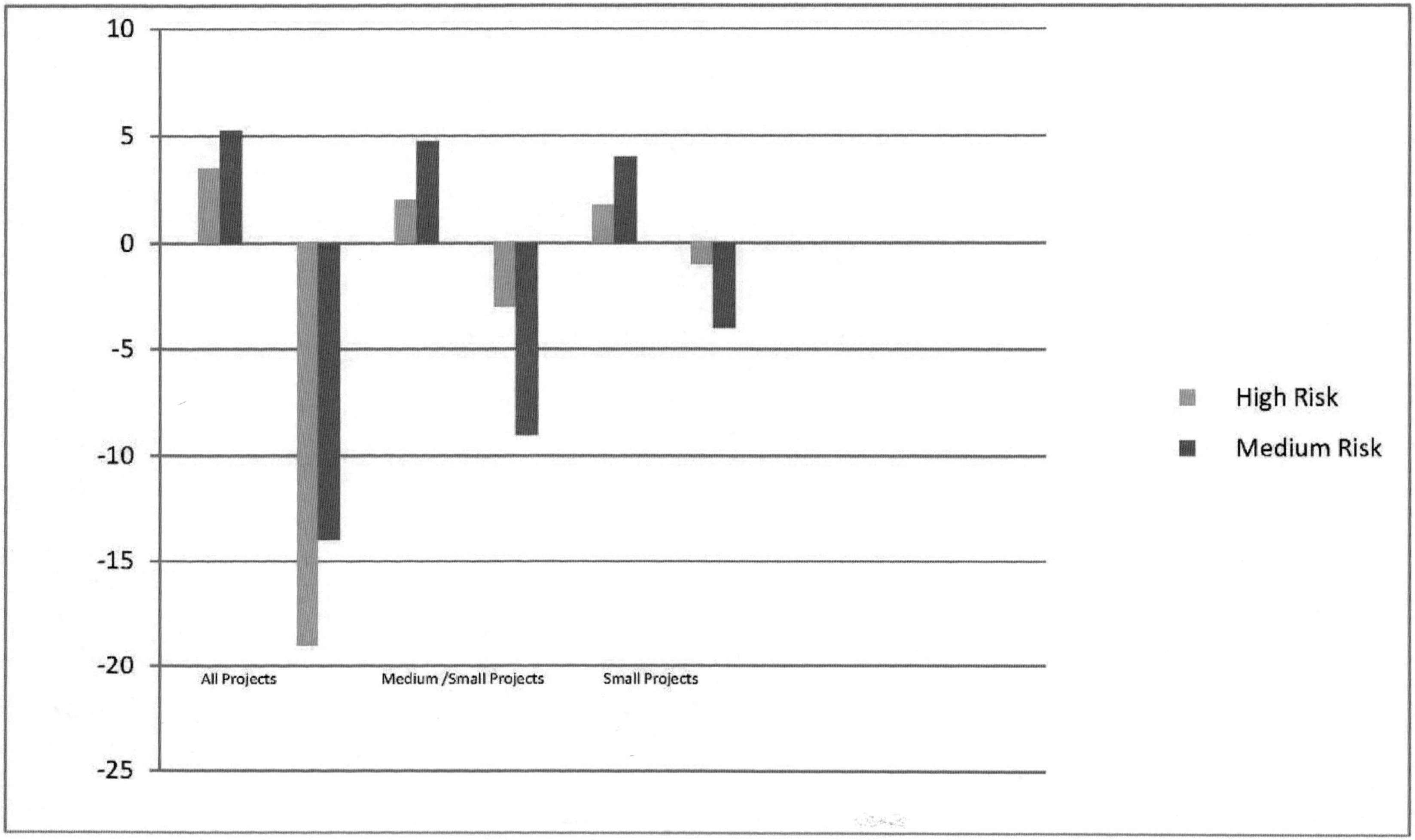

Figure 11: Discounted Cash Flow Analysis

The annual expected loss can be discounted to give the net present value of the safety modification over the remaining life of the plant.

Year	1	2	3	4	5	6	7	8	9	10
Loss Averted $ Millions	1	0	3.26	3.26	3.26	3.26	3.26	3.26	3.26	3.26
Project Costs $ Millions	1.1	1.1	0	0	0					
Cash Flow $ Millions	-1.1	-1.1	3.26	3.26	3.26					

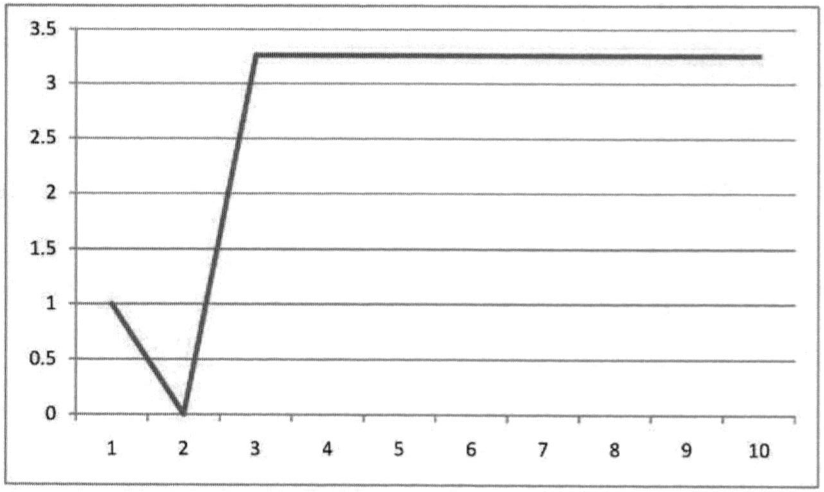

Figure 12: Ranked by Size of Risk Averted

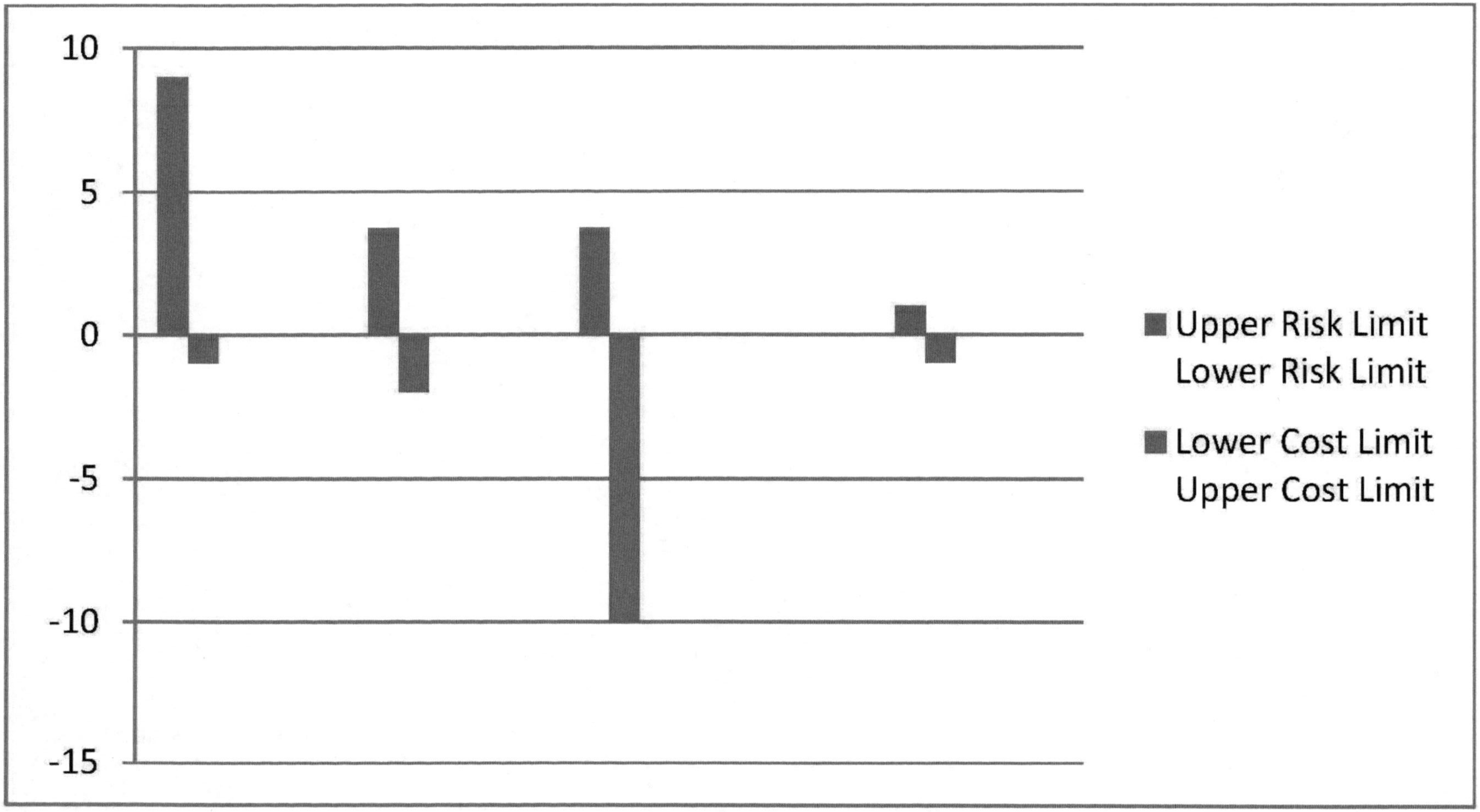

Figure 13: Sample / Cost Benefit Summary Table

HAZOP Rec. No.	Description of Project	Cost of Implementin g ($MM)	Loss Potential ($MM)	Fatality Potential	C/B Ratio	Action
2	Install Emerg Vent Hdrs & Scrubber System	5.10	100.0	11	0.1	Implement
21	Install ROVs on all Rxr & tank outlets w/ SSD	2.00	10.0	5	0.2	Implement
5	Reduce and reroute piping	0.45	1.0	1	0.5	Implement
13	Install independent HP alarms on Rxrs & Stills	1.20	1.0	1	1.2	Re-analyze
7	Install deluge systems in all Class I tk Farms	5.00	4.0	1	1.3	Re-analyze
10	Install independent HHL on all tanks	3.00	1.0	1	3.0	Re-analyze

Figure 14: Ranked by Fatality Index

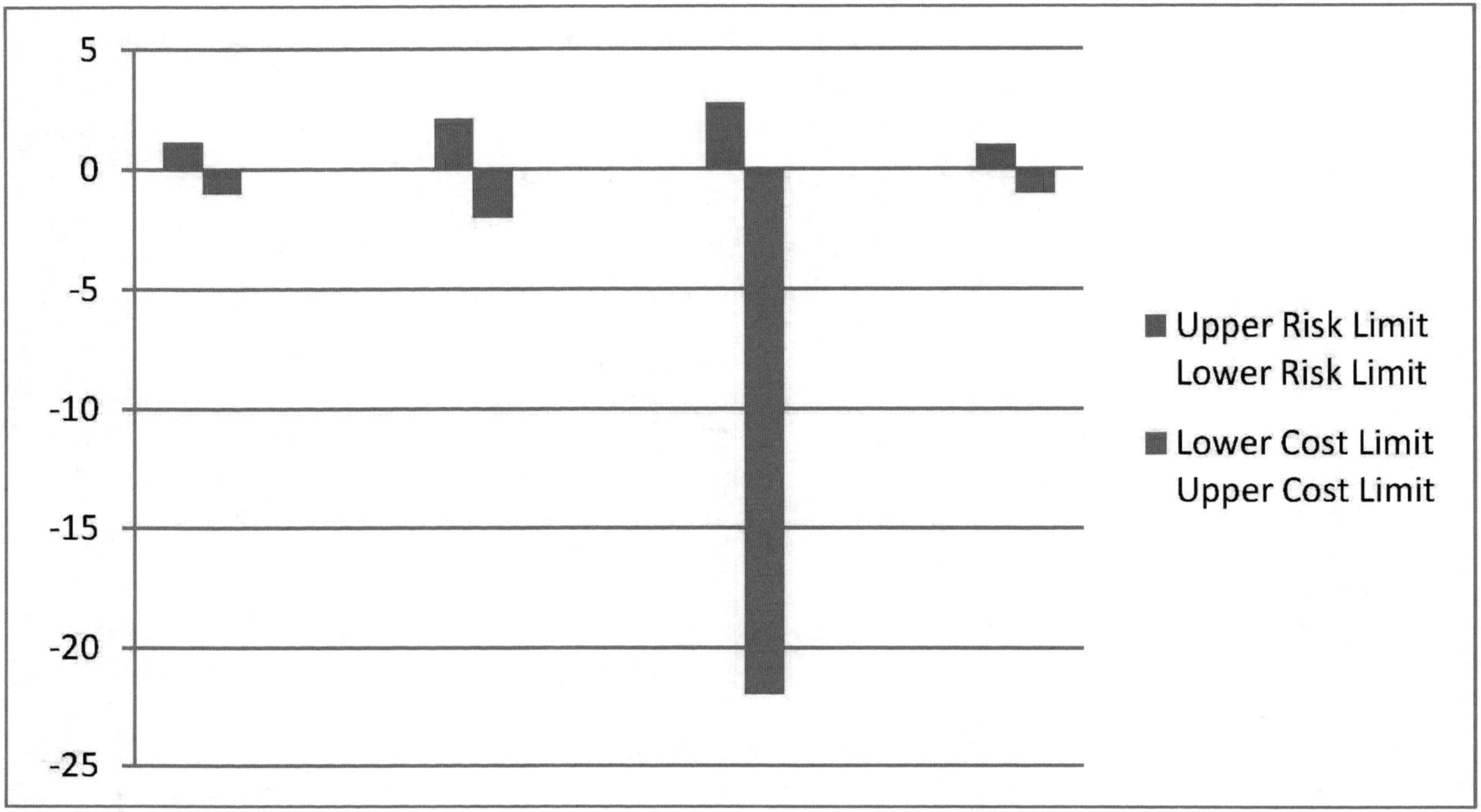

Implementation of PSM in Capital Projects

Charles Soczek
Principal Consultant
DuPont Sustainable Solutions
4417 Lancaster Pike, BMP 13/1108 Wilmington, DE 19805
charles.a.soczek@usa.dupont.com

Keywords: Process Safety Management, Capital Projects, DuPont

Abstract

It is critical to embed Process Safety Management (PSM) in the capital project system to help achieve a safe start-up. This is the foundation for assuring safe operations throughout the life of the process, especially in the construction of facilities for high-hazard processes.

Today's PSM system, practiced by most companies, is Deming-based and specifically a plan-do-check-act system that strives for continuous improvements. The capital project process is very different, with specific starting and ending points that are driven by cost and time. The result is that all required tasks for each PSM element are also time-driven, with the objective of having a full PSM program in place at the time of start-up. Additionally, capital projects must achieve designs that meet the balance of competing demands.

This paper explores some of these issues that are critical to achieving on-time start-up, design operating rates, and safe operation with optimum investment and life cycle cost.

Introduction

There are three types of changes that occur in high-hazard processes that address PSM-related issues: 1) facility and technology changes in existing operating units, 2) small projects, and 3) major projects.

Facility and technology changes are managed with the PSM management of change process. They are relatively low-cost changes and modifications to existing facilities that are charged to operating cost, investment cost, or a blend of the two. These changes require minimal design. However, PSM requires consideration for a Process Hazards Review, changes in process documentation and operating procedures, the need for a Pre-startup Safety Review, and possibly other PSM elements depending on the change. These changes are implemented through the plant work order system.

Small projects are for higher cost changes, generally to existing facilities that are charged as capital investment. They require design, frequently completed with a local engineering group. Because they are completed on sites with an existing PSM program, not all the PSM elements are required tasks for the project team. Below are the elements that should be addressed:

- Developing of and storing process safety information
- Completing the PHAs, including impact on the site facility siting study
- Developing the operating and maintenance procedures as necessary
- Provide training as necessary to operators and mechanics
- Implementing a quality assurance process for equipment procurement and installation
- Developing or updating:
    - Mechanical integrity inspection and test program
    - Audit program
    - Emergency planning and response program as necessary
- Implementing Contractor Safety for Construction Safety
- Pre-startup safety review

Large capital projects can be implemented on either a Greenfield or Brownfield site. On a Brownfield site, where there is an existing PSM program in place, the PSM elements that need to be addressed are the same as a small project. For a Greenfield site, with no PSM in place, all of the PSM elements need to be addressed. Therefore, in addition to the small project requirements, large capital projects require the following:

- Incident investigation
- Management of change (facility and technology)
- Management of change – personnel (a DuPont requirement, not currently an OSHA 1910.119 requirement)

Continuous Improvement PSM Process versus a Linear Process

DuPont implements PSM in high hazard operations according to a continuous improvement plan, based on the Deming Plan-Do-Check-Act system. For capital projects, all of the required PSM tasks must be completed by turnover and start-up. Figure 1 contains a diagram showing the timing sequence for PSM project activities.

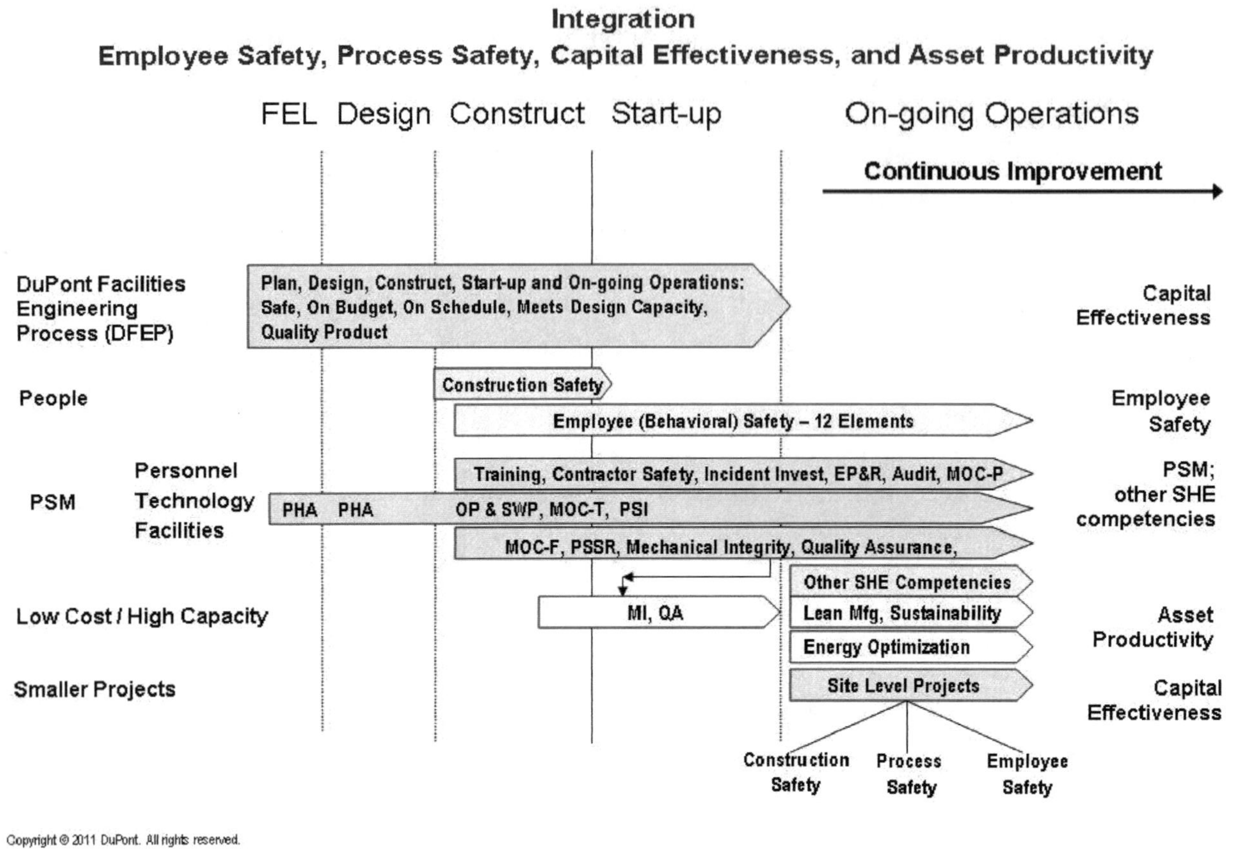

Figure 1. Timing of PSM activities (for more detailed information, see Appendix A).

In a DuPont capital project, business objectives are set during front-end loading (FEL). This is done in a business objectives letter that is issued when the need for a project is recognized and is completed prior to any gate reviews. This letter sets financial objectives and identifies key issues such as investment, volume, and schedule and sets safety, health, and environmental strategies. A key issue is setting project priorities. Project priorities in a DuPont capital project include consideration of:

1. Safety and environment
2. Investment/internal rate of return/investment intensity
3. Schedule
4. Manufacturing cost
5. Product quality
6. Proprietary information protection
7. Capacity
8. Reliability
9. Flexibility

In DuPont projects, safety and environmental protection are always number one priorities. In order to focus on these priorities, plus meet the other objectives, Process Safety considerations must begin early in the conceptual phase of the project. For the project referenced above, additional guidance emphasized design for zero leaks and a requirement to minimize toxic chemical inventories.

While the business objectives describe what needs to be accomplished, the project objectives detail how the project will be accomplished. The draft project objectives letter is prepared in the FEL. Following this step, DuPont completes 20-30% design, developing control strategies to manage hazards. This includes inherently safer processing technologies, process control, and safety interlocks. In this way, the project is adequately scoped to assure safe operation of the facilities prior to beginning production design.

The process hazards analysis (PHA) is the starting point for PSM-related activities. It is the responsibility of the project team to achieve a balance among priorities in order to meet the overall project objectives. An integrated project team at DuPont is assembled and includes members from the operating department, DuPont Engineering, the design contractor, and construction contractor.

In DuPont this is a staged process starting with a screening review, a pre-authorization review, a detailed review, and the pre-startup safety review. At the conclusion of the project, a project safety report is prepared that compiles all the PHA related activities.

Screening Review

The objective of the screening review is to identify potential acute hazards, any unknowns with potential process safety implications, and any other relevant concerns, and to develop broad recommended changes in scope to significantly reduce these hazards. A key activity in this review is to identify and consider the use of inherently safer technology. Identifying hazards early in the process and developing the scope to manage the risks of these hazards is a critical step in achieving a balance between capital cost, operability, and life cycle cost. In some cases, implementation of inherently safer processing strategies can add to the capital cost, but have a significant positive impact on operability and life cycle cost.

There are examples of projects that began the PHA reviews near completion of production design. This meant that implementing ISP strategies would add significant cost and delays in start-up. When the PHAs were completed, there were fewer options to manage risks. The result was a large number of safety interlocks, operability issues, and high maintenance costs to meet current IEC and ISA safety interlock design, inspection, and test requirements.

Pre-Authorization Review

The objectives of this study are to review the screening PHA, review any changes in scope or design intent since the screening PHA, confirm all process-related hazards and hazardous events have been identified, help ensure that the consequences of the identified events have been assessed, and determine if the current project scope is adequate to control all of the hazards. This assures that the project is adequately funded to control hazards.

Detailed Process Hazards Analysis

The detailed PHA includes a review of the earlier PHAs and an in-depth, systematic analysis of the process to identify and evaluate:

- All process-related hazards and hazardous events
- Potential routes to the hazardous events
- Adequacy of safeguards included in the process and facility design

Lastly, it is important to develop recommendations for further process hazards elimination or control. This review is done soon after the design drawings are released.

Final Project Safety Report

This report may be either a compilation of all the project PHAs (i.e., the screening, pre-authorization, and detailed PHAs) and related documentation for all other process safety related topics addressed during the project engineering phase, or it may provide a summary of the work completed in these other reviews and the key findings and recommendations, along with a reference to the complete individual reports. This report is prepared prior to the pre-start-up safety review (PSSR).

Below is a brief overview of some key PSM elements that apply directly to a capital project. **Note:** Not all the requirements of OSHA 1910.119 and EPA RMP are described.

Pre-Startup Safety Review

The PSSR provides a final checkpoint for new equipment to confirm that all appropriate elements of PSM have been addressed satisfactorily and that the facility is safe to start up. For capital projects, planning for success with the pre-start-up review begins with assembling a competent project team. Effective project management, including careful planning and coordination of all PSM elements, is important to achieve a smooth start-up.

Process Safety Information

The process safety information package provides a description of the chemical process or operation. It provides the foundation for identifying and understanding the hazards involved, which are the first steps in the PSM effort. The package consists of the following parts: hazards of materials, process design basis, equipment design basis.

Process safety information is critical to help ensure safe operation throughout the life of the facility. This information is needed by the PHA teams so they understand the impact of proposed changes and can perform cyclical reviews. Unfortunately, this information is very costly to reproduce if lost; therefore, permanent storage systems must be established. It is essential that the process safety information requirements need to be built into design and procurement packages.

Mechanical Integrity/Quality Assurance

Mechanical integrity (MI) covers the life of the facility from the initial installation through dismantlement. It focuses on maintaining and continuously improving a system's integrity to contain hazardous substances and control the mechanical hazards throughout the life of the facility. It addresses such topics as the following:

- Maintenance procedures
- Training and performance of maintenance personnel
- Quality assurance (QA)/quality control (QC) procedures
- Equipment tests and inspections, including predictive and preventive maintenance
- Repairs and changes
- Reliability engineering

The development of maintenance procedures and the equipment test and inspection program is a large effort, and the project team is normally responsible for developing these procedures.

In DuPont, quality assurance applies to project materials, whereas quality control applies to maintenance materials. The project team is responsible for QA, and activities may include correct specification of project materials, vendor inspections, inspections upon arrival, correct storage and preservation of project materials prior to use, and final inspection at installation.

Construction Safety

Many companies use contractors for construction of facilities. In DuPont, whether completed by site employees or by contract employees, it is essential that all tasks are completed safely and in accordance with established procedures and/or safe work practices and are consistent with the principles and essential features of PSM. For construction activities, particularly near or within an operating unit, additional activities such as a constructability analysis are required. This may include, for example, erection of physical barriers between operating units and construction activities.

Operating Procedures

Operating procedures are prepared by the project team during the construction phase of the project after drawings have been released. They should contain all the requirements listed in OSHA 1910.119. In the case where operating procedures are supplied by a vendor, they should be reviewed by the project team to help ensure operability with the actual unit under construction. These procedures are the basis for operator training.

Training

The project team is responsible for ensuring that all operating and maintenance personnel are properly trained. Initial training should include:

- An overview of the process, emphasizing specific safety, health, and environmental hazards
- Emergency procedures
- Operating procedures
- Safe practices applicable to the employees' job tasks
- Additional training such as basic skills and job task training

Management of Employee and Process Safety

In manufacturing processes that contain high hazards, it is important to build a strong safety culture. This is done by implementing both solid employee safety and process safety programs. Failure to implement either one of these has a negative influence on safety culture and may result in poor safety performance. For a Greenfield site, implementation of an employee program begins when operating, maintenance, and technical personnel are identified and the operating organization is established. DuPont has developed a 12-element program for management of employee safety.

Management Commitment

Management commitment is the basic component of a successful safety management system. This commitment must exist at the top and then cascade to all levels of the organization, including partners and contractors. In addition, safety is an organizational value that must be considered as important as any other business decision-making parameter such as quality, productivity and cost.

Overall, management must demonstrate commitment and respect for safety through "example, attitudes, behavior, and follow up." In addition, management must be consistent in all they say/speak/write and what they do. When done properly, this "felt leadership" will create an environment of trust that will be the foundation of safety improvement.

Policy and Principles

The organization shall establish and implement safety policy and principles that are based on the corporate values, mission, and vision. The safety policy and principles are effectively communicated through the line organization, and their understanding and adherence by members of the organization are continuously evaluated. In short, the actions of all members of the organization shall be guided by the policy and principles.

Integrated Organizational Structure

An integrated organizational structure facilitates efficient management of safety through involvement of all levels of the organization, and is applied to everyone from the head offices and administrative areas to the operational sites. Typically, a management committee headed by the highest level of the organization should lead the overall safety effort by developing and improving safety management systems.

In addition, line management has accountability for planning, developing, implementing, controlling, evaluating, and validating the safety management system, and for involving all people in their area of responsibility.

The management committee will sponsor a network of subcommittees throughout the organization. Subcommittees are responsible for guiding and supporting the implementation and continuous improvement of safety programs and procedures. Subcommittees are multifunction teams led by line managers and often include safety specialists, experts, and operational representatives.

Line Management Accountability and Responsibility

An effective safety organization is achieved when roles and responsibilities are clearly defined and implemented. The Safety Management System is based on line responsibility and states that leadership should be actively engaged and committed to safety, taking responsibility for their performance and the performance of those reporting to them. This commitment should be reflected in the people management processes (HR), which should consider both individual and organizational Safety performance in career and development plans.

Goals Objectives and Plans

Like any other important management system, Safety Management Systems achieve excellence through continuous improvement concepts, establishing challenging goals and objectives, and developing action plans accordingly. These goals set the overall direction of the system, and the objectives define the targets that must be met. In establishing challenging goals and objectives, the organization develops safety programs and management system improvements and drives safety performance.

Safety Personnel

Safety professionals are responsible for providing support to the line organization on regulatory and technical issues and on the Safety Management System improvement process. Usually, safety professionals:

- Serve as an advisor to leadership;
- Support and coordinate the safety management model;
- Influence, advise, and assist the entire line organization on safety management and program implementation;
- Participate in field audits to be used as reference for trend analysis;
- Periodically audit the safety performance and management model;
- Interpret safety regulations and interact with external agencies.

Procedures and Performance Standards

High operational standards are achieved by high-quality operating procedures, use of best practices, and best-in-class installations. Procedures should be written, disseminated widely, followed, and reinforced.

Operating procedures should also provide a clear understanding of the parameters of operation and the risks to people, community, and environment. Processes must exist to continuously improve and check for adherence to those procedures in order to guarantee operational discipline. In addition, installations are continually checked for improvement opportunities in safety.

Training and Development

The Safety Management System helps ensure that people are prepared for their jobs physically, mentally, and technically. Selecting people and training are basic principles that increase personal capability and awareness of risk when dealing with operational hazards. In addition, proper selection and training help to improve personal and team performance. A people development process including initial orientation certification for special activities, professional progression, and retraining should be in place.

Effective Communication

Information related to safety needs to be clearly, objectively, and promptly communicated in a way that enables all levels of the organization and the local external community to interact effectively to improve safety performance. Communication processes are based on a two-way approach throughout different levels of the organization, ensuring that both employee understanding and active leadership participation is achieved. Often, these processes use different communication vehicles for different objectives.

Motivation and Awareness

Safety motivation and awareness occur when the management system includes programs and processes that continuously promote safety and connect individual contributions to an effective results-oriented system. Leadership participation and initiatives are designed to help ensure that rules and procedures are followed and that people are motivated to follow them.

Audits and Observations

A continuous improvement system is based heavily on safety audits and inspections with effective leadership participation at all levels of the organization. Processes exist to enter, report, and monitor implementation effectiveness and to complete periodic, critical analysis for all audit conclusions and recommendations. Use of the overall safety management audit and the three audit levels (first, second and third party) are consistently used as a reference guide.

Incident Investigations

The learning process from undesirable events occurs in a structured way, uses the line responsibility concept, and is treated as a priority by the whole organization. The accidents and incidents investigation process includes activities from classification to analysis of all possible causes, as well as follow up on the effectiveness of the actions taken. Support of investigations methodologies exists to guarantee high quality results.

Conclusion

PSM must be embedded into capital projects to help achieve a safe start-up. This is the foundation for assuring safe operations throughout the life of the process, especially in the construction of facilities for high-hazard processes.

The PSM system practiced by most companies today is a Deming-based, plan-do-check-act system that strives for continuous improvements. The capital project process is very different. To meet the competing priorities in a project, PSM considerations need to start very early in a capital project, with specific starting and ending points that are driven by cost and time. The objective is to have a full PSM program in place at the time of start-up.

Appendix A – Timing of PSM Activities

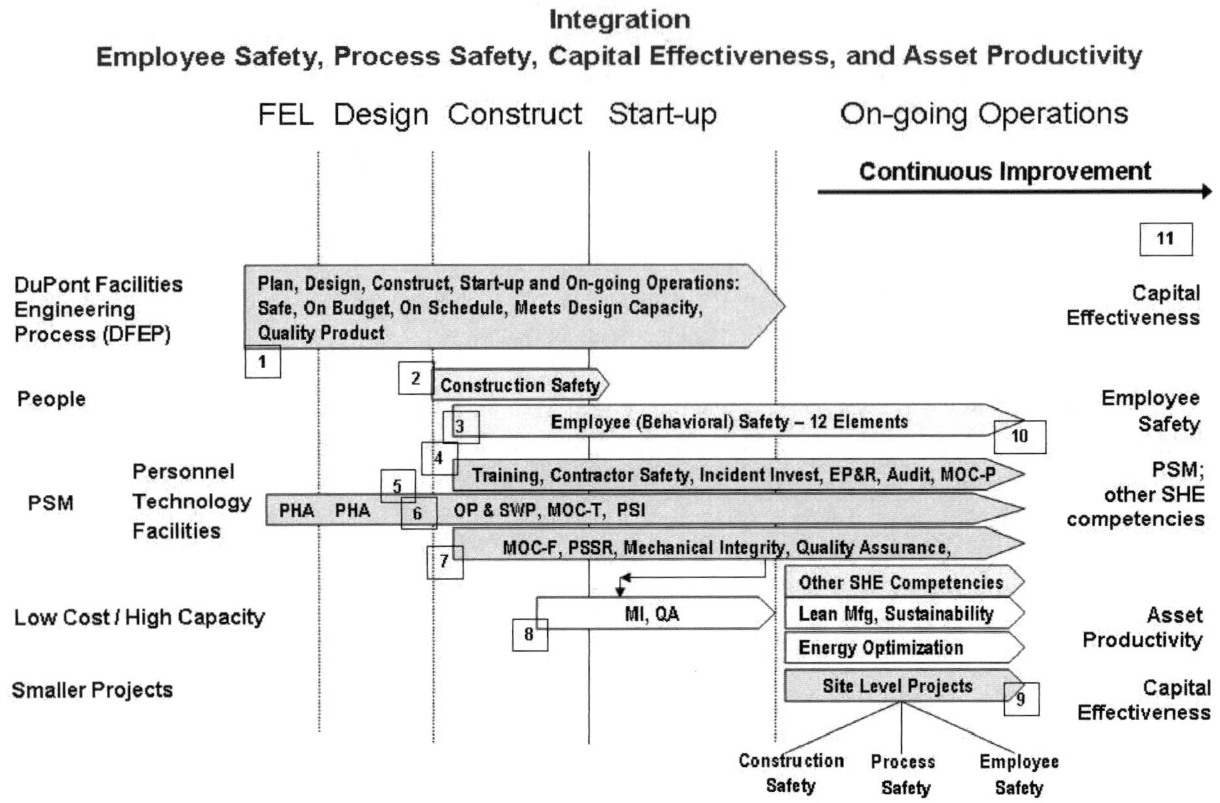

Notes:

1. The DFEP relies on a strong front end loading to assure the project meets budget, SHE, capacity, quality, and manufacturing cost objectives. In addition, it makes effective use of best practices and value improving processes and sound contracting strategies. The project team is composed of DuPont Engineering and Manufacturing representatives, plus others as design, construction, and technology representatives.
2. Construction Safety is targeted at assuring safety in capital projects during construction. It is very similar to contractor safety, which is part of Process Safety Management and targeted at ongoing maintenance and other activities in an existing facility.
3. Employee safety is based on 12 elements and targeted at the entire organization to assure full participation and commitment to building a strong safety culture. It builds the operating discipline necessary for continued safe operation of the facility and is a foundation for Process Safety Management activities. In a capital project, the activities must begin when the operating team is assembled, and activities continue throughout the life of the process.
4. Process Safety Management focuses on three areas: Personnel, Technology, and Facilities. There are six areas in the Personnel part of PSM. Certain activities, such as Training,

Emergency Planning and Response, and Contractor Safety must be complete by start-up. Also, at start-up, the plant converts from the project Incident Investigation system to the plant system. After start-up the plant needs to implement a first- and second-party audit system and a Management of Change/Personnel system.

5. There are 14 elements of Process Safety Management in three major areas: Personnel, Technology, and Facilities. Four elements are common with Employee safety: Audits, Training, Incident Investigation, and High standards of Performance, which includes Operating Procedures and Safe Work Practices. For Greenfield Project, the project team and site operations team are responsible for developing a comprehensive PSM program plus building a safety culture and having it in place for start-up. For existing sites, the project team is responsible for integrating project PSM activities into the site program.

6. The PHA process begins very early in the project with a screening review, followed by a pre-authorization review. The objective of these reviews is to assure project funds are adequate to manage process related hazards. A detailed PHA is done at the completion of the design step. For ongoing operations a base line review (detailed Review if available) is necessary, followed by periodic updates. Changes of Technology or Facilities could trigger additional PHAs.

7. The Facilities part of PSM covers four areas: Management of Change/Facilities (or Subtle Change), Pre-start-up Safety Review, Mechanical Integrity, and Quality Assurance. The project team is responsible for the conversion of the project change system to the plant change system. The PSSR is done upon project completion and turn-over to the site operations team. A major task of the project team is to address project requirements for Mechanical Integrity including Maintenance Procedures, Training and Performance of Personnel, Quality control Procedures, Equipment tests and Inspections, Repairs and Changes, and Reliability Engineering. A Quality Control program, aimed at parts and supplies for plant maintenance, and a Quality Assurance program, aimed at equipment for plant projects, must be developed.

8. The Mechanical Integrity and Quality Assurance Program developed under PSM are targeted to assure containment of hazardous materials and control of risks from high energy and other mechanical equipment. This program provides the basis for further equipment improvements targeted at reducing maintenance costs, replacement costs, capacity increases, and other sustainability efforts. This is done through the Asset Productivity programs as Lean Manufacture and Sustainability.

9. Smaller or site level projects are done to implement site driven improvements in safety, health, and environmental, process safety, cost reduction, capacity improvement, and replacement. A streamlined version of the DuPont Facilities Engineering Process is normally used. Site level projects require he same elements of FEL, design, construct, and start-up as the larger projects. They must also include Construction Safety and the project elements of PSM and integrate them into the site system. The training required for Employee Safety depends on the experience of the operating group.

10. The project team is responsible for development of the Process Safety Information package. Key areas are Hazards of Materials, Process Design Basis, Equipment Design Basis, and Process Technology Documentation. For the project team, pulling this information together is a substantial effort.

11. The typical area where DuPont works is in on-going operations, where the facility is built and in operation. The site wants to make improvements in safety culture, PSM, capacity, quality, reduce cost, etc., or add a new facility. These can be done through programs and projects. The management system in place uses the continuous improvement process of envision, plan, implement, and check.

Systematic Workforce Involvement in Process Safety

Cory D. Shelton
BP
501 Westlake Park Blvd
Houston, TX 77079
Cory.Shelton@bp.com

Keywords: Workforce involvement, process safety, behavioral safety, management systems

Abstract

Workforce involvement (WI) is considered to be an essential element in promoting effective process safety management. This paper describes some proven methods of workforce involvement that are considered to have further strengthened process safety. The potential scope, organization, process, and means to evaluate a workforce involvement program are described. Critical success factors and pitfalls are also illustrated. This paper presents a systematic approach to periodically assessing and improving workforce involvement in process safety.

1. Introduction

The importance of employee involvement and engagement is not limited to process safety, but rather seen as essential elements for success in numerous business sectors. One Department of Energy report[1] addressing organizations committed to high reliability gave several examples of the importance of employee involvement in numerous sectors and noted that new management skills might be called for:

> "These new ideas, these new experiences, these new realities may offer an answer to the question 'why'; why culture, why engagement, why involvement. Perhaps in the not too distant future senior managers will be equally skilled in designing organizations for people as they are today in finance or technical project management. *Perhaps in the not too distant future designing organizations to support human performance will be a core competency heralded as a distinctive strategic advantage. Perhaps in the not too distant future excellence might become the norm, rather than the exception.*" [Emphasis added.]

The Institute for Safe Medication Practices, a federally certified Patient Safety Organization listed several characteristics of a *high reliability organization* (HRO). Under the rubric of "feedback loops" they wrote[2]:

> "HROs have established cross-departmental, meaningful feedback systems that keep the workforce informed about safety, errors, and causal trends-not just in sophisticated charts and graphs, but also in rich stories about safety that can be

recalled as a constant reminder and sound rationale to keep safety a priority. *Likewise, leaders are visible in work areas to hold discussions with the workforce to learn firsthand about the barriers to safe work, and also to share strategic goals, build trust, and demonstrate safety priorities.*" [Emphasis added.]

Closer to our topic of process safety in the chemical and petroleum industries, the AICHE states that promoting *workforce involvement* (WI) is one of the key elements of committing to process safety:[3]

"Organizations often have a similar opportunity to tap more closely into the knowledge and expertise of workers. Operators and maintenance personnel may have a deeper and more authentic understanding of a particular problem, and its solution, than does the engineer...The contract maintenance employee who happens to notice a sound or a smell that is different from what it was on the first two days on the job may, on that third day, may have a more important role to play in the safety of the facility than any 30-year veteran company employee."

Process safety management is a vital component of industrial sites where hazardous substances, if released, might lead to fire, explosion, or toxic exposures.

In general: the larger the organization and/or site, the more difficult it may be for members of the workforce to have an active voice in raising concerns over potential gaps in barriers of protection and providing insights into how to improve those barriers.

Numerous books and articles[4, 5, 6, 7, 8, 9, 10, 11, 12] have been written on the topic of workforce involvement relative to process safety. This article summarizes relevant literature on the scope, organization, and process of workforce involvement. It also highlights the traditional approach to WI in process safety and a rationale for increasing this involvement. An initial assessment of both the current level of workforce involvement in a business as well as readiness can form the basis for improvement steps. Examples of successful WI in process safety are provided as well as some potential pitfalls in implementation. Finally, the paper finishes with conclusions on sustainability.

2. Legal Issues (a brief aside)

Experience across industry shows that there are many tangible benefits of active workforce involvement in developing a robust process safety management program (see below). Workforce involvement is also required by law in some countries, including the United States and the United Kingdom. In the US, for example[7,9]:

"Section 304 of the Clean Air Act Amendments states that employers are to consult with their employees and their representatives regarding the employers efforts in the development and implementation of the process safety management program elements and hazard assessments. Section 304 also requires employers to train and educate their employees and to inform affected employees of the findings from incident investigations required by the process safety management program. Many employers, under their safety and health programs, have already established means and methods to keep employees and their representatives informed about relevant safety and health issues and employers may be able to adapt these practices and procedures to meet their obligations under this standard. Employers who have not implemented an occupational safety and health program may wish to form a safety and health committee of employees and management representatives to help the employer meet the obligations specified by this standard. These committees can become a significant ally in helping the employer to implement and maintain an effective process safety management program for all employees."

In the UK, the *Health and Safety Executive* (HSE) provides guidance on the applicable legal requirements for workforce involvement.[6]

3. What is workforce involvement?

Reduced to its essence, workforce involvement may be considered the end objective of a system designed to enable the active participation of employees and contractor personnel in the design, development, implementation, and continuous improvement of a risk-based process safety system. And just as there are generally considered to be levels in the evolution of safety culture, such as those described in the Hudson "ladder" as rungs (i.e. pathological, reactive, calculative, proactive, and generative)[13], so too are there levels of workforce involvement.

The UK HSE has defined three possible levels of workforce involvement illustrated in Fig. 1. These include one way, two-way, and joint involvement. BP aims to reach "Level 3," which features managers and workers jointly involved in making decisions about how to create and maintain a safe workplace. Just as the "generative safety culture" represents the ultimate evolution of an organization's safety culture, Level 3 represents the highest level of workforce involvement.

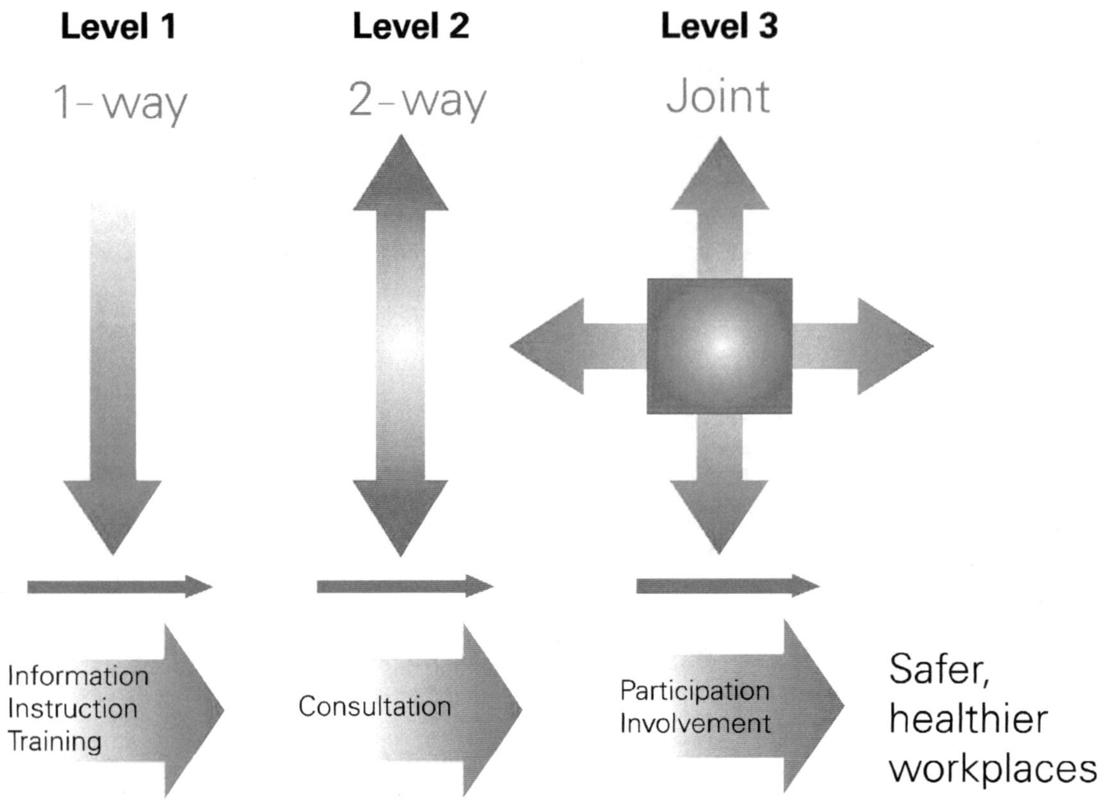

Fig. 1. *Levels of workforce involvement (source: UK HSE).*

Three key elements of a workforce involvement program are *scope, organization, and process.* These are described below.

Figure 2 identifies a series of components that define the possible *scope* of a workforce involvement program. Such a program should include all the components for a "Level 3" program, and should seek to meet all applicable legal and regulatory requirements regarding process safety management.

A complete discussion of all of the components in Fig. 2 is beyond the scope of this paper. Fortunately, each of these components has been well studied and there is an extensive body of available literature on each component. For example, the drafting and modification of *standard operating procedures* (SOPs) for a variety of industries has been extensively researched,[3, 14, 15, 16, 17, 18] including the publication of numerous informative case studies.[19, 20]

Potential Components of Workforce Involvement	
Review of accident/near miss data	Workers have the opportunity to learn from mishap experience. Workers can profit from shared lessons learned from mishaps and can offer useful suggestions for safety improvements.
Setting health and safety goals	Workers can help to establish health and safety goals for the facility or business unit. Goals mutually agreed upon are more likely to be attained.
Champions	Individuals that can be used to share best practice information throughout the facility or business unit. Champions can be recruited from a range of positions including operators, supervisors, engineers, and managers.
Hazard spotting	Workers can participate in a formal "hazard spotting" program where workplace hazards are identified and reported in periodic safety meetings.
Participation in investigation teams	Workers can contribute to accident or near miss investigation teams. Often their direct experience with the facilities and processes enables them to offer unique insights on error- or violation-producing conditions.
Participation in joint labor-management committees and other advisory or specific purpose committees	Workers can make useful contributions and also help publicise the work of these groups.
Review and modification of standard operating procedures (SOPs)	Workers have direct, day-to-day experience with SOPs and can offer constructive suggestions for their modification and improvement.
Management of change	Experience shows that process, procedural, or organisational changes often have potential safety impacts. Workers can be involved in management of change activities.
Process hazard analysis	Workers can be included in multi-disciplinary teams that survey facilities and processes to identify and reduce possible hazards.
Equipment design and review	Workers can provide useful input to the design or selection of equipment.
Development/assessment of safety training programs	Workers can provide constructive critique of safety training programs and help in the development of improved training programs.
Providing programs and presentations at workshops and safety and health meetings	Workers can bring unique real-world perspectives.
Auditing	Workers (perhaps with special training) can be effective members of safety audit teams.
Communication	Workers can help ensure effective communication of key safety messages. Posters on bulletin boards, hazard warning signs, newsletters, and other media can benefit from worker review and design.

Fig. 2. *Potential components of workforce involvement (© BP International).*

4. Organization for Workforce Involvement

Scope (discussed above) is one important dimension of a workforce involvement program. *Organization* is another. Several organizational mechanisms including the following may be used to facilitate worker involvement.

Including workers:[5, 6]

- As designated safety representatives on formal health and safety committees,
- In investigation teams or task forces with specific safety-related charters,
- As instructors and trainers in formal workshops and informal "tool box" or "tail gate" talks and meetings, and
- As designated "champions" with the responsibility of sharing "best practices" information with others in the organization.

The above list is illustrative (see http://www.hse.gov.uk/involvement/factorstoconsider.htm for other factors to consider). The specific organizational form is considered less important than the respect and support accorded participating workers and the demonstrated willingness of management to make prompt changes based on their recommendations. Though the organizational form is considered less important than the quality of the effort, *some form is necessary*. *Ad Hoc* efforts seldom have staying power and may not be perceived to be genuine by the workforce.

5. Process

Figure 3 shows a series of steps that the UK HSE has identified in the creation of an involved workforce program, which generally follow the "plan-do-check-act" paradigm:

- Get a clear, written management commitment,
- Establish specific roles for management and workers for the particular facility or business unit,
- Develop a written plan that describes the organization and defines the responsibilities of all parties,
- Implement these plans, and
- Periodically evaluate the program, identify lessons learned, and make changes as necessary.

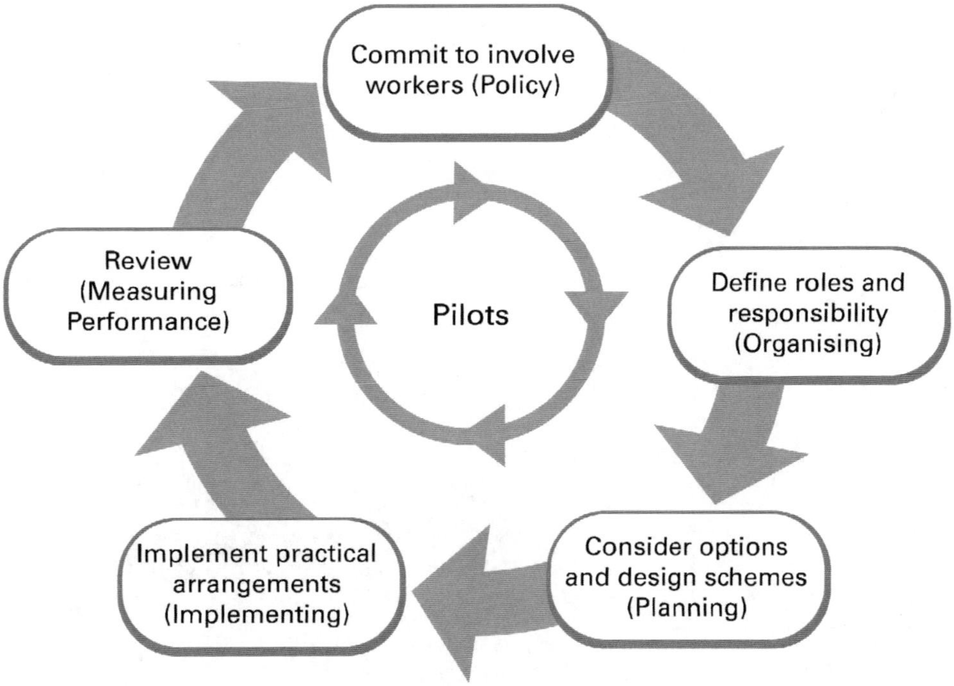

Fig. 3. *Steps involved in creating an involved workforce (Source: HSE).*

6. Safety culture and readiness?

In order to assess an organization's readiness for workforce involvement in process safety, it is considered prudent to assess the environment, climate and context of the organization. The culture or "the way we do things around here and why," is an important consideration in the nature and effectiveness of changes to workforce involvement. To illustrate, consider where your organization fits on the Safety Culture ladder depicted on Figure 4.

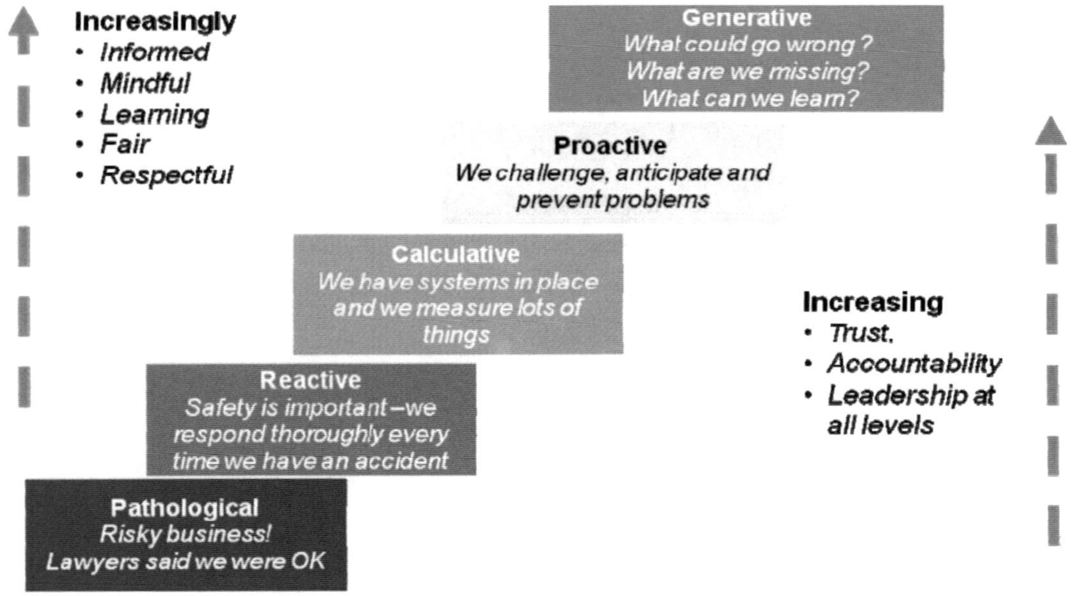

Fig. 4. *The safety culture ladder.*

This figure shows the familiar "safety ladder," used to describe the evolution of safety culture. While this model might be considered overly simplistic, it does provide a useful set of benchmarks for developing detailed plans and actions. For example, one feature of "moving up the ladder" is that there is increasing clarity of expectations, accountability, support, and trust.

Moving from "reactive" through "calculative" to "proactive" on the ladder requires a significant leadership involvement and support at all levels. Being an effective role model, actively integrating safety into business decisions, deferring to expertise, allocating appropriate resources, setting effective goals and measures, actively measuring real performance, and supporting active learning are among the many leadership behaviors that would drive the transition to a proactive culture.

How does trust figure in the safety culture ladder? At levels of "calculative" and below, suspicion and lack of trust may result if the behavior of leaders is not congruent with their statements such as "safety is our number one priority." For most organizations, worker trust cannot be taken for granted, it has to be earned. The demonstrated readiness to make and sustain safety related changes based on worker input helps to promote an atmosphere of trust. Lack of worker trust is certainly a challenge, but not an excuse for failure to further strengthen process safety. Trust can be built. For example, in circumstances where there is a lack of management-workforce trust, a carefully implemented workforce involvement activity could potentially boost and begin to restore that trust. In that situation, it may be good to pilot a pin-pointed activity.

What would an organization's climate or culture need to be in order for WI activities to be successful? Among other factors, there would need to be a foundation of active leadership involvement and support.

One of the differences between process and personal safety is that actions to further strengthen process safety can potentially involve engineering modifications with cost implications and may affect site production. These short-term consequences should be carefully analyzed against the potentially larger, longer-term impacts associated with process safety incidents, low morale, distrust and reduction in engaging the discretionary performance of the workforce. Therefore, as WI efforts identify process safety issues, management should carefully consider the long-term (as well as short-term) impacts of their decisions and actions. To state the obvious, any reliance on "luck" as a process safety strategy is fundamentally flawed.

6.1 Rewards (another brief digression)

How do you motivate and improve workforce involvement in process safety? There are differences of opinion on the subject,[21] but in my view a common trap is for leaders to think that cash or other tangible rewards are necessary to stimulate this performance. In respect of safety, if not managed appropriately, such rewards can lead to unintended consequences such as not reporting incidents. This is particularly true of the low-level process safety incidents, and of underlying precursor conditions and behaviors.

If not through cash or tangible rewards, then how can superior performance be recognized? In general, the most underused leadership tool is personal, one-to-one feedback on the desired behaviors of their people. The most effective leaders, managers, and supervisors, *reinforce people while they are performing.* The proper balance between positive reinforcement and critical feedback is a minimum of 4:1 and can be as high as 10:1 (Daniels).[22]

7. Management Systems Approach

Incorporating workforce involvement activities in a company's management system can be an important tool in driving further improvements in process safety. For example, at BP, the systems which guide operations are contained in our *Operating Management System* (OMS) and in our OMS essentials (requirements), which describe several forms of workforce involvement. It is useful to step back and consider workforce involvement as a whole (in order to highlight gaps), as well as the quality of both required and optional workforce involvement activities. Another feature of embedding involvement activities into your company's management system is that almost all such systems incorporate a form of the continuous improvement cycle.

8. Assessing Workforce Involvement

BP has developed an assessment tool to identify and evaluate the quality of proven workforce involvement activities based upon the wide body of research and guidance. Low injury and incident rates are the basic test of any safety program, but these are lagging indicators and other tools are also appropriate. The BP assessment tool permits assessors to define the various components of the workforce involvement program (see below) and scores each on the Level 1 – 3 scale suggested by the UK HSE. Examples of process safety focused workforce involvement include the components identified in Fig. 2:

1. Review of accident/near miss data: Workforce involvement in reviewing company data (from local and other sites) and industry data can result in moving from "lessons learned" to "learning lessons." Several government agencies provide relevant data on incidents. For example, the US Chemical Safety and Hazard Investigation Board ("CSB") (see http://www.csb.gov/) publishes useful incident reports. It also produces videos, which add interest. The National Transportation Safety Board (see http://www.ntsb.gov/Publictn/P_Acc.htm) publishes useful reports on pipeline accidents. In the UK the HSE also publishes useful reports.

2. Setting health and safety goals: Workforce involvement in goal setting for process safety goals can provide feedback on the utility and impact of such goals. Activity or achievement goals could include measures such as the percentage of procedures updated or the number of process safety concerns raised. The UK HSE published a useful document on developing process safety indicators[27] and workers could participate in the selection of the indicators themselves as well as specific goals for these indicators.

3. Champions: Workforce involvement thru the "champions" approach can provide a focus and role for individuals as well as cross-organizational conduits of information and concerns. Champions can have process safety specific duties such as bringing process safety concerns or ideas from their teams to committee meetings or periodically conducting process safety team discussions. The potential of champions is now being realized in many safety contexts, such as medicine (see Soo *et al.*,[28] for a useful discussion).

4. Hazard spotting: It may be useful to involve the workforce in conducting hazard spotting activities which focus entirely on process safety. This could potentially serve to deepen and challenge their current practices, as well as raise awareness about risk management.

5. Participation in investigation teams: Workforce involvement in investigating the underlying causes of process safety related incidents could potentially enhance the technical understanding of what went wrong and strengthen the communication of the potential consequences to members of the workforce involved in the incident. The perceptions of administrative justice following incidents may be inaccurate or incomplete which could result in a cycle of mistrust and a reduction in communication of concerns.

6. Participation in joint labor-management committees and other advisory or specific purpose committees: Workforce involvement participation in various committees can help to confirm that the perspective of the workforce is represented and help to test decisions and identify unintended consequences at an early stage. Perceptions of the decision making process can also be improved.

7. Review and modification of standard operating procedures (SOPs): Workforce involvement in the development and revision of SOPs can play a very important role in confirming that procedures are practical and relevant, and address process safety. SOPs drive consistent performance of process safety critical activities such as start-up, shutdown, and response to upset conditions, as well as energy isolation and permitting. Therefore, workforce involvement in SOPs is a major process safety workforce involvement activity. See Hopkins[20] and Bourrier[29] for illustrations of the potential benefits of worker involvement in drafting and revising SOPs.

8. Management of change: Typically, proposed process or equipment changes are routed thru various designated authorities to confirm that proposed changes are safe, well planned and do not compromise other aspects of plant operation. Workforce involvement can potentially provide a unique opportunity to tap the experience and insights of workers.

9. Process hazard analysis/Equipment design and review: Workforce involvement in process hazard analysis and equipment design and review can provide a mechanism for the end-user to provide input to design, operability, and controls to confirm that the new plants or modifications to existing ones can be operated safely. Small changes to design and controls can potentially make significant differences to the human/machine interface, and workforce involvement can help drive risk reduction.

10. Development / assessment of safety training programs: Active workforce involvement in development/assessment of safety training programs can potentially provide "end-user" perspective to improve program effectiveness.

11. Providing programs and presentations at workshops and safety and health meetings: Workforce involvement in such activities can tap into the creativity and vast knowledge of the workforce. Because workers face the process safety energy sources on a daily basis, their "take" on how to manage process safety can potentially assist with "keeping process safety real" and raise their own awareness as well as that of their peers.

12. Auditing: Auditing plays a very important role in process safety management. Active workforce involvement in auditing can help increase transparency and awareness and coupled with management action should serve to build trust and accountability.

13. Communication: Process safety communications can potentially benefit from workforce input and detailed examples. This can also provide an opportunity for management to express appreciation for stories of defect elimination and to value process safety risk reduction.

14. Behavioral safety: Workforce involvement in behavioral safety is quite common in industrial plants. However, behavioral safety efforts typically do not focus on process safety since process safety is on the whole more technical and less visible, and because process safety incidents tend to occur less frequently. Though some (e.g., Hopkins[20]) suggest that behavioral safety programs are of limited utility for improving process safety, incorporating process safety into training, checklists and performance measures has merit.

AIChE[3] has developed a number of possible measures of effectiveness of a workforce involvement program.

9. Workforce involvement benefits

One of the primary motivations for having a workforce involvement program is that it is considered to be a positive, inclusive activity, which can help to drive risk reduction and further strengthen operating behaviors. In some countries, it is also a legal and regulatory requirement. The UK HSE has identified the following benefits of a good workforce improvement program (supported by data):[25]

- Lower accident rates,
- More positive health and safety climate,
- Better control of workplace risks, and greater awareness of workplace risks.

(The HSE also publishes a series of successful case studies to illustrate the above points.) Further, there is anecdotal evidence to suggest that companies with successful workforce involvement programs enjoy higher employee morale and reduced turnover.

10. Workforce involvement pitfalls:

The list of possible reasons why a workforce involvement program may fail to achieve its objectives is long indeed. Here are a few potential pitfalls for your consideration:

- Program does not have genuine management commitment *at all levels*—lack of commitment is readily detected by the workforce and reduces the effectiveness of the WI program,
- Program is perceived as a "flavor of the month" that will not last,
- Program is not well planned and people are not clear about roles and responsibilities,[4]
- Managers are inconsistent in the scope or timing of the program leading to worker confusion,
- Resources allocated are insufficient to allow full participation,[4]
- Program seen as method for blaming workers for failures in process safety,[12]
- Some program elements, such as behavior-based safety, seen as intrusive and resulting in employee concerns about spying and reporting on colleagues,

- Lack of effective communication on program success (and failures) to all audiences,[4]
- Program was too ambitious initially—it is all well and good to have aspirations for a Level 3 program of full scope; however targets that are unrealistic and set too high can invite failure and disillusionment. See Ordóñez *et al.*[23] for an interesting discussion of pitfalls with goals.
- Failure to sustain program momentum and to reinforce the need for continuous improvement. Sometimes a program may be successful initially yet loses momentum as success is taken for granted and/or perceived limits to improvement are taken to be correct. Sustaining momentum is a major challenge for many programs and books have been written on the subject.[24] (A Google search on the phrase sustaining momentum generated over 1.3 million web site addresses!)

11. Conclusion

Systematic workforce involvement – particularly integrating it into management systems - is considered to be a key component of building a strong process safety culture. Involvement of the workforce can potentially help affirm management's trust in workers and create a virtuous cycle of positive reinforcement. This reinforcement can positively impact overall performance and can extend to the other aspects of the business (such as cost, efficiency, morale, and quality). Workforce involvement can also help foster positive relationships in teams, businesses and organizations as well as reach into customer and societal realms. Periodic assessment and adjustment is important in sustaining and building from strength to strength.

In the early 1920s the National Research Council conducted a classic series of experiments conducted at AT&T's Western Electric's Hawthorne plant located in Cicero, Illinois. The intent of the first of these studies was to investigate the effects of environmental variables, initially lighting, on the productivity of female workers. Researchers were puzzled by the results which apparently indicated that either reduced or improved lighting could result in increased productivity—the real incentive was that management seemed to care about the worker—and the worker responded. In the end, these classic experiments undercut prevailing theories of "scientific management" at the time and showed that behavior is not merely physiological, but also psychological.[26] We seemed destined to replicate the nearly 100-year-old Hawthorne studies periodically. Promoting the active involvement of workers (with their reservoirs of talent and experience) in designing a robust safety program is the right thing to do and likely to offer tangible benefits.

11. References

[1] US Department of Energy, "Employee involvement, engagement, culture—why?" http://hsshpi.wordpress.com/2010/03/11/employee-involvement-engagement-culture%E2%80%93-why/ (accessed Dec. 11, 2010).

[2] Institute for Safe Medication Practices, "High reliability organizations (HROs): What they know that we don't (part 1), 2005, http://www.ismp.org/newsletters/acutecare/articles/20050714.asp (accessed Dec. 11, 2010).

[3] American Institute of Chemical Engineers, Center for Chemical Process Safety, *Guidelines for Risk Based Process Safety*. Wiley-Interscience, New York, NY, 2007.

[4] Pennie, D., Leach, P., Wright, M., *Effective workforce involvement in Health and Safety, A Guide*, Greenstreet Berman Ltd., London, UK, 2007. See also Energy Institute, *Guidance on Effective Workforce Involvement in Health and Safety*, Energy Institute, London, UK http://www.energyinstpubs.org.uk/tfiles/1292082198/798.pdf (accessed Dec. 11, 2010).

[5] Health and Safety Executive, "Worker Involvement," undated, http://www.hse.gov.uk/involvement/. See also "Play your part! How offshore workers can help improve health and safety, http://www.hse.gov.uk/pubns/indg421.pdf (both accessed Dec. 11, 2010). See also Risk Support Limited, Optimising Hazard Management by Workforce Engagement and Supervision, prepared by the UK Health and Safety Executive, 2008, http://www.hse.gov.uk/research/rrpdf/rr637.pdf (accessed Dec. 11, 2010).

[6] Health and Safety Executive, "Consulting employees on health and safety, a brief guide to the law," http://www.hse.gov.uk/pubns/indg232.pdf (accessed Dec. 11, 2010).

[7] United States Department of Labor, "VII. Safety and Health Programs and Employee Involvement," http://www.dol.gov/_sec/media/reports/dunlop/section7.htm (accessed Dec. 11, 2010).

[8] Sepeda, A. L., "Understanding process safety management, " Chemical Engineering Progress, August, 2010, available electronically at http://www.aiche.org/uploadedFiles/Publications/CEPMagazine/081026_Public.pdf (accessed Dec. 11, 2010).

[9] US Department of Labor, Occupational Safety and Health Administration (OSHA), "Appendix C to §1910.119—Compliance Guidelines and Recommendations for Process Safety Management," http://www.osha.gov/pls/oshaweb/owadisp.show_document?p_table=STANDARDS&p_id=9763. See also "Process Safety Management of Highly Hazardous Chemicals (1926.64)," http://www.osha.gov/doc/outreachtraining/htmlfiles/psm.html (accessed Dec. 11, 2010).

[10] Vogus, T. J., and Welbourne, T. M., "Structuring for high reliability: HR practices and mindful processes in reliability-seeking organizations," Journal of Organizational Behavior 24, 877-903, 2003, http://owen.vanderbilt.edu/vanderbilt/data/research/1586full.pdf (accessed Dec. 11, 2010).

[11] Hudson, P., "Applying the lessons of high risk industries to health care," *BMJ Quality & Safety*, Vol. 12, 2003, 7-12.

[12] Step Change, "Changing Minds, a Practical Guide for Behavioral Change in the Oil and Gas Industry," http://stepchangeinsafety.net/ResourceFiles/Changing%20Minds%20Guide.PDF (accessed Dec. 11, 2010).

[13] Hudson, P., "Safety Management and Safety Culture: the Long, Hard and Winding Road, http://www.caa.lv/UserFiles/File/SMS/Read%20first%20quick%20overview/Hudson%20Long%20Hard%20Winding%20Road.pdf (accessed Dec. 11, 2010).

[14] US Environmental Protection Agency, *Guidance for preparing standard operating procedures (SOPs),* EPA/600/B-07/001, April 2007, http://www.epa.gov/quality/qs-docs/g6-final.pdf (accessed Dec. 12, 2010).

[15] Federal Emergency Management Agency, *Developing Effective Standard Operating Procedures for Fire and EMS Departments,* FA-197/December 1999, http://www.usfa.dhs.gov/downloads/pdf/publications/fa-197-508.pdf (accessed Dec. 12, 2010).

[16] International Atomic Energy Agency, *IAEA Safety Standards; Conduct of Operations at Nuclear Power plants, Safety Guide No. NS-G-2.14*, 2008, http://www-pub.iaea.org/MTCD/publications/PDF/Pub1339_web.pdf (accessed Dec. 12, 2010).

[17] Reason, J., *Managing the Risks of Organizational Accidents*, Ashgate, Burlington, VT, 1977.

[18] Dekker, S., *The Field Guide to Understanding Human Error,* Ashgate, Burlington, VT, 2006.

[19] Skriver, J., Flin, R. H., "Decision making in offshore emergencies: are standard operating procedures the solution?" SPE Health, Safety, and Environment in Oil and Gas Exploration and Production Conference, 1996, New Orleans, LA.

[20] Hopkins, A., *Lessons from Longford; The Esso Gas Plant Explosion,* CCH Australia Limited, Sydney, NSW, 2000. See also Hopkins, A. (2008). *Failure to Learn. The BP Texas City Refinery Disaster,* CCH a Wolters Kluwer, Australia, 2008.

[21] Ozer, I., "Motivation Minute effectively rewarding safety and tangible rewards," (attributed to *Occupational Health & Safety*, 2006), http://mcdowell-incentives.com/effectively-rewarding-with-tangible-rewards (accessed Dec. 12, 2010).

[22] Daniels, A. B., *Bringing out the Best in People: How to apply the astonishing power of positive reinforcement,* McGraw-Hill, 2000.

[23] Ordóñez, L. D., Schweitzer, M. E., Galinsky, A. D., and Bazerman, M. H., "Goals Gone Wild: The Systematic Side Effects of Over-Prescribing Goal-setting," http://opimweb.wharton.upenn.edu/documents/research/goals_gone_wild.pdf (accessed Dec. 12, 2010).

[24] Senge, P., Kleiner, A., Roberts, C., Ross, R., Roth, G., and Smith, B., *The Dance of Change: The Challenges of Sustaining Momentum in Learning Organizations*, Doubleday, Garden City, NY, 1999.

[25] UK Health and Safety Executive, available electronically at http://www.hse.gov.uk/involvement/factsfigures.htm (accessed Dec. 12, 2010).

[26] Management Guru, "The Hawthorne Experiments: Management Takes a New Direction, http://www.mgmtguru.com/mgt301/301_Lecture1Page10.htm (accessed Dec. 12, 2010).

[27] Health and Safety Executive, *Developing Process Safety Indicators,* UK Health and Safety Executive, 2006, http://www.hse.gov.uk/pubns/priced/hsg254.pdf (accessed Dec. 13, 2010).

[28] Soo, S., Berta, W., and Baker, G. R., "Role of champions in the implementation of patient safety practice change," *Healthcare Quarterly*, 12, 2009, http://www.longwoods.com/content/20979 (accessed Dec. 13, 2010).

[29] Bourrier, M., "Elements for designing a self-correcting organisation: examples from nuclear plants, in Hale, A., and Baram, M., Eds. *Safety Management: The Challenge of Change,* Oxford, Pergamon.

A Key Factor in Hazardous Processes: Safety Culture

Charles Soczek
Principal Consultant
DuPont Sustainable Solutions
4417 Lancaster Pike, BMP 13/1108 Wilmington, DE 19805
charles.a.soczek@usa.dupont.com

Keywords: Safety Culture, Risk Management, Process Safety Management, Bradley Curve, DuPont

Abstract

Safety culture is increasingly recognized as a key contributor to the safe operation of hazardous processes. This paper describes the four stages through which manufacturing companies travel in their journey to a world-class safety culture and applies them to Process Safety Management (PSM).

The foundation for this journey is illustrated in a model that describes the four stages of culture strength: reactive, dependent, independent, and interdependent. The model has been calibrated against performance and is based on years of data from both DuPont facilities and external companies. Also described are useful tools that manufacturers employ to improve their safety culture and performance.

Introduction

The petroleum, petrochemical, and chemical industries in the United States continue to report some of the best safety performance when compared to other industries and to the U.S. average. While injuries and illness cases with days away from work per 100 employees average 1.3, the petroleum, petrochemical, and chemical industries average 8% or less of the U.S. industry average (U.S. Dept. of Labor, Bureau of Labor Statistics, 2009).

Despite this impressive performance, these industries continue to be plagued by high-profile incidents worldwide. As a result, the U.S. Chemical Safety Board (CSB) is considering additional regulations to improve safety. One example is a requirement that companies develop and submit a "Safety Case."

The argument can be made that existing regulations under OSHA PSM 1910.119 and the Environmental Protection Agency's Risk Management Plan are sufficient to prevent incidents if fully and properly implemented. Incident investigations continue to point to deficiencies in the implementation of Process Safety Management. The CSB in its special-emphasis program targeting U.S. refineries has shown that although companies have programs in place, their implementation can be poor, and this performance can result in significant incidents.

Likewise, it is increasingly recognized that a root cause for poor PSM implementation is a weak safety culture. This was cited in the Baker Panel Report, The BP U.S. Refineries Independent Safety Review Panel, and in other studies as a key contributor to poor PSM performance. As stated in the Baker Panel report, "a positive safety culture is important for good safety performance." The report discussed several elements of safety culture, including process safety leadership, employee empowerment, resources and positioning of process safety leadership, and incorporation of process safety into management decision-making. Implementation of these and other elements such as operating discipline are important to help a company achieve superior performance.

This paper discusses 1) a survey that measures relative safety cultural strength and how it relates to performance, and 2) a process that can be used to develop a site program to improve safety culture.

Discussion

DuPont has developed a 24-question survey that is used to determine a site's Relative Cultural Strength. The survey has been used for eleven years in 41 countries across 52 industries, including energy, food, paper, chemicals, manufacturing, transportation, and mining. There have been over 500,000 responses to the survey. Results can be sorted and analyzed in a number of formats. Figure 1 contains an example of survey results.

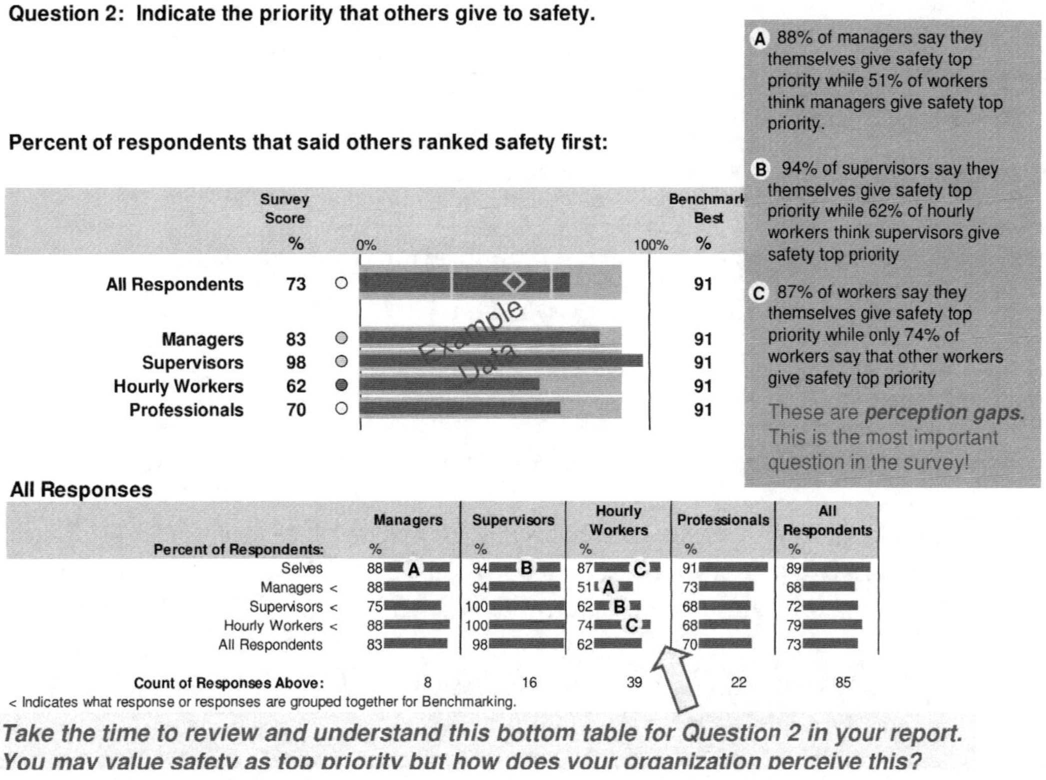

Figure 1. Relative Cultural Strength Survey: Safety Perceptions

In this example, you will notice that there is a perception gap among hourly workers, supervisors, and managers.

Figure 2 contains a summary of all the questions and the average of all respondents in three categories – Leadership, Structure, and Process and Actions.

Survey Results by Job Category

A: Managers B: Supervisors C: Hourly Workers D: Professionals

	A	B	C	D	Leadership
Q1	◍	◍	○	●	Priority individuals give to safety
Q2	◍	○	●	●	Priority respondents think others give to safety
Q3	●	◍	●	●	Belief that injuries can be prevented
Q6	●	●	●	●	Extent that safety is built in
Q7a	○	○	●	●	Presence of safety values
Q7b	●	●	●	●	Influence of safety values
Q9a	●	●	●	●	Involvement in safety activities
Q14	●	●	●	●	Extent safety rules are enforced
Q19	●	●	●	●	Recognition for safety achievements
	A	**B**	**C**	**D**	**Structure**
Q4	◍	●	●	●	Effect of a drive for safety on business performance
Q5	●	○	●	●	Level of safety where the cost-benefit break-point occurs
Q8	◍	●	●	●	Extent line management is held accountable for safety
Q13a	●	●	●	●	Quality of safety rules
Q13b	◍	○	◍	◍	Extent that safety rules are obeyed
Q21	◍	◍	●	●	Knowledge of safety performance
Q22	●	●	○	●	Rating of the safety organization
Q23	◍	○	○	○	Rating of the safety department
Q24	●	●	●	●	Satisfaction with the safety performance of the organization
	A	**B**	**C**	**D**	**Processes and Actions**
Q10	◍	◍	○	●	Extent individuals feel empowered to take action in safety
Q11	●	●	●	●	Extent of safety training
Q12a	●	●	●	●	Frequency of safety meetings
Q12b	◍	◍	●	◍	Safety meeting attendance
Q12c	○	◍	●	●	Quality and effectiveness of safety meetings
Q15	●	●	◍	○	Thoroughness in investigation of injuries and incidents
Q16a	●	●	●	●	Extent of involvement in safety audits
Q16b	●	●	●	●	Quality of safety audits
Q17	○	*	○	○	Rating of modified duty and return-to-work systems
Q18	●	●	●	●	The presence of off-the-job safety programs
Q20	◍	●	◍	○	Rating of the safety of facilities and equipment

Figure 2. Relative Culture Strength Survey: Sample Company Survey

This report is color-coded with survey scores averaged across all respondents. Red indicates scores lower than 70, yellow represents scores between 70 and 80, and green is for scores 80 and higher. Overall, these scores are used to determine RCS. The formula is as follows:

$$RCS = \frac{\text{Site Average Score} - \text{Benchmark Worst} * 100}{\text{Benchmark Best} - \text{Benchmark Worst}}$$

The scores range from 0 to 100%. "Benchmark Best" companies are determined based on:

- Total Recordable Rate (TRR) less than 1 over a 5-year period
- No fatalities over a 5-year period
- Lost Workday Frequency Rate less than 0.25 over a 5-year period
- Site population greater than 200 employees
- Verification by DuPont field assessment
- 10 companies, with at least one from each region of the world: North America; Latin America; Europe, Middle East and Africa; Asia/Pacific.

When the Relative Culture Strength has been correlated to Total Recordable Rate, we have found the relationship illustrated in Figure 3. Known as the Bradley Curve, this representation is a Six Sigma box plot of a database correlating RCS as a strong predictor of safety performance.

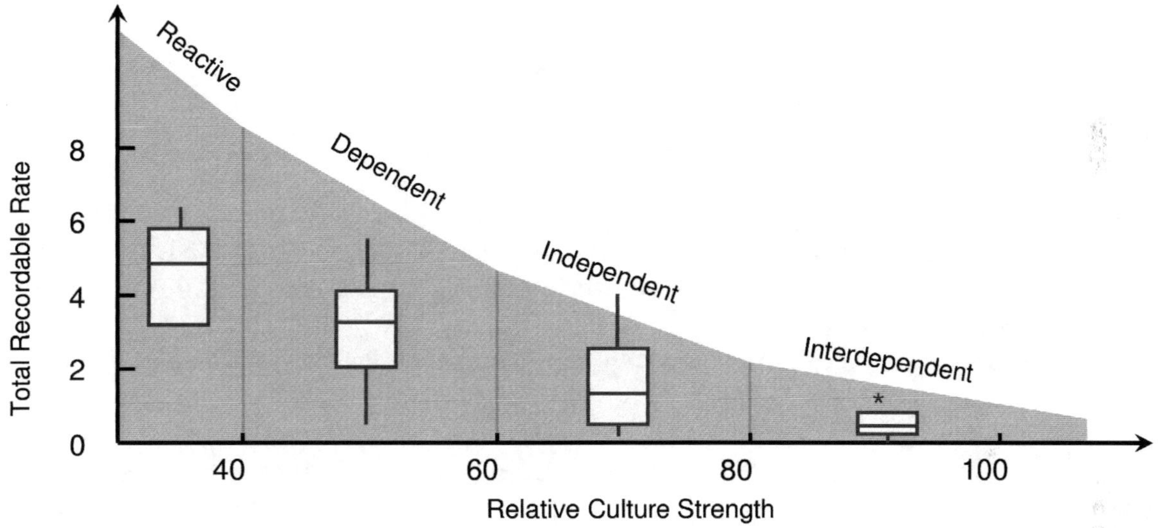

Figure 3. The Bradley Curve

The four stages of cultural strength depicted in the Bradley Curve are described in Table 1.

Table 1. Cultural Strength Stages of the Bradley Curve	
Stage	**Characteristics**
Reactive	• Safety by natural instinct • Compliance is the goal • Delegated to safety manager • Lack of management involvement
Dependent	• Supervisor control, emphasis, goals • Management commitment • Condition of employment • Fear/discipline

Table 1. Cultural Strength Stages of the Bradley Curve	
Stage	**Characteristics**
	• Rules/procedures • Value all people • Training
Independent	• Personal knowledge, commitment, standards • Internalization • Personal value • Care for self • Practice, habits • Individual recognition
Interdependent	• Help others conform • Others' keeper • Networking contributor • Care for others • Organizational pride

Unfortunately, many companies operate in the Reactive stage. As a result, their safety culture and performance is weak, and it will remain that way unless they move toward the Dependent stage. In this stage, organizations set rules and enforce them rigorously. People follow the rules because if they don't, they will loose their job. Also, in this stage, we often find that safety performance stagnates, and in particular, off-the-job performance may be significantly worse relative to on-the-job performance.

As organizations continue to move to the right on the Bradley Curve, further improvements are possible but are usually tied to management commitment and programs that enhance the characteristics of stronger safety cultures. Overall, the Interdependent stage is the goal achieved by the best performing companies, but the challenge is often in sustaining the Independent and Interdependent stages. Constant vigilance is required to keep an organization performing at a high level.

The Safety Perception Survey can be used with the TRR data to place a company on the Bradley Curve, and in a statistical sense, there is a strong correlation between the two. The challenge is then to identify specific actions that can be implemented to improve the safety cultural score and achieve better safety performance.

A petroleum, petrochemical, or chemical manufacturing site often has a complex mix of employees with a wide range of education, experience, skills, and expertise. To achieve good safety performance, sites must implement employee safety and Process Safety Management programs. Table 2 lists typical employee and process safety activities:

Table 2. Employee and Process Safety Activities		
• Process Hazards Analysis • Process Safety Information • Management of Change • Mechanical Integrity • Quality Assurance • Audits • Inspections and Tests • Operating Procedures and Safe Work Practices • Emergency Response	• Goals, Plans and Employee Safety and Process Safety Metrics • Housekeeping • Safety Meetings • Motivation - Operating Discipline • Safety and PSM staffing • Safety and PSM budgets	• Leadership by Example • Safety Communications • Safety meetings • Central Safety or Responsible Care Meetings • Incident Investigation • Off-the-job Safety • Management of Change – Personnel • Monthly Safety Program

The activities in Table 2 are examples of the many programs and cultural elements (e.g., rituals, totems) that sites use to improve performance. It's important to be aware, however, that many employees on a site may see only part of the safety effort; in fact, in many cases, no one sees all of it. For example, an operator may see operating procedures, operating discipline, safety meetings, housekeeping, the monthly safety program, and management audits. A technical engineer will receive PHAs, process safety requirements and change management information. An office worker may be included in safety meetings, safety communications, and off-the-job safety programs. And site leadership will participate in central safety meetings, budgeting, leadership audits, and setting goals plans and metrics.

Sites that have a good safety program and are continuously working to improve their safety culture recognize that employees have different views and experiences related to employee and process safety. As a result, to improve safety culture and safety performance, the safety-related efforts must change and evolve to reach all site personnel in some way many times throughout the year. At a typical site, this may mean dozens of efforts, some lasting for only a month or two, some for a year or more. The key is to have everyone on the site engaged in some way in the safety improvement effort at all times, and to understand that their actions are part of the safety improvement process. Sites that do this successfully will improve their safety culture and performance.

Here are some examples of programs implemented at various DuPont sites to help drive operating discipline and improve safety culture and performance. Each site developed a unique set of programs and other efforts that met the current site culture and performance.

- Goal-setting and continual communication and discussion around the goals
- Safety meetings and attendance tracking, making a special effort to get everyone to a safety meeting at least once per month
- Training and audits
- Periodic town meetings, connecting safety to business issues
- A "Safety Points to You" program emphasizing wage-roll participation, where employees earn points and rewards for participation

- Training of exempt personnel on safety behavior auditing and using them for field audits
- A "Safety Pulse" program measuring current safety performance and attitudes

Conclusion

Safety culture can be measured and related to employee safety and PSM performance. The Relative Culture Strength metric developed by DuPont allows an organization to take qualitative safety perception survey data and change it to quantifiable data to measure a company's progress on its journey toward world-class safety.

Integrating Relative Culture Strength into the Bradley Curve provides the ability to benchmark a company's safety culture. This knowledge can provide an insight into the overall safety process, which is different from other traditional indicators and key to building an organization that demonstrates world-class safety performance. To sustain this performance, specific programs must be put into place to reinforce and improve the safety culture and facilitate the engagement of all employees.

References

(1) U.S. Dept. of Labor, Bureau of Labor Statistics, 2009

(58c) Exploring the Emerging Trends and Practices for Process Safety

Mónica Grillo[1], Alfred Arias[2], Ana Crespo[3], Jordi Cuadros[4], Juli Sempere[1] and Rosa Nomen[1], (1) Chemical Engineering Department, Institut Químic de Sarri IQS URL, Barcelona, Spain, (2)Training and Education Department, Dow Chemical, Barcelona, Spain, (3)Associació Seguretat i Vida, Barcelona, Spain, (4)Statistical Deparment, Institut Químic de Sarri IQS URL, Barcelona, Spain

To explore what are the emerging trends and practices for process safety it is necessary to approach companies and examine their practices for safety. It is desirable to engage with a number of companies working in the same industrial sector in order to study the safety program style, as this is the key which influences the response of employees towards safety actions. This is ultimately the reflection of the safety culture of an organization. To be able to identify consistencies or differences in implementation of practices, it is necessary to compare vital statistics between companies presenting good and bad performances of safety. A suitable way to commence this activity is to review their respective accident rates. In order to evaluate and breakdown the safety programs being applied by each of the companies, it is necessary to utilise a comparative tool which penetrates the companies' safety programs style and content. Considering the approach above, a tool has been created which consists of the measurement of 5 components describing the "safety culture" of an organisation. The tool has the capability to score each of the safety components. This enables analysis to recognise the weaker aspects of the safety program and/or management. This means that those (previously unknown?) barriers to success can be identified to act as a basis for a continuous improvement plan. It is necessary, therefore, that a management system platform be sufficiently flexible to accommodate changes or additions discovered through analysis and measurement of the management system. This forms the basis of implementation of Continuous Improvement, and may follow the traditional Shewhart Cycle, i.e. Plan, Do, Check, Act. This investigation was made in cooperation with the Chemical and Statistical Departments of the Institut Químic de Sarri , Universitat Ramon Llull in Spain with the participation of AsSevi Association and a number of chemical companies located in the Tarragona City, Barcelona. The tool has been derived from an original concept proposed by Sheffick et al., 2007, Distributed Safety Management, (GDS for its abbreviation in Spanish), which establishes those aspects to which attention should be posed in order to achieve excellence in safety. The tool was developed to focus on the human aspects of the approach of Health and Safety in the workplace. It determines the degree to which human aspects interact with the management system. The five safety components comprising the tool are: ï Commitment of the managers towards safety. ï Participation of the workers in the process safety. ï Responsibility of the workers regarding safety aspects. ï Zero tolerance regarding safety rules. ï Manager's consistency with the safety rules.

Validation was made upon developing the tool whereby survey results from the various chemical plants aligned with their actual statistics of accidents reported to the Spanish Labour Ministry.

Lifecycle methodology for effective safety training

Shay Segev
Center for Loss Prevention in the Process Industries Ltd.
POB 175 Kibbutz Beit Kama 85325 Israel
ssegev@cflppi.com

Keywords: training, effective, instructional design, evaluation, objectives definition

Abstract

This paper suggests a methodology for achieving effective safety training by a diligent analysis of each of the following steps in the safety training life cycle:

- **Objectives definition (need analysis)**: understanding why training is needed and what would be considered a success. Whenever possible, the objectives should be in terms of actions the participants need to perform.
- **Decision if training is the appropriate solution**
- **Decision regarding the instructional approach**- i.e.: behaviorist, cognitive or constructive.
- Decision regarding the most appropriate tool for delivering the training.
- Development of the training materials.
- Delivering the training.
- Evaluation of the training effectiveness.
- **Extraction of the lessons** from the evaluation and implementing them in the training program.

1. Why effective safety training is a must in today's business environment (Introduction)

Safety training has always been an essential tool for minimizing the likelihood and the consequences of accidents and losses. However, in today's business environment there is also a necessity that the safety training will be as effective as reasonably practical due to the following reasons:

- The investment in safety training is usually significant (most of the training cost is usually associated with its participants' time). As with any other business investment, one of the key parameters for defining a success is effectiveness (i.e: achieving more with less).
- There is zero tolerance by the authorities and the public for catastrophic accidents. In addition, in case of a severe loss, the lean manufacturing and the just in time supply chain, may jeopardize the company's business continuity, which may result in loss of customers and market share. Ineffective training may contribute to a false sense of security by the management and the on-site safety professionals. For example, following an ineffective operator's training for handling of emergency situations, the management

may mistakenly believe that operators are competent to tackle abnormal /emergency situations. Should a triggering event occurs, this gives rise to an error in the form of unsafe act or decision. This in turn combines with an unforgiving environment that does not support recovery, may result in a severe accident.

- Understanding the risks and the basis of safety is the foundation of almost every operation (e.g: design, operation, maintenance etc.) that is carried out in the CPI (Chemical Process Industry). Without such an understanding, an accident or loss is just a matter of time.
- Effective safety training helps to brand safety training as an inherent, important and challenging duty of every role (rather than a burden and a lip service).

Unfortunately, ineffective safety training is still common in the CPI . One of the reasons for that inefficiency, in my opinion, is the lack of a formal methodology for analyzing the various steps in the safety training life cycle. The objective of this paper is to suggest such a methodology.

2. The lifecycle steps for an effective safety training:

2.1 Step 1- Objectives definition:

Before starting to design a safety training, the need for the training and its detailed objectives should be clearly stated (this step is known in the literature as "needs assessment"). Whenever possible, the objectives should be measurable (this will help to establish the training effectiveness in step #2.7 of this model much easier). This step is highly important as its will help to ensure that the training will address the need and meet the objectives.
Safety training usually has one or more of the following objectives:

2.1.1 Ensure safety knowledge retention in the organization

This includes retention of the insights learned from safety incidents within and outside of the organization, training of newcomers to the organization / new managers etc..
For example: a new lab manager was hired following a successful career in the university. There is a need to bring him quickly up to speed with all the safety and safety related practices, asthe 'common practice' in most of the leading companies is that each manager in the company is also the "safety manager" (i.e: in charge) of the unit which he manages. Therefore, without having a good understanding of all the required safety practices and the lessons learned from previous incidents in similar activities inside and outside the organization, he is not able to adequately manage the safety in his unit.

2.1.2 Address performance and competency gaps:

Bridging the gap between the expected performance and the existing performance.
The training has to be directed at the underlying causes of an error /gaps, for example:
o A JSA study concluded that operators stand in the "line of fire" when replacing steam valves and hence can be injured in case the line is not properly drained. The training objective is that the trainees will understand the “line of fire” concept and implement it in their routine work (this should be verified by future field audits and JSA studies).
o An audit found that PAPRs (Powered Air Purifying Respirators) in one of the departments are contaminated as operators don't clean and wrap the PAPRs in bags after use. The training objective is that the operator will understand the importance of cleaning and implement it in his routine job.

2.1.3 Address lack of skill / knowledge how to perform a task in a safe manner:
Training to address the lack of skill or knowledge. For example: how to safely remove contaminated gloves, new process safety engineer that is not familiar with the Hazop methodology and doesn't know how to lead an Hazop study, training regarding the safety aspects of new processes and new equipment etc.

2.1.4 Develop or upgrade the participants' mental model of the process:
A mental model is an internal representation in each of the training participants' long term memory which is used as a basis for solving ill structured problems.
This objective is usually required in situations in which the operator is needed to respond to a wide range of abnormal /emergency situations that can't be predicted in advance.

2.1.5 Fulfill mandatory requirements:
Training that is mandated by an authority: e.g: training regarding the MSDS of new materials (in some countries).

2.2 Step 2: Define the target audience for the training

Training should be tailored for each target audience (managers, engineers, operators, maintenance, etc..)
The concept of "one fits all" (i.e: the same training for the various target audiences) often results in inefficient training, as each audience has different tasks, needs, abilities, strength and weaknesses that needs to be addressed during the training development.

2.3 Decision if training is the appropriate solution

Training is sometimes perceived by some managers as the "cure for all diseases". However, this is not the case. In this step, the training designer should check whether any of the primary causes for ineffectiveness of safety training listed below exist, and if they exist, the situation should be thoroughly analyzed (usually the best solution would not consist of training):

2.3.1 The tools necessary to implement the training are not there:
These are the cases where flawed environment /poor management/ poor design or lack of commitment for implementation makes the training irrelevant, for example:

- An operator attended a course for conducting internal safety audits but due to lack of commitment of his line manager, there is no implementation and he doesn't do any audits after the course.
- An engineer attended a course on static electricity. However, as most of the piping in his plant is made from non-conductive materials (due to corrosion consideration) and there is no budget for significant changes, he is not able to implement what he learned.

2.3.2 Improper incentives:
What are the consequences of doing the job in an unsafe manner? Does safety behavior / implementation of the training insights result in additional and nasty work assignments?
For example: what is the incentive of an employee who participated in an incidents debriefing course to start making detailed debriefing for near miss incidents in his plant?

2.3.3 Lack of motivation:
Rossett (1) claims that **motivation=value*confidence**:
o **Value**: How much and whether the topic is valued
o **Confidence**: The amount of confidence the employee has that he or she can master the skill or knowledge. For example: sizing of emergency vents may be perceived by the course participants as extremely complicated (especially if the instructor starts with detailed differential equations, resulting in low confidence in their ability to implement the methodology and hence low motivation).

2.3.4 "Inert" training materials:
The training is based on "inert" materials (i.e: materials that are irrelevant to the work activities or that the materials that the training participants don't see how they can implement them in their real life activities ,e.g: training that focus on instructions like "you should pay attention when…:", "you should be careful").

To summarize this step, the main question the training designer should answer is: If your life depended on it, could you meet the performance objectives without the training? If the answer is yes, training is not the optimal solution.

2.4 Decision regarding the instructional approach

The most suitable instructional approach should be specified for each of the training objectives. The main instructional approaches are:

2.4.1 The behaviorist approach:
The behaviorist approach_suggests that learning occurs due to reinforcement or weakening of reactions or of associations between stimulus to reaction (i.e: the learning occurs by acquiring and reinforcing the required action /reaction). This approach is effective when we want to implement a set of routine actions that the trainee will perform (e.g: safe wearing and removal of gloves, gowning and de-gowning, use of PPE etc.) and when we want to achieve a quick automatic action (for example in training of emergency response teams to use fire extinguishers, oscillating monitors etc.).
An effective training in this approach is when the trainee does the required activity /response automatically without having to think about it, as the thinking process will damage the performance.

2.4.2 The cognitive approach:
The cognitive approach suggests that knowledge can be transferred directly from the trainer / book / e-learning kit to the trainee. This approach is effective when new knowledge should be acquired for the understanding of concepts, processes and phenomena (for example: training of a new procedure, training of new process that had already been thoroughly analyzed, training of material hazards from an MSDS etc.)
An effective training in this approach is when the trainee understands the structured knowledge that has been presented to him (i.e.: can repeat it and understand how to implement it in his working environment).

2.4.3 The constructivistic approach:
The constructivistic approach suggests that learning occurs when the trainee actively constructs and reconstructs the knowledge (i.e: construct or refine their mental model of the subject). This approach is effective in training aimed for the solution of complex, ill structured problems (e.g: training to solve complex process abnormal situations, training for development on inherently safer process / solutions etc.).

2.5 Decision regarding the most appropriate tool for delivering the training
In this step it should be decided which is the most cost effective tool to delivering the training, using the required instructional approach. As in case of illness it is possible to give the API (Active Pharmaceutical Ingredient) in several types of medicines (tablet, injection, inhalation etc.), it is also possible to deliver the training using the required instructional approach in several formats (e.g: traditional/ classroom, e-learning, guided project etc...) and it is the duty of the training designer to use the most cost effective format in the given situation.

2.6 Development of the training materials
The development of the training materials should:

2.6.1 Address the training objectives:
The training material should be developed to meet the training objectives defined earlier. Any material that does not support the objective achievement should not be included (as it would be an inert knowledge that would only increase to the cognitive loading).

2.6.2 Address real life situations and real life dilemmas:
The training materials should be developed in such a way that they will address **real life situations and real life dilemmas** (the problem is usually not the lack of knowledge, but the lack of understanding how to apply that knowledge). The vast use of real life dilemmas, lessons learned from accidents and real life situations will make the application of the knowledge much easier.
For example: in a confined space e-learning which we developed, we asked the trainees: what would you do if you see your friend lying unconscious at the bottom of the reactor due to nitrogen asphyxiation, assuming you don't have an SCBA near you (i.e.: would you try to hold your breath like you do in the swimming pool and enter the tank without SCBA to save your friend's life)? After they decide what they would do, they saw the CSB video on the Valero accident and got an explanation as to why they can hold their breath in the swimming pool but are unlikely to succeed in holding their breath in an inerted reactor.

2.6.3 Address the current level of the participants knowledge, experience and skills:
An additional point that should be addressed in this step is the variation of the different starting points of the target audience (different employees have different knowledge levels, experience and skill).

2. 7 The training delivery design

The design of the training delivery should address the following points:

2.7.1 Trainees' engagement and active participation

Treating the training participants as a sponge that should only absorb knowledge is a common mistake that usually results in ineffective training. Most people **learn by doing** much more than they learn by passive listening. Therefore, engaging activities, such as role play and simulation should be embedded as much as possible in the training. As an example, a study done by Pirani and Reynolds , indicated that the most successful campaign to encourage the use of PPE involved the use of role playing.

2.7.2 The instructor

The best instructor is often the team's direct manager as:

- He / She will be much more committed to the long term implementation of the training objectives and will be perceived by the team as such.
- He/ she is more authentic and can use more authentic real life examples and scenarios.
- During the training preparation, he/ she masters the subject and becomes a knowledge source to other employees about it.

Other options for good instructors include: a charismatic peer / team member or an external expert with a relevant experience.

2.7.3 Training duration and frequency:

In order to increase retention and decrease the cognitive load it is usually worth to consider the option to spread the training (for example training of 10 minutes every day are more effective than one full day in a year), although this is not always practical.

2.8 Evaluation of the training effectiveness

The suggested method for evaluating the training effectiveness is based on the Kirkpatrick Model (1959) 2, which consists of four levels (level 1 is the most basic and level 4 the most comprehensive). Therefore, the difficulty and cost of conducting the evaluation increases with each level:

2.8.1 Level 1 – Reaction :

How did participants react?
The training effectiveness will be evaluated using:
completed participant feedback questionnaire
informal comments from participants
focus group sessions with participants

2.8.2 Level 2 - Learning :

To what extent did participants improve knowledge and skills and change their attitudes as a result of the training?
The training effectiveness (i.e: the "increase in knowledge") is evaluated using:
Pre- and post-test scores
On-the-job assessments
Interviews

2.8.3 Level 3 - Behavior :
To what extent did participants change their behavior back in the workplace as a result of the training?
The training effectiveness (i.e.: the "Behavior change"-the extent of the applied learning back on the job) is evaluated using:
Observations and interviews over time
On-the-job observation
Reports from peers and participant's manager

2.8.4 Level 4 –Results:
What organizational benefits resulted from the training?
The training effectiveness (i.e.: the Change in the business results and environment) is evaluated using the routine KPIs such as:
Increase of the near-miss reports
Reduction of the OSHA index / LWD
The main challenge in this level is to establish the relation between the training and the change in the KPIs.

2.9 Extraction of the Lessons learned and their implementation in the training program
In order to achieve continuous improvement, it is important to conduct a formal debriefing, with the participation of the instructors, after the evaluation of each training. The debriefing should focus not only on understanding the root causes of the problems and deficiencies identified, but also on understanding the root causes of success. The debriefing should include a table with assignments, responsibilities and due date and there should be a person in charge to follow up the implementation status

3. Conclusion
This paper has outlined a structured and systematic methodology for achieving effective and results driven safety training. The author of this paper believes that this methodology can be easily refined and applied for achieving more effective safety training throughout the CPI..

4. References

[1] Rossett, A.,"Needs assessment". In: Anglin, G.J. (editor), InstructionalTechnology, Englewood, Co, Libraries Unlimited Inc.:183-196, 1995

[2] Kirkpatrick, D. L.,"Evaluating Training Programs", 2nd ed., Berrett Koehler, San Francisco, 1959

[3] Kirkpatrick, D. L. (comp.) ,"Another Look at Evaluating Training Programs", ASTD, Alexandria, USA,1998

[4] Semple A., "Learning theories and their influence on the development and use of education", *Australian Science Teachers Journal;46, 3; Education Module p. 21,* Sep 2000

[5] CCPS (Center for Chemical Process Safety), " Guidelines for Preventing Human Error in Process Safety ,1994

[6] Pirani, M & Reynolds, J., "Gearing Up for Safety", Personnel Management, 25-29, February,1976.

Assessment of Human elements to avoid accidents and failures in task perform, cognitive and intuitive schemes

Ávila, Salvador
Universidade Federal da Bahia, UFBA
Rua Aristides Novis, Federação s n, Salvador Bahia
avilasalva@gmail.com

Keywords: task analyzes; human factors; human elements; human reliability

Abstract

Analysis of the activity of the operator in the task of the industry requires the establishment of definitions regarding the Human Reliability, Influence Environments and, affective, cognitive and intuitive functions connected with the job. These explanations depend on the operational and tactical point of view and can be embedded in complex systems which hamper research into the nature of human error. The discussion on human errors is old by production management but not effective. The main efforts so far have been undertaken in the search for solutions based on equipment, process technology and management. These issues bring solutions not effective with regard to human error and can result in inappropriate projects that fail to achieve the status of "human error-proof."

The analysis of the causes of human errors in the work is highly complex due to the fluctuation of the behavior resulting of environments influence. The operator who performs the tasks of the job has variable behavior due to several factors, among them: non-availability of suitable tools and environments to accomplish the task; changes in the balance between personal and organizational goals, between individual and organizational values (commitment); and due to inherent issues in human formation for the job, emerging with motivational aspects, often, of unconscious dimension.

The complexity of this analysis involves: environmental aspects; types of cognitive processing that lead to specific decision models; forms of planning, executing and controlling the task; levels of knowledge and ability to implement and task analysis including verifying the demand for training for skills training; commitment to task; verification of the statement of causality techniques for failure; and verification of mentoring in the implementation of projects by providing task interface between man, the equipment and process.

The PADOP methodology (Standards and Procedures in Operation) is proposed to analyze the task including: the environments influence; the type of cognitive processing with decision models; and verification about the efficacy and effectiveness in achieving the ideal state. Some important aspects of the give importance to PADOP utilization: • clear identification of human and technical default state to be reached; • Inventory Analysis of knowledge based on: actual demands (causation problem) and commitment in the implementation, as written; • Level of cooperation in Teams (written and verbal communication); • Characteristic type of profile on the creativity and standardization in performing the task; • Type of mental map constructed for the procedure by the operator: time, clarity, sequence, team work; • Logic roles in performing the task and availability of tools and information; • Interpret procedures taking into account cultural differences; and measure results to determine the efficacy and effectiveness.

The discussion of human behavior in the operation of industries prioritizing environment aspects and type of personalities and cultural relation become possible to construct cognitive, intuitive and affective schemes. These schemes have particularities depending of type of technology, organizational movements, type of routine operation, failure warning indicating different types of root cause as: difference between politics and practice, cultural traits, inadequate task planning, inadequate design criteria, risk management based on individual guilty, and economic biases.

1. Introduction

The Chemical Industry has not been engaged enough in the discussion on human factors in the causal relationship of major events such as accidents. Thus Programs that decrease Process Losses and Risk Management cannot be effective, not finding the root cause of problems and not achieving the best overall result.

Human errors can result in accidents and events with rising costs of production and impacting on the company's image. Some studies were presented in the [9] API770 about the importance of human factors to avoid accidents, where, in the petrochemical and petroleum refining industries (BÜTIKOFER, 1986) inform that over 80% of the causes of human errors are inappropriate designs and mistakes among professional production (operation and maintenance). There are indications [10] that knowledge is insufficient and poorly designed procedures also contribute to initiate human error. Statistical studies (Garrison, 1984, OSHA, 1990; Coco, 1998) indicate the high cost of hazard mitigation in the chemical industry, the losses exceed $ 2 billion on 30 accidents, indicating losses of 70 million dollars per accident. Sebzali and Wang [12] indicate that human errors contribute significantly to accidents reaching 80% of the causes (Fleming and Lardner, 1999) and about 60% contribute to fires in refining industries (Uehara and Hasegawa, 1986).

Human error can lead to accidents and, or incidents. New concepts and techniques are proposed by Avila [1] [2] to treat the problem with applications made in the chemical process industries. In the proposed work, Ávila discusses methods of how to analyze the human factors to avoid operational failures and that can in its unfolding, turn into accidents or major events. Methods to analyze faults use intuitive and cognitive schemas in the routine.

Figure 1 presents the flow of issues to be discussed. General aspects are presented on the factors that influence human factors as social and organizational culture, aspects about task planning, and others. For each technology there are specific interactions with workers depending on this automation level. Some researchers discuss the requirements and limitations for analyzing the human factors including the planning task. Thus, the task analysis includes analysis of the environments and how the cognitive apparatus performs during the production decision. In continuity, psychological functions at the task analysis are detailed valuing operation for the different organizational movements. Indicating that, at every moment, different parts of the cognitive apparatus are activated. Thus, leisure, skills acquisition, emergency situations, the organizational integration, and social meetings activate different psychological functions in different orders. Some of the routine cases are presented to illustrate the method with the interpretations of agents around the fault (environments, leadership, skill and decision). The work closes with an analysis of some of these agents that cause the appearance of human error and the factors leading to the accident showing different points where interventions can be performed.

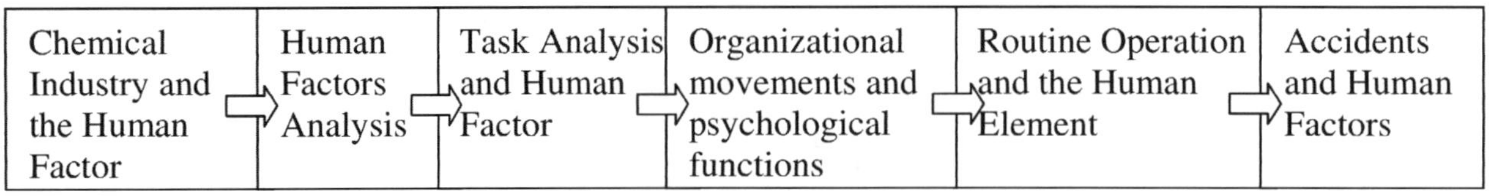

Figure 1 – Flow of the topics covered

2. Chemical Industry and the Human Factor

The effort of selection and competency development differs depending on the type of technology in the industry. This depends on the level of automation, the risk with product handling and scale of the plant. The complexity of social processes which relationship(s) affect the performance of worker should be considered. Some specific aspects of the chemical industry are discussed in the executive function of the production in industrial sites.

In Oil Exploration, the platforms are inhabited by the need for availability and quick action in case of failure. The rituals of entrance and exit during the boarding of workers are critical and the possibility of human error can be reduced with greater communication within the team including contractors. In Petroleum Refining there is an increase of volume processed, high level of automation coexisting with other plants with lowest levels of automation. Some refineries have differentiated commitments of the team leading to very different results and can cause environmental impacts such as CO_2 emissions, overuse of water, oil contamination to the sea, and atmosphere releases. Both for exploration and refining, the oil industry has the need to maintain the positive image trying to avoid the low yield and occurrence of accidents and incidents.

The Chemical Industry has been concerned with the reduction of major events such as accidents and great incidents, but cannot treat small deviations leading to low productivity due to existence of hidden areas in the failure investigation. With technology oriented around the renewable resources and with sustainability challenges affecting global competition, these industries have turned to green chemistry. Managers of these industries know how to build corporative management programs, but they do not establish robust programs for routine management. In these industries there is relative ignorance about energy efficiency in the management of routine leading to higher fuel consumption and inefficiencies in cooling systems and combustion in furnaces and boilers. Entrepreneurs are worried, though, to acquire equipment that meet market demand, with technologies that require large size of equipment leading to high investment, debottlenecking and consecutive loss of efficiency in energy utilization.

The mining and pulp industry has high potential for degradation of nature. The theme of sustainability is part of the strategic definitions in the mining and metallurgy industry. Due technological intensity of thermal energy using, the metallurgical industry needs to establish management programs to control the consumption of thermal energy leading to improve the controls performed by the operation. Another important detail is the size of equipment involved in operations with solids, it is important to maintain the reliability of large machines and additional care as the safety aspects of logistics activities.

Current technologies in the plastics industry are based on high levels of automation with integration of various machines to "mass production" and high level of flexibility requiring many manual operations to exchange templates and setup to operation. Thus, human and operational issues depend on what is the focus of product and what type of technology is adopted. The movement of raw materials and finished products has become increasingly automated as well as the integration of manufacturing with the control of many machines in the unique supervisory. The blend technology for each type of plastic and green plastics require additional knowledge from the operation team.

The storage and transport of liquids and gases can be made through pipelines, trucks, ships and trains, where special care are took to avoid environmental accidents, to avoid impact on cities and communities. The machines used to promote transportation of fluids by pipeline are pumps and compressors and pressure and temperature are controlled to ensure operation exit without leaks or accidents. The control network involving large pipeline network in large geographic regions requires intensive instrumentation for the maintenance of automatic operations. In the case of ships, docking operations and ship movements can be critical and depend on designs of ports and the level of adaptation to new situations outlets. The interconnection between vessels with pipelines and tanks should be careful with extra care in the manuals operations. The transport of solids requires the use of conveyor belts or channels to facilitate the flow of large quantities of material to ports and ships. Although there is the possibility of failure of such equipment, the probability is low for the occurrence of fire or explosion. In the case of chemicals, the transportation is done by road ways that have low quality requiring special attention by the truckers' drivers.

Organizational and human characteristics influence on the quality of work of each operator depending on industrial segment. The features are ranked in Table 1 as very influential (red), little influence (yellow) or no influence (green). The psychological functions related to cognition are discussed by Reason and Dekker to investigate the origin of human error in the workplace and here are treated in each market segment at industry and service.

Table 1 – Aspects for Human Reliability Analysis and Operations by type of technology

Environmental and Human factors	Oleum	Chemicals	Plastics	El. Energy	Logistic
COGNITION					
Mental Map to decision					
Omission at action					
Flexibility on procedure					
CULTURE					
Organizational Cultural Influence					
Safety Cultural Influence					
Technical cultural aspects					
LEADERSHIP AND TEAMS					
Competition and Cooperation					
EXTERNAL ENVIRONMENT					
Physical Stressors					
Psychological Stressors					
SOCIETY AND ECONOMY					
Image					
Globalization Impacts					
Safety and contracted					

The construction of specific mental map is demanded for each complex processes and tasks. When blocks of processes are connected by principal pipeline processes or secondary pipelines of utilities with recycling pipes this generates the need for more complex mental map that often are not available in the team. One way to resolve these difficulties is to build redundant systems for process safety and facilities where the commitment of perception of the signs must be maintained to avoid cognitive traps. The industry of manufacturing and logistics operation has simple operation, but with integration of various operations it can be complicated. The chemical process industries and petroleum refining require greater complexity in their controls and then require special competences to these works.

The perception function is used in all activities performed at industry routine, but, due to the complexity of cases, there may be a lack of signs demanding the greater presence of operator in the operations place, at industry field. Industrial facilities and services are focused on developing automatic controls without actions to improve human reliability in the field. Lost productivity is the result of poor visibility of signals, thus hindering the action on the root cause. The omission is a phenomenon resulting from fuzzy rules, when appointments are not established, and powers of perception are not developed. This behavior deserves to be the main cause of accidents in the continuous process industry.

Every procedure requires some flexibility in action due to the social relations in the workplace. There is a need to adapt operations to enable the application of the perception and attention competences to completing the cognitive process at decision making. The rules that lead to the best script are not "dictatorial" in form and allow changes to the route.

The lack of knowledge about local culture elements that influence the behavior of teamwork leads to the establishment of rules that expire and situations of low trust in relation to management systems. Thus, changes in organizational culture, safety culture and technical culture must be well planned and controlled so as not to be considered as random and promoters of human error.

The organizational movements can change the stability of the psychological functions and build social or individual ills of human behavior in the workplace. The conflict between policies and practices can lead to disruption of the psychological contract and the improper application of skills leading to reduced productivity and influencing the competitiveness of industry and services in latent form. These phenomena (or disease of management system) are hard to identify, leading to a desire to study the social relations in the workplace.

Organizational activities can allow different safety patterns for different tasks. It depends on the organizational situation that can cause anomalous behavior, deviations in the task or in the function, thus causing the incidents or accidents, and that, depending on the type of safety culture, may have a higher or lower probability of occurrence.

The changes in human behavior caused by social and technical aspects can change the technical culture (TC) influencing the safety state and the productivity at work. Every industry and service has its own characteristics related to the type of TC installed. It is important to analyze the operator speech from shift events to enable the identification of their TC. When knowing about the TC characteristics, it can avoid or treat "vices" from management systems, "sick" organizational systems and "high complexity" computer systems, reducing then, the possibility of inappropriate behavior in the workers. The "excessive" preoccupation of the operator with reports required by management system will hinder meeting the real demands of process and equipment control. Some companies in its politics, even without meaning to, encourage the selfish behavior, and others, do not encourage the commitment, these situations impacting differently on the TC.

Strategies to promote cooperation and competition are adopted to integrate and motivate the team, respectively. Cooperation within the team allows the exchange of information and competition between people leads to the challenge to achieve new challenges. Each type of strategy has its advantages and disadvantages then, are not always being applied. It is important to understand the factors that motivate the harmonization of teams before changing social environments through the use of these strategies.

In cooking, it is known how to mix different flavors giving to a result that satisfies many people, in music, particularly in clusters that add different sounds; seek the same goal to unite these sound movements in search of the perfect cadence of the willing people. It also thinks similar when doing orchestration of groups or teams in industries in general: different seasonings or people with different characteristics joining different qualities and establishing built-in safeguards with regard to defects allowing, then, to achieve high productivity. The reverse situation is true, when the leaders do not know how to harmonize the teams in the midst of economic process of sustainability.

It is important to examine the stressors that cause reduced performance of human behavior as the risk of product, high temperatures, the possibility of mechanical impact at work (physical stressors), or even, the pressure of time and centralized management (psychological stressors). The risk of accident can be affected by these stressors resulting in employee dissatisfaction that affects the quality of the task.

The risk and reward analysis of performing the task may generate conflicts between organizational values and individual ones, resulting in different situations of routine communication: active in seeking solutions, or passive communication living with low standards of safety in execution and control of the task. It is therefore important to analyze the type of task, tool implemented, used information and equipment. This knowledge prevents task deviations and the events tha can cause the accident. The standards need to be reviewed from the perspective of the execution team. These aspects are important in industry, especially on oil facilities due to public company with stable employment. Excess temperature involves the metallurgic works and in case of hydroelectric operations, excessive pressure. The possibility of mechanical impact is intense in logistic services and in the mining industry.

The psychological stressors can be mitigated by management style in the activity or by the social relations arising from the actions of informal leaders. It is therefore important for each type of facility to select and prepare appropriate leaders. Time pressure is an important factor that can generate negative impacts on shut-downs and assemblies for new designs in the industrial area.

Social and economic movements were analyzed by Rasmussen and are pressures that cause changes in the law, management systems and practices of routine. Sometimes, social pressure does not expect the amendment of laws and neither of management systems and act directly on the routine of production. The velocities which social and economic changes are taking place affect social relations in the company and hinder the action of the managers (emerging movements without visibility). The image is essential for the maintenance of business affecting: the stock price, the financial loans by banks and funding institutions, the social pressures for new facilities (public hearings) and, indirectly, the civil and job lawsuit. This image comes from workers within the institution by "active" accepting by employees of organizational values, thus, confirming the psychological contract established the entry rite.

Economic pressure is directly related to: reduction of production costs, increase of sales and increase of profits, due to appropriate features of the product. When the company cash flow is positive, there are pressures leading to psychological stressors from the leadership and organization. This situation also creates more harmony to the team, but if there is not stress in the work, there is no minimum level of stress to keep the team's performance, then, it can cause discomfort to achieve high productivity.

In the relationship between contractors and companies, there is the challenge of maintaining the commitment with safety and organizational goals of the contractor. Issues related to maintaining social relations in the workplace should include performance analysis of support functions involved with the work. Thus, the sense of justice and the treatment of interfaces are important to avoid conflicts between functions of the operation and maintenance services.

3. Human Factors Analysis

To analyze the activity of the operator in the task of industry definitions should be established in the context of human reliability and environmental influences and, how the affective, cognitive and intuitive functions are linked at workstation. These concepts depend on the kind of decision (operational or tactical), if it is entwined in complex systems, and what arc the difficulties for human error investigation. There are types of human errors studied by Reason that add environment, task, and technology characteristics. These errors are designated here as nucleation agent and if identified allow preventive action taking to increase Human Reliability.

The discussion on human errors is old in the production administration and adopted ineffective management programs by: they are top down propositions, do not meet the specific characteristics of socio-technical systems, and due to not considering the complexity of industrial processes. Management systems imported from large multinational companies do not internalize easily in the local technical culture, and do not get the same results.

Most efforts to prevent human errors have been made in developing facilities and devices, or safety barriers, installed in the equipment and in the process control technology. Human factors at the current level of automation technology have failed to prevent stoppages of production systems with time losses, accidents or even losses at industrial process. Some projects resulting from these efforts with multiple redundancies to prevent human error have not reached their goals demonstrating the need to understand the decision processes and information processing of worker.

Reports published in the industrial (nuclear industry) and transportation services (train, ship and Airspace) present the following conclusion: 'The models for predicting human error while performing the task are unreliable due to variations in human behavior based environmental changes'. [8]

Statistical studies indicate that the worker is involved with 85% or more of the process losses related to accidents and incidents. The human contribution may be much greater when considers that many latent failures contribute greatly to reducing the capacity of equipment and of people, causing the loss of competitiveness in the industry.

The analysis of human errors causes at work can be complex due to the possibility of human behavior fluctuation resulting from new conditions of physical, legal and socio-economic environments factors. The operator who performs the task at work can have variations in their behavior due to: non-availability of appropriate tools for the job, changes in the analysis between individual and the company values, changing commitments; character and personality traits that impact on labor (motivational aspects including unconscious dimensions); and inadequate relation between real risk and perceived risk by the operation can cause discomfort and insecurity in the job.

4. Task Analysis and Human Factor

Paul Sharratt (UMIST, 2006) commented that research on representative indices of organizational climates and safety have low correlation with the rate of incidents and accidents indicating the need of methods to study the human factors in complex systems and that accept the influence of organizational and social environments. Also, it is noted that the adoption of Management, Health, Safety and Environment Systems based on successful experiences (such as DUPONT), does not represent, in the oil industry, ensuring reduction of negative consequences of accidents and incidents to the Community and Nature. Inter-and multidisciplinary groups should be formed for the development of work in search of greater human reliability in technical systems.

Human error ([4], [7]) may cause failure in the task in relation to the planned pattern. There are several reasons that can lead to failure happening, among these: the overconfidence due to operating experience indicating that the pattern set is not the best way to operate, skipping steps, and changing patterns. Another possibility is due to no consensus in planning and writing when the writing task does not represent the performed task. Lees shows the need for knowledge to plan the task and the need for ability to correct parameters, change steps, measure and control the final state of the process after the completion of the task.

The analysis of these complex issues is through the technique PADOP ([1], [2]) Standards and Procedures in Operation (Figure 2) at three different stages: (1) Identification of environments (IAT), (2) Analysis of Cognitive Processing and Task Execution (PCET) and (3) Analysis of Efficiency and Effectiveness in the Task. The use of PADOP is important because it identifies the patterns of job including environments fulfillment. This technique measures and analyzes the inventory of knowledge, verifies the level of team cooperation, considers the type of mental map in decision-making, examines the roles in the task perform, and evaluates whether the interpretation of the procedures is correct on the group of workers.

(1) The step of identifying the environments intends to check the intensity of physical and psychological stressors of the environment surrounding the job. Intending to detect the presence of social movements emerging in response to reports of activities related to the company, quantity and quality of product, or the type of processing performed. The influence of organizational movements are studied, that with the mergers and "confusion" organizational. Changes in management profile generating motion protection in small groups in the workplace. The modification at environmental or occupational legislation can be studied to change tasks in preventive way. The influence of economic results and high scale of production can lead to dissatisfaction that generate a loss of quality in the cognitive system. At this stage of PADOP, it measures levels of knowledge and applied skills required by technology checking with knowledge offered by the team. From this analysis it is possible to define the training program, reviewing the allocation of workers into the job, define and establish educational programs on the profiles settings management. In PADOP, the potential imagination and creativity are measured too. With these innovation characteristics is possible to design the task giving tools for man's comfort during operation of the equipment and process.

(2) The types of cognitive processing schemes installed in the workplace or in the individuals are result of decision models with risk situations that vary with environmental stressors. In PADOP, assertiveness and greater reliability is searched to perform the tasks. Maybe there is need for social flexibility without impacting on the necessary indisputable requirements for safety systems. The control task is established to promote the revision of procedures ensuring that writing is the same as it is practiced or, best practices in terms of quality, safety and cost. The task is fulfilled properly when the sequence of activities follows a causal connection for the processing of materials with value added in the final product. The kind of mental map of the worker must allow the taking of corrective and preventive actions on the failure. It is important to examine the technical culture (TC) through the speech of the operator to correct problems of omissions in the task activities.

(3) To check for efficacy and effectiveness of task in achieving the ideal scheduled state lets you draw the corrective and preventive actions and makes it possible to track which items are to be reviewed in the task

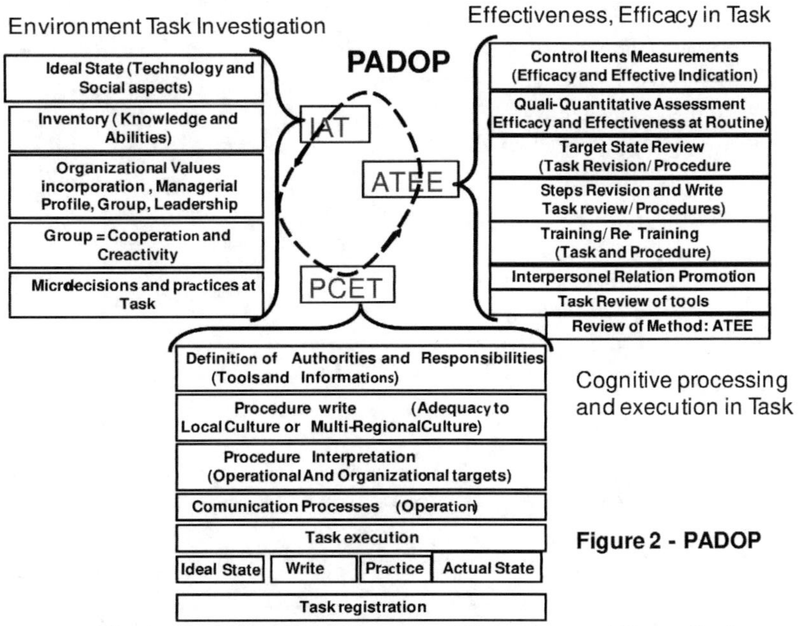

Figure 2 – PADOP (Patterns and Procedures at Operation)

5. Organizational movements and psychological functions

The movements of the individual (in the Company and in the Society) need to enable perception, cognition, and memory, and depends on emotional stability. Every move, as shown in Figure 3, uses psychological functions in order to achieve certain ends that can be conscious (mostly) or unconscious.

In Free Thinking move the leisure moments, with contemplation and no requirements about the utility to Company or the Society, involving functions such as: perception, memory, emotion and imagination.

The individual demands on the psychic survival (need for social integration) and physical survival (maintenance of body stability with safety and nutrition) indicate the movement of Survival. In Enterprise, the struggle for survival involves feuds, powers and influences that make the routine saves and full of exercises, so named as survival and maintenance of power in employment relationships. The psychological functions involved are: perception, memory, emotion, cognition, imagination, communication / language and motor action.

On Experience and concept movement, intends to build the knowledge base to facilitate professional growth. Without this exercise, the operators with skills will have difficulty bridging the understanding of problems in industrial processes. The experience is engaged in the conceptual basis for bringing power to the job. The tasks involved in this case are: perception, cognition, emotion / memory / imagination, action, verbal / motor, perception, cognition, memory, skills and knowledge.

On Movement of think, decide and act. The moment of decision is different from the process of forming the conceptual basis and memorization of tasks. In the decision some dynamic skills are built. The moment of decision requires some knowledge about the environment. So to act is necessary to think and decide. In this case the related functions are: perception, cognition, memory, imagination, cognition, knowledge, verbal action, action and motor skills.

In Figure 3, the individual uses psychological functions mentioned in the various moments in the organization. Thus, the formation of the knowledge base takes place just prior to the organizational as well as the survival instinct. Leisure (free thinking) emerges and is maintained with the need for social inclusion, almost always in the moment of rest. After learning about the job task, the use of knowledge to decide, learn how to perceive and take impulsive actions. At maturity, at work, it establishes a maximum level of learning ability and having reached the full condition of the use of intuition...

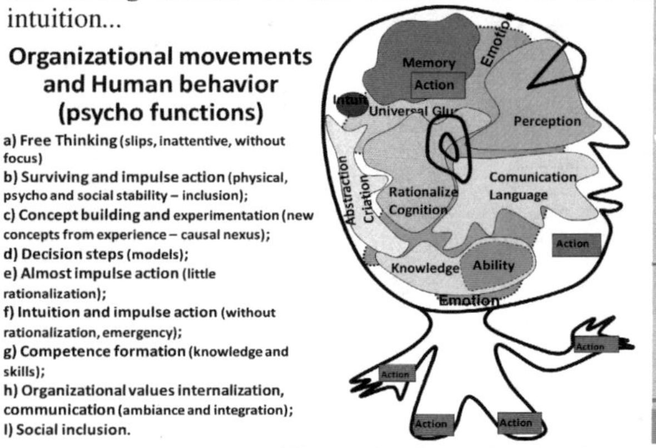

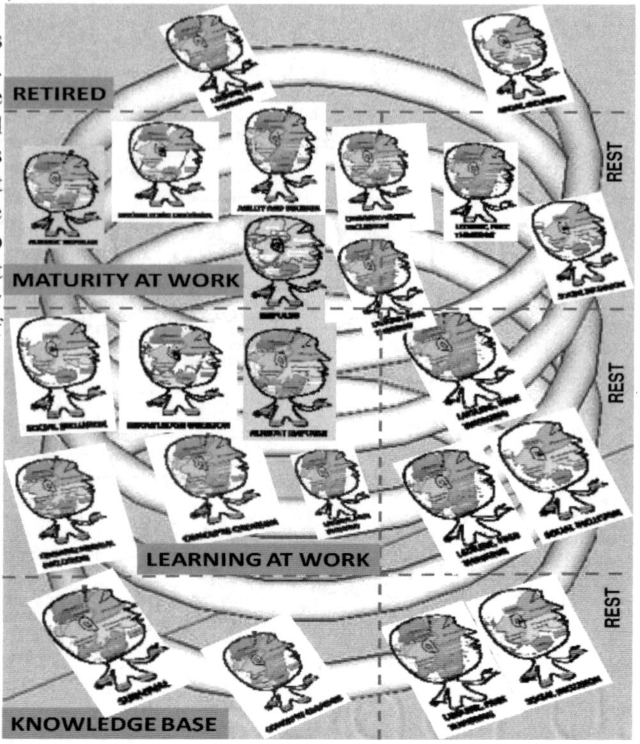

Figure 3 – Organizational movements and Psychological Functions

On Movement of almost impulse, compare and act. In this case, though cognition is not the most important function, involves quick actions based on experience with impulsive actions. Thus the functions involved are: perception, memory, emotion (balance), cognition, imagination, cognition, motor activity, skill, communication and language.

In impulsive action, the habituation through repetition of emergency tasks allows the application of practices when the situation will be real (fire, major leaks or explosions). The tasks involved are: intuition, emotion (balance), skill, action, verbal, and motor action.

In the course of training skills and build competences, creation, thinking, decision and action are important psychological functions. The operator is involved with the formation of competence beginning on the abstraction through the rationalization and thence to the action. It is more complex and has many physiological functions such as perception, abstraction, and emotion (balance), memory, imagination, cognition, skills, verbal action and motor action.

In the movement of Organizational Inclusion, organizational values are integrated into the body of work, at least the intent, and must have identification with the organizational practices of the directors and managers. The organizational inclusion involves entry ritual presenting the vision, mission and company values to employees, as well as, security aspects of the processes and people. The tasks involved are: internal and external perception, memory, emotion, imagination, cognition, verbal action / motor, communication / language, skill, and knowledge.

On Social Inclusion movement: the value of man through work allows social and financial activation, then, has tasks running in the middle of productive means that this worker have market value and balancing the emotional side. The emotional tie of the group may be the result of feeling of social inclusion. The financial balance is conducive environment for health even though many workers with depressive traits.

6. Routine Operation and the Human Element

Some occurrences or cases of routine operation are discussed in regard to aspects of leadership, influence of environments, and possibilities of derivative human error occur [4] [7] [2]. Some actions have been taken by the organization, operator or engineer or were perceived by the action of the organization, operator or engineer, all on the production line. We sought to examine the following aspects: environmental analysis, leadership, skill, human error, and the decision task. These actions were classified as in Table 2: right (green), adequate but not effective (yellow), inadequate for that situation (orange) or inappropriate (red).

A) **Director** (ex-operator and ex-engineer) actions centered on the departure of industrial plants inhibiting the activities of shift team;
B) Inappropriate behavior in **waste disposal** (burying) and cooperation in the best possible solution;
C) Overview financier director's decision regarding investment in measuring flow causing oscillations in the reaction and, team's practical solutions using "**alchemy**";
D) Mapping perception abilities (**ear for tuberculosis**) to identify loss of vacuum in the plant;
E) Publishing of the report with the **hypothesis** of operational event of fluctuations in the vacuum distillation causing the appearance of benzene contamination in the effluent and causing the operation team **in search of root cause**;
F) Lack of cooperation and **selfishness at shift change** causing false signals and delaying the resolution of problems for the next shift (educational campaign: pass the ball round to neighbor);
G) **Operational Solution**: quick perception of losses in the effluent through the application of polymer plates and verification of attack of the solvent, making it possible to adjust the process and investigate the root cause;
H) The interpretation of **routine signs** allows it's re-planning in causality anatomy, and test for changes of procedures, process parameters, operating standards and behavior patterns;
I) Respecting forms of **regional behavior** allow change of patterns with no great efforts of persuasion (negotiation of temperature change with the operator panel before writing the final rules);
J) **Inspection of the feet** in changing room for employees to identify contamination by direct contact with chemicals, inadequate standards in the use of PPE (invasion of personal space in an educational campaign);
K) Behavior of natural ownership of specific industrial areas by operators to maintain **organization and cleanliness**, without requiring management programs "top-down";
L) The **fear of punishment** caused by contamination inhibits the operation in reporting events to the HS&E. The unfolding of this communication generates records and investigations that may affect promotions leading to the great omissions of investigations into this area;
M) Inadequate conditions of operation since the **start of the industrial plant** in alternative pumps due to ignorance about the root cause of problems, the case of o-ring;
N) **Change Manager** ➔ Changes in **rituals** for social inclusion ➔ causing occupational instability in labor relations ➔ illnesses, absenteeism, accidents and labor issues;
O) During sampling for environmental control of waste, **improper operation of alignments** distorts the analytical result indicating "false" normality in the process;
P) Large amount of solvent loss to the effluent with consecutive errors by and consciousness group omission, **social vices**;
Q) Director authorizes the solvent evaporation (using improper purging of heat exchanger) at the end of organic wastewater system to prevent high levels of contaminants in environmental sampling by the agency. The sacrifice of **human health by organizational health**;

R) Operator with **lack of control** in **transfer** activity of liquid effluent (lack of commitment or omission on the divided attention) and loss of control by sending oil to the sea;

S) The remedial action to address the hole in the reactor (caused by acid gases in the process) are inadequate, using no appropriated inspection method, under pressure from the Board, creating redundancy in the operation of inspection and the resulting report, not accepted by the management of production, but reported to management of technology. The record has generated unease at the managerial level and has caused an impact on the outcome of the technical culture **ethical conflict**;

T) TDI tank released by the operation after recognition of signals of several instruments (deviation generating multiple false signals) added to the lack of skill and knowledge of operators' **→release of improper service.** Maintenance team performs the opening of tank using techniques not suitable generating tank explosion and death of workers;

U) **Truck failed** added to the inappropriate actions of reactivation (spark) plus the presence of toluene in the dike and dam plus the overflow due to rain, coupled with the valve locked in the bottom of the dam caused the deaths of two workers;

V) Service inside height tower with improper action of operation at control **panel** beginning start-up with water almost causing an accident in workers (**construction site**);

Table 2 – Cases of Routine operation and human and organizational elements

Lt	Failure agent	Environment	Leadership	Skill	Human error	Task decision
A	**Director**					
B	**Waste disposal**					
C	**Alchemy**					
D	**Ear for tuberculosis**					
E	**Hypothesis of root cause**					
F	**Selfishness at shift change**					
G	**Operational Solution**					
H	**Routine signs**					
I	**Regional behavior**					
J	**Inspection of the feet**					
K	**Organization and cleanliness**					
L	**Fear of punishment**					
M	**Start of the industrial plant**					
N	**Change manager, rituals**					
O	**Improper alignments**					
P	**Social vices**					
Q	**Human by organizat. health**					
R	**Lack of control in transfer**					
S	**Ethical conflict**					
T	**Release of improper service**					
U	**Truck failed**					
V	**Panel, construction site**					

7. Conclusion: Human Factors that promote accident/ incident

The root cause of the problems may directly involve human factors or environmental situations in which industry is involved. It can be noted from Figure 4 that the following factors are listed as primary causes: policy and practice; cultural traits; task planning; design criteria, risk management, and economic bias. Also in this Figure 4 is presented that, depending on the amount of fault energy can be classified as major events such as accidents or incidents, events with medium energy as the several technical, human and social failures, and events with low energy such as signs of failure or deviations that are greater in number, but which are not investigated adequately with the omission of caused by the possibility of punishment.

Investigation of failure and decision making

Often the decision to be made is about or inside complex processes, tasks or complex social relations, lacking concepts to support this decision. At the same time, the leaders cannot create new concepts and search for external knowledge base limiting the alternatives to the decision, and lowering their quality. In other situations there may not be skilled enough to decide due to high staff turnover and low valuation of execution to experienced operators who can read signs of abnormality in routine operation. On the other hand, to investigate problems we look for profiles that can build causation map, favoring more clearly decision. Vices can be established in the decision when using the customary answer by management and automatically assume a single direction (and known) for granted, thus being able to reach the best place in the company's goals. Creativity is an important feature for investigation of failure, without which, there will be not innovative and dynamic ideas to work on complex and new processes.

Leaderships and group formation

Leadership is not structured for the challenges of cultural change and sustainability issues in the industry. In many situations the doubt is installed and the sense of fairness lacks in decision making. Another characteristic of poor leadership is the lack of feedback on the quality of routine tasks. These positions of leadership and societal changes of mass culture indicate that groups have less cooperation currently generating truncated information and worsening the aggregation of the teams.

Task Efficiency

As a result of globalization, jobs are filled by people from different regions and from different generations and there may be errors in written and oral communication based on language errors. This may cause bias or bias the formation of "cultural pots" or castles. In an attempt to avoid human error seek the installation of barriers to prevent the accident is ineffective at complex process where multiple causes are not looked by unprepared team. Another phenomenon indicate that when installing several safety barriers it could provoke cognitive laziness in operation functions giving other possibilities to the accidents. This type of behavior occurs in routine tasks and in emergency situations.

Tasks due to poorly planned or due to unprepared formulator or due to safety policies end up causing distrust in the team and motivating, consciously skip of steps. It can cause task failure where lack of planning stages or steps in excess affecting the behavior of omission in the routine. This omission is enhanced when the tools for the job are not available or were not mentioned in the task.

Safety and Emergency

It is important to study the behavior of the team when they suffer extreme stress in the task, generating low performance. When the team is fortunate to have on their staff, people with high level of intuition based on experience and knowledge, you can take agreed actions without causing chain reactions or in worst situation causing accidents, affecting communities and the environment.

Organizational Culture

The positives and negatives aspects generated on the practice of the operation depend on corporate policies and trust level between the company and the employee. Non-acceptance of organizational values can be caused by conflicts between policies and managerial practices, then, staff team become fragmented and don't feel inside organization.

Traces of social and individual behavior

The current society is produced by globalization and suffers influence from mass culture. This society may be producing incompatible personalities with the needs of the working environment by generating mock group. This situation occurs when the unit of the teams cannot be established, do not form genuine leaders and power prevails over collegiality, result: the task group loses efficiency. Often, this fuzzy situation affects strategic issues where, corporate leaders insist on breaking cultural barriers causing disrespect to population stereotypes.
It is important that in the operation team there is the presence of different profiles that seek the same goals with specific points of view. The generation of not fighting (arguments) in the discussion may result in a team situation "warm" which adopts the first solution is probably not toward the root cause. The overconfidence resulting from self-centeredness, dash of mass culture eventually causes difficulty in accepting the standards of the task. It may happen that the reverse is the acceptance of planned procedures, which are inadequate, with no questions or propositions.

Individuals in the workplace inside globalization and with trends of automation and greater risks end up suffering more psychological stress at work which may, depending on the profile at each station, generate quality loss of cognitive function. This leads to memory errors and inattention as a result of environmental or worker profile.

Contracts and Interfaces

The production team involves operation, maintenance, laboratory, process engineering, the project team, and many others, finally, various functions to enable productive activity. The low commitment caused by contract poorly formulated with the maintenance has resulted in corrective actions and not solving problems acting over root causes. Thus, the improper maintenance of systems is lowering the patterns of the production team as a whole. The relationship between different functions must be transparent to avoid jealousy, noise and other communication failures that may cause suspicion and rework.

Risk Management

Ignorance of the risks in productive activity, or worse, the false knowledge of the risks generate a fictitious comfort that, when discovered, causes fear of activity due to the possibility of accident or incident with several impacts. This situation cause omission in making, in control and perform the task. The fear of losing power or position also creates situations of failure in communication with deviations hindering the root-cause identification due to low visibility of the factors involved. Thus, the lack of information by default can become active fault in a latent failure. The cognitive or physical stress due to repetitive stress or because of outside activities occupying the attention can affect the quality of cognitive processing is essential analysis.

8. References

[1] ÁVILA FILHO, S. Metodologia para minimizar efluentes na fonte a partir da investigação de anormalidades operacionais: caso da indústria química. 2004. 115 f. Dissertação (Mestrado Profissional Gerenciamento e Tecnologias Ambientais no Processo Produtivo) –Teclim, Universidade Federal da Bahia, Salvador, 2004.

[2] ÁVILA FILHO, S. Etiologia das Anormalidades operacionais na Indústria. 2010. 298 f. Thesys (Chemical Engineer Doctorate at Chemical Post-Graduation Program) – Universidade Federal do Rio de Janeiro, Rio de Janeiro, 2010.

[3] DEKKER, S. The field guide to human error investigations. Hampshire, Burlington: Ashgate Publishing Limited, 2002. 160 p.

[4] EMBREY, D. Preventing human error: developing a consensus led safety culture based on best practice. London: Human Reliability Associates Ltd. 14p. 2000.

[5] FRUTUOSO, P.F.M. Análise da Confiabilidade Humana (ACH) na AQR. In: 11o Congresso de Atuação Responsável, São Paulo: ABIQUIM. 2007. Palestra disponível em: <http://www.abiquim.org.br/atuacaoresponsavel/11cong/ segundo dia/Paulo_Frutuoso.pdf>. Acesso em: 16 de ago. 2008.

[6] HIRSCHBERG, S. Critical operation actions: human reliability modeling and data issues. 1998. Final task report by a group of experts of Nuclear Energy Agency (NEA) Committee on the Safety of Nuclear Installations (CSNI), Paris, France, 1998.

[7] LEES, F.P. Loss prevention in the process industries: hazard identification, assessment and control. Great Britain, GB, Butterworth-Heinemann. 2. ed. v. 1-3. 1996. 3.680 p.

[8] LEES, F.P. The hazard warnings structure of major hazards. Trans IchemE, UK, v. 60a, p. 211-221, 2005.

[9] LORENZO D.K. API770 – A manager's guide to reducing human errors, improving human in the process industries. API Publishing Services, Washington, 2001.

[10] RASMUSSEN, J. Risk management in a dynamic society: A modeling problem. London: Elsevier Safety Science. England. v. 27, n. 2/3, p. 183-213, 1997.

[11] REASON, J. Human error. Cambridge: Cambridge University Press, 2003. 302 p.

[12] SEBZALI, Y.M.; WANG, X.Z. Joint analysis of process and operator in chemical process operational safety. Journal of Loss Prevention in the Process Industries, Leeds, UK. n. 15, p. 555-564, 2002.

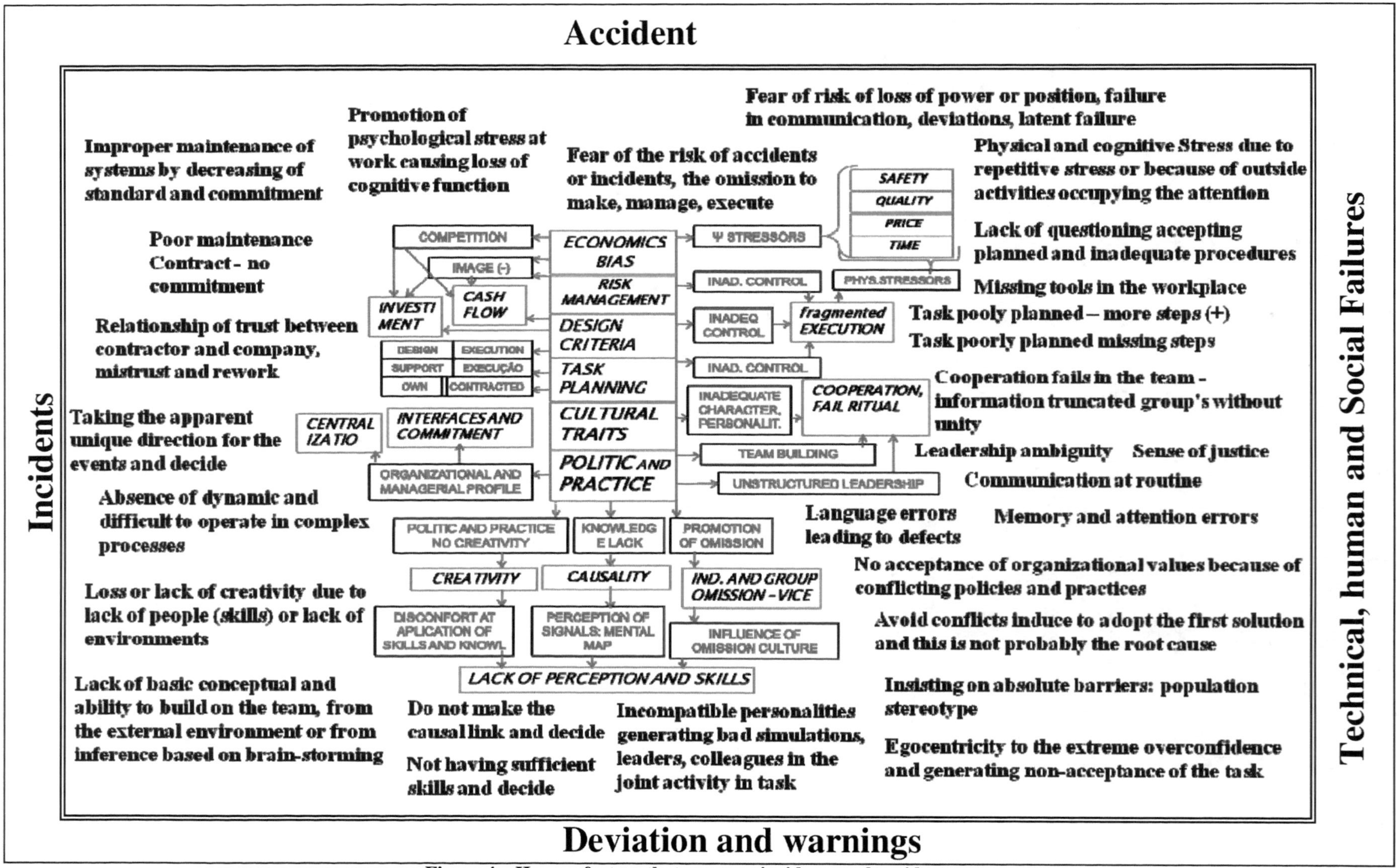

Figure 4 – Human factors that promote incidents and accidents

It's People, Stupid!
Human Factors in Incident Investigation

Michael P. Broadribb
Baker Engineering & Risk Consultants, Inc.
3330 Oakwell Court, Suite 100, San Antonio, Texas 78218-3024
mbroadribb@bakerrisk.com

Keywords: human factors, incident investigation, ergonomics, human error,

Abstract

People have inherent strengths and weaknesses which can affect their performance. Issues such as fatigue, emotional stress, and motivation can adversely affect performance. Their performance is also influenced by factors external to the individual, such as poor equipment design, inadequate training, excessive workload, and the work environment.

Most incidents in the process and related industries involve human factors, often as a key causal factor. Understanding these performance shaping factors is essential for conducting good incident investigations. However, besides those involved in the actual incident, human factors can affect other aspects of the incident investigation process, especially when considering other stakeholders.

The content of this paper will raise awareness of how human factors can affect the success of the incident investigation process, and help others plan for improved investigations.

1. Introduction

People have inherent strengths and weaknesses which can affect their performance. Issues such as fatigue, stress, and motivation can adversely affect performance. Their performance is also influenced by factors external to the individual, such as poor equipment design, inadequate training, excessive workload, and the work environment.

Most incidents in the process and related industries involve human factors, often as a key causal factor. Historically companies, and agencies that investigate incidents, have tended to overlook human factors causes almost entirely. Material deficiencies, such as equipment malfunction, can normally be identified easily. The real difficulty is to answer **why** the deficiency occurred, and invariably the answer is related to human behavior or performance.

Understanding these performance-shaping factors is essential for conducting good incident investigations in order to develop meaningful recommendations for maximizing human reliability and minimizing future failures.

However, besides those involved in the actual incident, human factors can affect other aspects of the incident investigation process. Table 1 lists some of the internal and external stakeholders who have an interest in the investigation. These stakeholders will ask questions of the investigation team, and seek to exert pressure and influence the investigation. By understanding human factors from the standpoint of all stakeholders in the investigation, it is possible to plan for improved investigations.

Table 1. Stakeholders in the Investigation

Persons directly involved in the incident
Other workforce members (including contractors)
Union
Witnesses
Operating company
Owners
Regulatory agencies
Local government
Media
Local community
Plaintiff attorneys
Insurance companies
Other plants with similar process
Disgruntled former employees
Investigation team

Human factors can affect the success of the incident investigation process, especially for major, catastrophic incidents, but many of the issues apply to lesser incidents also.

2. Persons Involved in the Incident

2.1 Managers are not Human

It often seems that managers must not be human because incident investigation reports rarely indicate that they make mistakes [1]. The same is true of the plant design engineers. Although no investigation should seek to apportion blame, historically accidents have been blamed on someone, such as an operator or technician, at the bottom of the organization. These individuals can't blame someone below him/her.

Fortunately this attitude is changing. We need to realize that most incidents are due to management failings. Those individuals **directly** involved in an incident should include managers (and supervisors) and design engineers, and not just the operators and technicians. Management and organizational factors invariably have a huge influence on incidents, either directly or indirectly through their impact on the behaviors of the workforce.

The most significant management issue is allowing economic pressures to compromise safety by not ensuring an adequate balance between production, cost and safety. Managers should recognize this as a potential systemic hazard, especially in operations with low margins or in highly competitive environments. Other typical managerial failings include failure to provide adequate training, instruction, supervision, equipment, or methods of working. Managers need training, instruction and supervision like the rest of us. Unfortunately executives rarely provide this training, instruction and supervision to middle managers, and often do not realize they could do more to prevent incidents.

The most significant issue with design engineers is that they fail to provide operations personnel with the ability to recognize and recover from unsafe conditions. It often seems that design engineers are setting operators up to fail. Invariably the recommendations from process hazards analyses (PHAs) rely on managing the identified hazards by operating procedures rather than implementing inherently safer measures to eliminate or prevent the hazard. Design engineers also need training, instruction and supervision, but rarely receive it.

Safety training for executives is probably more important than safety training for those at the bottom of the organization. Those individuals at the bottom of the organization generally follow the approach of those at the top. So, those at the top of the company or site set the culture of the organization.

2.2 Culture

Culture is the combination of an organization's values and behaviors that determine the manner in which production operations including process safety are managed – sometimes referred to as "How we do things around here." The process safety culture of an organization determines how its process risks are controlled. As noted above, the organization's leadership plays a huge part in establishing the culture. Essential features of a sound culture include:

- Establishing and enforcing high standards of performance
- Maintaining a sense of vulnerability
- Ensuring open and effective communication
- Providing timely response to issues and concerns

Cultural deficiencies are often a contributory factor or underlying cause of management system failures that lead to process safety incidents. Investigation teams therefore need to analyze the company's culture, and especially the site's local culture, in order to identify any cultural deficiencies. Some typical cultural deficiencies are listed in Table 2.

Table 2. Examples of Process Safety Culture Deficiencies [3]

Absence of vision statement/policy on process safety

Unclear / conflicting roles & responsibilities, accountabilities, authorities

- unclear/conflicting reporting relationship,
- unclear/conflicting assignment of responsibility,
- improper/insufficient delegation of authority,
- inadequate shared responsibility / empowerment for process safety at all levels of organization

Enforcement of management system

- no or inadequate enforcement of policies, standards, procedures,
- non-compliance not corrected/confronted,
- tolerance for willful violations,
- inadequate reinforcement of proper behavior

Maintain sense of vulnerability

- staff do not understand hazards/consequences of deviation

Open and effective communications

- input/bad news unwelcome,
- management not accessible to workforce,
- managers/supervisors do not listen & act

Timely response to issues/problems

- recommendations from PHA/incident investigation/audit not resolved,
- actions not implemented

Embed lessons learned

- repeat incidents,
- findings from PHA/incident investigation/audit not communicated to workforce,
- inadequate knowledge sharing with internal/external groups,
- inadequate learning from others

Case Study: An example of the underlying cultural issues that led to a major incident is the 2005 explosion at a Texas City refinery. A fractionation tower was overfilled and overheated over a four-hour period, and the resulting hydrocarbon vapor cloud explosion destroyed nearby temporary buildings, causing 15 fatalities and injuries to over 170 workers. The underlying cultural issues identified in the investigation [2] involved poor morale and motivation, absence of management focus on process safety, poor site communication, reduced workforce capability, poor hazard identification skills, poor understanding of process safety, failure to learn from outside experience, and absence of a system to provide early warning of process safety issues.

2.3 Behavior

During the course of an investigation, the investigation team will interview a variety of witnesses and seek assistance from other members of the workforce while gathering and analyzing evidence. In doing so, the team is likely to observe a range of human behaviors from caring and cooperative to a total lack of concern and even deliberately misleading. Sometimes the relationship between peers or a supervisor and worker can be a contributory factor in an incident, as can a lack of alertness due to distraction.

Case Study: Following a serious incident, the operators' log book was removed by an individual without informing company management or a government agency also investigating. Although when subsequently challenged, the individual claimed this was for safe keeping, it is unknown if the log was changed in any way. Investigators need to be alert, and, if something is missing or does not appear right, then suspect the worst of human behaviors and investigate accordingly.

Case Study: An incident involved a control board operator, who was an introvert and had few, if any, friends at the workplace. Other members of the crew apparently played jokes at his expense. One day the board operator closed a valve in error, while another crew member monitoring the process understood the error but intentionally delayed communication of the error to the board operator. By the time the crew member rudely informed the board operator of his error, it was too late to prevent the incident. It was also found that the board operator spent considerable time on non-work related telephone calls while the process was out of control.

To better understand incidents, investigation teams should attempt to identify the underlying human behaviors that contribute to incidents (see section 10 below).

2.4 Communication

A lack of communication or some deficiency of communication is often a key factor in incidents. Examples of communication deficiencies include:

- Communication not received or late
- Incorrect information/ instructions
- Information not understood (long/complex message, non-standard terminology, repeat back not used, etc.)
- Background noise/interference
- Language barrier

Common areas for poor communication exist wherever horizontal or vertical interfaces occur in the organization(s). Examples are communication with / between:

- different organizations including contractors (horizontal)
- work groups/departments (horizontal)
- shifts (horizontal)
- peers (horizontal)
- manager and supervisor (vertical)

- supervisor and individual (vertical)

Case Study: An investigation of a drilling incident found that the drilling team drilled into the gas cap of the oil reservoir with too little mud weight, resulting in loss of integrity of the well casing. Gas and well fluids flowed uncontrolled to the surface. It would have been easy to stop the investigation at this point and recommend greater care when specifying the mud weight. However, further analysis of why the incident occurred identified that three groups were involved; the well design team, the drilling team, and the reservoir management team. Over a period of several months during which the well was designed and the drilling rig mobilized, the reservoir management team increased gas reinjection into the reservoir to raise reservoir pressure in order to assist oil recovery. They did this without informing the other teams.

3. Witnesses

3.1 Hierarchy of Evidence Reliability

Eyewitnesses can provide valuable information, but can also be the least reliable source of evidence. Investigators need to interview witnesses quickly before the witnesses talk to each other or to their friends and family. It is well established that a witness's perception and recollection of events is likely to degrade with time, and/or become influenced and changed by conversation with others. Schedule interviews so that contact between witnesses is minimized.

The investigation team should place primary credence upon the physical and documentary evidence collected, and the use of witness testimonies needs to be carefully considered. In line with investigation guidance [4], the verbal testimony of an individual should be given less importance unless it has been possible to corroborate that evidence through other means, such as testimony from another witness or physical evidence. Sometimes follow-up interviews are conducted to gather additional evidence or to resolve inconsistencies in initial testimonies. Given that early witness interviews are likely to be more reliable than later testimony; these should be valued more highly. The hierarchy in Table 3 represents a weighting that can be applied to the evidence gathered.

Table 3. Hierarchy of Evidence Credibility (highest reliability listed first)

DCS / PI data
Paper / electronic documents
Multiple witness statements (where consistent)
Visual inspection (possibly disturbed by emergency response)
Process sample analysis (possibly changed by delay in sampling)
Process equipment testing (possibly damaged/changed by incident)
Single uncorroborated witness statement

3.2 Interview Techniques

The first interview of an eyewitness is the easiest. Simply ask the witness to recount in his/her own words what they saw, heard or smelled. The interviewer needs to listen and not interrupt to allow the witness to finish their story. Witnesses can often have a poor recollection of time or sequence of particular events in an incident. Only when the witness has finished their account should the interviewer seek clarification by repeating back what they heard, and asking open questions seeking information on gaps or inconsistencies.

If two investigation team members are present, one should act as the primary interviewer, and only at the end should the second team member ask any additional questions. The interviewer(s) should maintain eye contact with a positive expression to help maintain rapport, and not react verbally or non-verbally to any specific statement by the witness. On no account should the interviewer comment on other witness statements in front of the witness, and all statements should be treated as confidential.

Most witnesses want to help the investigation and will cooperate with the interview. Occasionally it may be necessary to interview an uncooperative witness who is intentionally withholding or modifying evidence. The interviewer should emphasize that the purpose of the investigation is to focus on finding the causes, not finding who is to blame. The interviewer cannot promise to eliminate punishment for actions discovered, especially if malicious or criminal intent is involved. However the interviewer should make a strong effort to convey that the purpose is not to find the guilty.

Later interviews of the witness are likely to use increasingly closed questions to confirm individual facts. However, as noted above, care should be exercised as the witness will almost certainly have had the opportunity to speak to colleagues and friends about the incident by the time of a second or later interview.

The choice of interview location and room layout should be carefully selected to put the witness at ease. Normally a quiet location away from interruption by site management or plant operations is preferred. Sometimes a neutral location may be appropriate, such as a union hall or witness's home, especially if the witness has been traumatized by the incident. The room should be arranged so that the interviewer and witness sit side by side rather than across a table or desk. A second interviewer or note taker should sit beside the primary interviewer so that the witness can see both without having to turn around.

Occasionally there will be no eyewitness of the incident, but valuable evidence can be gained by interviewing first responders on the scene and others, such as the preceding shift operators and technicians.

Further guidance on witness interviews is available in the CCPS guidelines on incident investigation [4] and [5].

4. Workforce

4.1 Managers always know Best!

When the investigation team arrives on site, the local manager usually welcomes the team and often thinks he knows exactly what has happened and why. The investigation team should thank him for his input and promptly commence their investigation. In particular, the team should resist any temptation to jump to conclusions, and should consider **all** possible scenarios - not just the manager's pet theory.

Case Study: A manager reported that cattle had died as a result of drinking fluids alleged to have been dumped illegally at a remote site by a third party. The investigation team gathered evidence for a range of potential scenarios. In particular, by interviewing an operator and a contract technician and analyzing fluid samples, the team was able to determine that the fluids were generated by maintenance and production activities at the site. No third party was involved.

Case Study: Upon arriving at the site of a major incident, an investigator was informed by a local manager that data from the control room were useless as the instrument air to the pneumatic instruments had failed during the ensuing fire. Ignoring this advice, the investigator studied the data and was able to exactly determine all process parameters at the time of the incident, which ultimately confirmed a different scenario from that being supported by local management. Although the numerous control instruments were not synchronized, the air failure provided a perfect reference as good as an exact time. Determining conditions at the exact time of an incident is valuable in identifying any precursors.

The moral of these case studies is that investigation teams should not believe anything until they have seen it for themselves, and generally that operators, technicians, and contract workers know what really happened. Managers only think that they know what happened.

4.2 Impact on Workforce

An incident, especially those resulting in serious injury/fatality, can have a significant impact on workers at the site. Many will know the victim(s) and/or may have witnessed a distressing experience. Different people will respond in different ways. Some may be traumatized, even appearing almost in a zombie-like trance. Others will be angry and/or distraught. The way that some individuals deal with the experience is to be helpful, and they may start cleaning up the incident site without instruction. This housekeeping can destroy valuable evidence, so investigation teams need to give precise instructions to secure the site and preserve evidence. Investigators should also be prepared for witnesses to break down in interviews while trying to recall the events.

Case study: A major incident involving a pressure vessel resulted in several fatalities to the operations team, but there were no eyewitnesses. Upon arriving on site, the investigation team interviewed a first responder who discovered the victims' bodies. This individual was distraught as one of the victims was a personal friend. His method of handling the distress was by talking at length "to get the experience off of his chest." The interview lasted over two hours with the interviewers listening sympathetically and barely saying a word. However valuable information for the investigation was obtained merely by listening to what the responder had to say.

Investigation teams need to consider the impact of a serious incident on members of the workforce and be prepared to take a flexible approach, especially when conducting interviews.

4.3 Impact on those in Authority

The impact of a major incident on those in a position of authority, such as managers and superintendents, can be even more distressing than on other members of the workforce. Some individuals responsible for operations take it hard that such an incident happened on their watch. Symptoms may include not thinking properly and/or erratic behavior.

Case study: A loss of well control resulted in a major release of gas and well fluids at the surface close to offices and production facilities. Fortunately there was no ignition but a spark at any time could have had catastrophic consequences. The site manager tried to handle this major incident locally without evacuating non-essential personnel and without requesting corporate support. After a couple of days the corporate crisis center became aware of the seriousness of the incident, and an executive then ordered the evacuation. Significant external resources were required to regain well control.

4.4 Contractors

Many production sites will invite one or more representatives of the contract company to participate in the investigation if a contractor(s) suffered injury in the incident. This can be helpful to the investigation as cooperation between the parties will often facilitate the evidence gathering and analysis. However sometimes the contract company will decline the invitation to participate, possibly for legal liability reasons, and may conduct a separate investigation. Under these circumstances the production site's investigation may be denied access to contractor witnesses.

5. Agencies

5.1 Multiple Agencies

When the investigation team arrives, the site may be under the control of the local Fire Marshal. Sometimes the emergency response may be ongoing, and, while all personnel may have been accounted for, efforts to make the site safe and prevent escalation may be continuing. It is important that the team establish a good relationship with the Fire Marshal as soon as possible, especially in order to preserve evidence and achieve any other immediate objectives that the team may have.

By the time that the initial emergency response is over and the site is rendered safe, other government agencies may have arrived. In the event of a major incident in the USA with one or more fatalities, it is not uncommon for federal agencies such as OSHA, EPA, CSB, and ATF to initially respond in addition to law enforcement and the coroner. Despite memoranda of understanding between the agencies, it is not unknown for disagreements to surface over which agency has overall control. Depending on the circumstances, various state and local government agencies may also respond.

5.2 Impact on Investigation

The presence of one or more government agencies on site can have a significant effect on the environment under which the investigation is conducted. Any activity which the investigation team wishes to undertake may be subject to prior agreement by the agencies. This is in addition to any requirement to operate within restraining orders imposed as a result of applications by plaintiff attorneys. This may require detailed written protocols to be agreed by all parties prior to process fluid samples being taken, debris being moved, metallurgical testing of failed equipment, and removal and testing by independent third parties of critical controls, valves and other process equipment. Establishing a good working relationship with the relevant agencies will be important for the investigation team to be able to progress their investigation in a timely manner.

Case study: Following an incident, several government agencies responded to the site. Some rivalry between the agencies arose, and concerns around harmful air emissions were used to delay access to the site of the incident by the company and other agencies for several days. Access was denied despite the appropriate personal protective equipment (PPE) being available. Some fragile, time sensitive evidence was lost as a result.

Other factors can influence the investigation environment. Some agencies may not have the appropriate technical disciplines required to properly investigate the incident but nevertheless are committed to investigating it anyway. In the event of a major incident to a company operating in a foreign country, the company and its investigation may be subject to additional surveillance, possibly by monitoring telephones or other means. Some agencies may not be interested in determining the root causes of the incident because their jurisdiction only allows them to cite existing regulations that have been violated whether these violations were material to the incident or not. Other agencies may not have any enforcement powers but may be able to use the power of the media to place pressure upon the company to respond to their proposals. Some agencies have limited resources and may be under pressure to complete their investigation, which can lead to jumping to conclusions without evaluating all possible scenarios.

Case study: An incident occurred where the agency personnel responding did not understand the technology of the process unit involved. None of the agency personnel were professional engineers. As a result the agency declared that certain instrumentation and controls were faulty and caused the incident, despite independent third party testing after the incident demonstrating that the instrumentation and controls were working as designed.

The investigation team will need to manage the investigation environment and all its competing pressures to the best of their ability.

6. Local Community

6.1 Human Behavior

When a serious incident occurs it may impact the local community, especially if an explosion or release of a toxic chemical occurs. Fires have less potential for offsite impacts unless emergency services close adjacent roads or clouds of smoke blow across residential areas. These incidents may cause damage to residential and commercial properties.

While some property owners may have valid claims for recompense, others may make frivolous claims in the hope that the company will settle. Sometimes companies will not have any evidence to the contrary with which to challenge claims, and may settle to keep the local community supportive of their operations. Often investigators will be primarily focused on the internal investigation and may ignore the offsite consequences.

Immediately following an incident, someone should take photographs of the surrounding neighborhood, especially if there is no apparent damage, as it may be advantageous later if claims are filed. If the emergency response is ongoing and the site is still under the control of the fire marshal, the investigation team may be able to undertake this themselves. Alternatively they may request someone else to take photographs. Irrespective of property damage claims, the presence or absence of broken windows, for example, can provide valuable data for determining the size and footprint of an explosion.

6.2 Communication

Any serious incident is likely to raise concerns within the local community, and the company may need to re-build trust with the community. Open communications through public meetings or other means are likely to be more successful in re-building trust than relying upon local media reporting.

While it is a company or site management responsibility to communicate with stakeholders, the investigation team will be best placed to provide information on the progress of the investigation and its findings. In a lengthy complex investigation, the team should plan on preparing several interim statements. One or more members of the investigation team may be called upon to participate in a public meeting, especially when findings and recommendations are ready.

7. Industry

When a major incident occurs, companies with similar operations will be interested in any lessons learned that can prevent a recurrence in one of their production sites. Like local communities, other industry will be looking for information on relevant findings and recommendations. In particular, an interim statement of provisional finding(s) may indicate temporary or even permanent safety measures to be considered.

Once again, the investigation team will be best placed to provide this information. These lessons learned may be communicated through industry associations, regional safety groups, joint venture partners, or other fora.

Case Study: During the ongoing investigation of a refinery explosion that destroyed nearby temporary buildings, the investigation team published an interim report on the internet after two months. This report identified provisional causal factors but not root causes. One of these causal factors addressed the proximity of the buildings to process hazards. As a result, other companies decided to immediately move temporary buildings within their production sites away from process units.

8. Media

Television channels and newspapers are likely to report on major incidents, especially where fatalities or offsite property damage occurs, but may ignore minor accidents. TV companies may mobilize outside broadcast vans, and sometimes helicopters, to broadcast live video of the emergency response. Reporters will be looking to interview employees and neighbors to include "sound bites" in upcoming newscasts and publications, in addition to any formal press release from the company.

After the initial interest, the media coverage may decline. However, where fatalities or significant offsite property damage has occurred, local media may continue to cover the story. It is not unknown for TV vans to scan cell phone messages, so investigation teams should consider how they communicate. It is also possible that reporters will try to contact investigation team members, in addition to employees or even disgruntled former employees, and team members should refuse to comment and refer the reporter to the company's media representative. The team should also ensure the security of their data and deliberations, especially when conducting the root cause analysis.

Even without comment, some rather creative stories or headlines can appear, probably with the intention of increasing readers or viewers. If a factual distortion is challenged, any damage has already occurred and may not be corrected. As a result companies need to be very careful what information they release in press statements, although regular interim statements may mitigate against some factual distortion.

9. Legal Representation

9.1 Impact on the Investigation

Different stakeholders in the incident may retain legal representation to help protect their interests. As a result, restraining orders may be placed upon the company and others to restrict activity, such as evidence gathering, that could potentially result in evidence alteration or removal. Sometimes these orders ensure that nothing happens until all parties can jointly determine the way forward or observe a particular activity.

The involvement of lawyers inevitably changes the investigation environment, and the investigation team will be constrained in what they can achieve in a certain timeframe. There will be a delay in drawing process samples and testing equipment and materials until such time as detailed written protocols covering the activity have been agreed by all parties. These delays can run to weeks or months. It is not unknown for process fluid samples to be totally useless by the time that protocols have been agreed. Investigators need to manage this environment and plan for a lengthy phase of evidence gathering and analysis. It may even be in the investigation team's interest to mobilize on site quickly and gather fragile, time-sensitive evidence *before* any restraining orders can be filed.

9.2 Lawsuits

The decision on whether to include a lawyer on the investigation team will depend upon the company's policy and the circumstances of the incident. There are advantages and disadvantages to having an attorney on the team, but a discussion of these pros and cons is beyond the scope of this paper.

The company will almost certainly want legal advice, even if there is no legal representation on the investigation team. This is especially true when there is the possibility of criminal and/or civil lawsuits. In that event, investigation team members may be called to give depositions and/or testimony, and will require legal support for preparation. Opposing counsel may be looking to attribute blame and discredit experts on the investigation team if the findings could harm their client.

10.Investigation Team

10.1 Objectivity

The investigation team needs to be aware of the demands of all the stakeholders discussed above and manage the investigation environment to the best of their ability, while remaining focused, independent, and objective on determining *what happened* and *why it happened.* The team should not believe anything they are told unless they are able to corroborate the facts, and should resist jumping to conclusions too early. Above all else the team should avoid apportioning blame.

10.2 Planning

The investigation team needs to plan for everything covered in this paper, which may require prior training before a serious incident occurs. The team should be selected from a cadre of trained individuals with the appropriate discipline expertise including root cause analysis (RCA) and human factors. Depending upon the scale and circumstances of the incident, it may be necessary to select one team member just to manage the evidence gathered.

10.3 Organized Labor

Union members, or hourly paid employees, can be valuable investigation team members and help establish a more open, trusting relationship with other hourly paid members of the workforce. With their assistance, it may be possible to gather evidence that would not necessarily have been readily available to the team.

10.4 Manage Emotions

In the event of a major incident resulting in fatality or serious injury, the investigation team is likely to be subject to similar emotional impact as that previously discussed for witnesses and the workforce. The team will need to manage their own emotions and this should be considered when selecting the team members.

The investigation team needs to interview cooperative and uncooperative witnesses, including heroes who may have saved others from serious injury and individuals who know/suspect that their actions caused the incident. Under these circumstances, experienced interviewers may need a flexible approach and occasionally have to act as the 'bad cop.'

10.5 Methodology

10.5.1 Checklists

The easiest techniques for investigation teams to use to identify human factors are undoubtedly checklists. The investigator does not need to be a psychologist or expert on human reliability to use the checklists, although teams may on occasion wish to engage the services of these disciplines to assist a major incident investigation.

One approach uses checklists based upon a model developed by Weigmann & Shappell [6], which in itself is an adaptation from James Reason's 'Swiss cheese' model [7]. The model depicted in Fig.1 comprises four categories of checklist:

- Organizational factors
  - top management defenses against an incident
- Unsafe supervision
  - supervisor defenses
- Preconditions for unsafe acts
  - substandard conditions (mental state, fatigue, illness, environment)
- Unsafe acts
  - skill, knowledge, perception

Further guidance on this approach is available [4]. CCPS has published several human factors checklists [5], including those for evaluating ergonomics/man-machine interfaces, capital projects and procedures. Other checklists are available that focus on the causes of human error, such as:

- Lack of knowledge or skill
- Distraction (external or internal)
- Limitations (physical or mental)
- Four common categories of human error:
    - omission
    - commission
    - out of sequence
    - wrong timing

Figure 1. Weigmann & Shappell Model

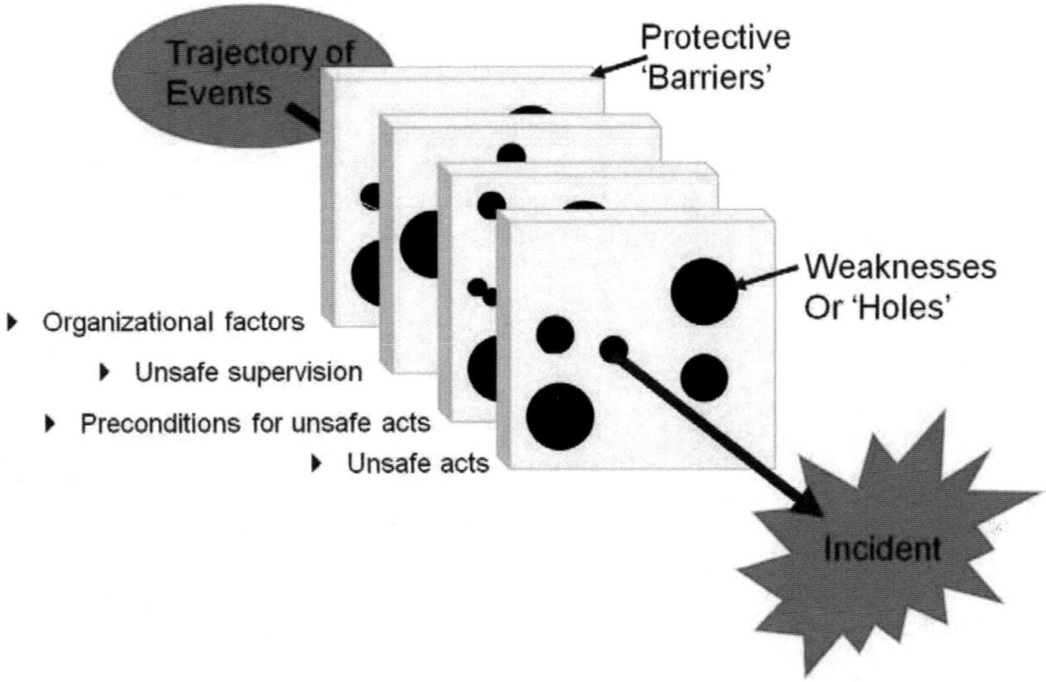

10.5.2 Intentional Behavior

A useful tool that investigation teams can use for analyzing intentional behavior that leads to an incident is Antecedent Behavior Consequence (ABC) Analysis [5, 8]. The antecedent sets the stage for the behavior, which is an action (say/do) that is observed by others. The consequence is what happens as a result of the behavior. If the consequence is negative (injury, reprimand, etc.), then the individual is likely to change their intended behavior in future. However if the consequence is positive (no injury, lack of reprimand, etc.) the individual is likely to repeat the behavior.

10.5.3 Unintentional Behavior

Where the behavior is unintended, a different analysis tool is required. Human Error Analysis, a.k.a. Human Reliability Analysis, covers a range of techniques based upon task analysis and a cognitive theory of control. Some of these techniques are:

Task analysis techniques
- Technique for Human Error Rate Prediction (THERP)
- Accident Sequence Evaluation Program (ASEP)
- Simplified Human Error Analysis Code (SHEAN)

Cognitive control based techniques
- Contextual Control Model (COCOM)
- Cognitive Reliability and Error Analysis Method (CREAM)

The techniques have origins in the nuclear power and air transport industries, and need to be applied by an expert. However there are checklists that non-experts can use that focus on human error.

11. Conclusion

Human factors are involved in most incidents that occur in the process industries, and can affect the conduct of incident investigations and their success. Investigation teams need to manage a variety of stakeholders' expectations, which also involve human factors, while at the same time remaining focused on identifying root causes and effective corrective actions to prevent recurrence.

12. References

[1] Kletz, Trevor A., "Learning from Accidents, 3rd ed.," Butterworth-Heinemann Ltd, Oxford, 2001.

[2] Broadribb, Michael P., "Lessons from Texas City - A Case History," paper presented at the AIChE 40th Annual Loss Prevention Symposium and CCPS 21st Annual International Conference, Orlando, Florida, April 2006.

[3] CCPS, "Guidelines for Risk Based Process Safety," New York, 2007.

[4] CCPS, "Guidelines for Investigating Chemical Process Incidents, 2nd ed.," New York, 2003.

[5] CCPS, "Human Factors Methods for Improving Performance in the Process Industries," New York, 2007

[6] Shappell, S.A., and Weigmann, D.A, "The Human Factors Analysis and Classification System - HFACS," FAA Office of Aviation Medicine, Federal Aviation Administration, Dept of Transportation, Washington D.C., 2000 (report #DOT/FAA/AM-00/7)

[7] Reason, James, "Human Error," New York, Cambridge University Press, 1990.

[8] Cooper, J.O., Heron, T.E., and Heward, W.L., "Applied Behavior Analysis, 2nd ed.," Upper Saddle River, NJ, Prentice Hall, 2007

Best Practices for an Evergreening Pressure Relief System Management Program

Mike Marshall
Marathon Oil Company, Findlay, OH
mtmarshall@marathonoil.com

Don Nguyen
Nitin Talreja
Siemens Energy, Inc., Houston, TX
don.nguyen@siemens.com
nitin.talreja@siemens.com

Keywords: Evergreening, PSM, Pressure Relief, process safety, pressure relief valve

Abstract

Most oil and gas refining and petrochemical organizations have a Process Safety Management (PSM) program that stems from the requirements outlined under OSHA's 29 CFR 1910.119. Some have a very extensive program that covers most of the 14 elements of PSM and is managed via an integrated platform while others follow an ad hoc approach that does not span all of the elements or is disjointed.

Maintaining accurate and current Process Safety Information (PSI) is a cumbersome and difficult task to accomplish due to the numerous sub-elements. One of the key sub-elements of PSI maintenance is the management of your pressure relief documentation. Historically this is an element that has not been given the attention it deserves. However, recent incidents in the industry have elevated the importance of maintaining accurate and current pressure relief systems documentation.

The ultimate goal is to allow all plant personnel to understand the basis of the current plant relief and flare header design and ensure that this current plant relief and flare header design provides adequate overpressure protection for all equipment. Thus, an evergreen program for relief systems documentation is created that is customized to fit the needs and resources of the operating company and has the ability to relate to the other PSI elements. As such with the support of Marathon Oil Company, Siemens has used its considerable experience in the pressure relief systems overpressure protection arena and, in conjunction with support from Marathon Oil Company, has developed a best practice program for pressure relief systems evergreening.

This presentation will identify the reasons and the benefits of having an evergreening program for pressure relief systems documentation as well as explore the interdependency of the key proposed evergreening elements listed below:

- Evergreening Workflow
- Resources and Processes
- Software and tools
- Key Performance Indicators
- Auditing
- Training

1. About Evergreening

Data integrity and control is the critical life blood of a Process Safety Management (PSM) program. Data and change management systems should be in place to ensure that data remains complete, accurate and up-to-date. In the petrochemical industry, inaccurate data can result in safety and environmental consequences as well as impacts to business continuity and productivity. It is therefore paramount that data be treated as a vital resource and that systems are developed to keep it evergreen.

2. Process Safety Management and Relief Systems

With promulgation of the Process Safety Management (PSM) Standard 29 CFR 1910.119, OSHA mandated that a management system comprised of several well-defined elements be established *"for preventing or minimizing the consequences of catastrophic releases of toxic, reactive, flammable, or explosive chemicals."* The Process Safety Information (PSI) element of the PSM standard states *"the employer shall document that equipment complies with recognized and generally accepted good engineering practices (RAGAGEP),"* with specific reference given to *"relief system design and design basis."* The PSI element of PSM is all about data integrity and control, and more.

At the core of a dynamic and properly integrated PSM program is an effectively administered Process Hazards Analysis (PHA) program. The integrity and value of a mature PHA program is entirely dependent on the PSM elements supporting it, namely PSI, Management of Change (MOC), procedures, training, and auditing.

These elements should be part of a holistic, fully integrated approach designed to be self-sustaining and continually improving. Without accurate PSI, without a robust and comprehensive MOC program, without proper procedures, training and auditing, a credible PHA cannot be properly conducted and the integrity of the resultant findings will be suspect. Rather than adding value, a flawed PHA process could result in hazardous consequences, especially as it relates to proper relief system design.

With consideration given to recent incidents and regulatory developments within the industry, it is clear that the integrity of relief system design warrants its own PSM emphasis program, a management system enhancement more aptly termed an "evergreening" pressure relief systems management program. The term "*evergreening*" is generally defined as "something that remains fresh, lively, interesting, or popular." In the context of PSM, evergreening would translate to keeping the elements of PSM complete, accurate and up-to-date.

Although the regulatory agencies such as OSHA and EPA do not explicitly use the term *evergreening* in their regulatory standards, they use equivalent terms such as *accurate, clear,* and *on-going.* For example, in the Appendix C compliance guidelines of 29 CFR 1910.119, OSHA uses the term *"complete and accurate"* in lieu of *"evergreen."* Following is an excerpt from Appendix C: *"Complete and accurate written information concerning process chemicals, process technology, and process equipment is essential to an effective process safety management program and to a process hazard analysis".*

Likewise, for the Mechanical Integrity element of 29 CFR 1910.119, OSHA uses the term *"on-going"* to describe the expectation to remain evergreen. The standard states the following: *"The employer shall establish and implement written procedures to maintain the on-going integrity of process equipment".*

Regardless of the preferred term, the regulatory agencies that oversee Process Safety Management expect programs, procedures, and process information to stay evergreen. It is safe to say that keeping process safety information *accurate*, ensuring operating procedures remain *clear* and mechanical integrity programs remain *on-going* supports the OSHA PSM intent and constitutes an evergreening process.

Recent enforcement actions demonstrate OSHA's expectation for operating plants to maintain an *evergreen* process safety management system. In 2007 and 2009, OSHA initiated a special enforcement initiative (OSHA's National Emphasis Program [NEP]) specific to refineries and chemical plants, respectively. Of the citations issued, many involved missing, inaccurate and incomplete process information as well as outdated relief system studies. Clearly, it is OSHA's expectation that operating plants have an *on-going* or *evergreen* integrity management program for process equipment associated with relief systems.

The abovementioned evergreening pressure relief systems management program is a natural extension of PSM in that it simply builds on elements already established in the OSHA PSM standard. Continuous improvement is at the core of any plan-do-check-act management system cycle, and what is being proposed here is consistent with that rationale. Accordingly, a relief system evergreening process should compliment and enhance existing PSM programs by accomplishing the following:

- Provide an evergreen record of process safety information (PSI) relative to the current relief system design basis and installation.
- Assist in the management of change (MOC) process.
- Support the process hazards analysis (PHA) program.
- Facilitate both internal and regulatory auditing.
- Ensure compliance with RAGAGEP developments.
- Provide for company and industry-wide standardization and continuous improvement based on best practices and metrics.

- Protect investments in relief system data and design assessment initiatives.

Moreover, in the interest of social responsibility, it is just good business to develop an evergreening system which not only enhances safety and environmental protection, but asset protection as well. Clearly, safety and environmental stewardship are of paramount importance, but asset protection, business continuity and public image also have vital significance in any business environment.

3. An Evergreening Pressure Relief Systems Management Program

A common misconception regarding an evergreening pressure relief systems management program might be that it simply applies to the management of documentation like relief device design bases and datasheets, with emphasis on data integrity and control. However, the concept behind this evergreening program is much more comprehensive, going well beyond basic documentation to focus on many other supportive areas including standards, practices, procedures, change management, training, study revalidation, auditing, metrics, risk management and mitigation.

All of these focus areas should be managed to work together in a proactive manner, promoting synergy, sustainability and continuous improvement. As previously mentioned, PSM was conceived out of a management system mentality of a continual plan-do-check-act cycle with continuous improvement at its core. Each element of a properly functioning PSM program employs this management system philosophy, all interconnected to form *"a whole greater than the sum of its parts."* The degree to which a management system achieves sustainability and continuous improvement is a measure of its integration profundity, effectiveness and ultimate success.

So, as for a relief evergreening program, the focus areas are similarly interrelated and dependent upon one another, and failure to properly manage any one of these elements compromises the integrity of the entire evergreening program. Even though it is quite evident that the focus areas are already elements of PSM programs across industry, opportunities for improving relief systems management may exist as evidenced by the breadth of recent OSHA NEP enforcement actions. Therein lies the justification for implementing a robust relief system evergreening program designed to augment any existing PSM program.

As previously mentioned the PHA program is at the core of a dynamic and properly functioning PSM program. Contrary to conventional thinking, the PHA process is not an effective tool for capturing relief system deficiencies. None of the hazard assessment methodologies commonly used in performing PHA's are effective in assessing the adequacy of relief system design, not HAZOPS, What-If/Checklists, FMEA, Fault-Tree, LOPA, etc. In fact, a PHA generally takes credit for proper relief system design as an inherent safeguard in mitigating or preventing loss of containment consequences.

If a PHA is going to take credit for relief system design as a safeguard, then the design must be accurate and up-to-date. Again, OSHA enforcement actions have revealed missing, inaccurate and incomplete process information as well as outdated relief system studies as an opportunity for improvement within the industry.

Accordingly, a facility-wide relief system analysis should be performed periodically and used to support the PHA process. This would not be unlike other hazard assessments which are commonly used to supplement or contribute to the implementation of a PHA, e.g., a facility siting analysis, human factors analysis, or site security vulnerability assessment. These studies are typically standalone hazard evaluations which compliment a PHA, and so too should be a facility-wide relief system analysis.

Flare/relief system modeling software like Siemens' Process Safety Pressure Protection Manager (PS PPM™) is typically used to perform a facility-wide relief system analysis, involving sizing calculations on multiple relief devices associated with inlet and outlet piping, and simultaneously accounting for header and flare stack hydraulics. Flare radiation effects and atmospheric dispersion modeling are also usually simulated. Such an analysis involves significant manpower expenditures, both by the consulting company and the operating company. An investment of this magnitude should involve the necessary controls to ensure continuing accuracy and data evergreening.

A facility-wide relief system analysis should be managed as process safety information (PSI) in order to properly support the PHA process. A PSI management system is a process, and as with any well-designed process, the control and integrity of data necessitates workflows which systematically delineate tasks from inception to completion, involving a series of checkpoints and controls along the way. From a PSM perspective, such checks are fundamental to PSM's management of change (MOC) and auditing elements.

As for the MOC element, comprehensive equipment-based checklists should be developed to manage any change impacting the relief system, and personnel should be trained extensively on what constitutes a relief system impact. Relief system specific auditing protocols should also be developed to provide for a higher level check of PSI accuracy and evergreening as well as the subsequent effect on the quality of PHA's.

Furthermore, metrics (or key performance indicators [KPI's]) should be utilized because *"if it can't be measured, it can't be managed."* The continued success of any program is contingent on developing smart KPI's and acting appropriately to promote continuous improvement.

Protocols need to be established which trigger the revalidation of existing pressure relief studies. The triggering events could be recommendations from the PHA team, recommendations from incident investigations, incidents that occurred in industry, or changes in design guidelines from industry or the corporate office. In addition to having an evergreening process, operating plants can also elect to revalidate their relief studies every 5 years in conjunction with PHA schedules. This revalidation is an additional layer of assurance necessary to ensure impact evaluations are conducted and PSI is updated to reflect process changes.

4. Evergreening Workflow

Maintaining an evergreening program involves a challenging yet manageable process. It requires commitment from all levels of management and employees, including contract employees. The program should include workflows and specific procedures to ensure changes which impact the existing relief system are evaluated, and associated PSI is appropriately tracked and updated. The workflow and procedures should interface and co-exist with the MOC system and should include steps for pressure relief valve (PRV) impact determination, evaluation, mitigation options, and documentation. Additionally, the evergreening program should also include elements for revalidation protocols, key performance indicators, audit protocols, and training requirements. Figure 1 is an example of a relief system evergreening workflow.

Figure 1: Evergreening workflow

5. Change Management Procedures

For a successful evergreening program, the program should also include clear procedures to manage changes that impact the existing relief system. These procedures should interface with the plant MOC process and include the following elements:

- Relief System Impact Determination
- Relief System Impact Evaluation
- Mitigation Options
- Risk Assessment and Management Plan
- Mitigation Approval Process
- Documentation
- Closeout

6. Relief System Impact Determination

After a MOC is initiated, a relief system impact determination is necessary to determine whether the projected change could result in an impact to the existing relief system. The MOC initiator and the process engineer responsible for the affected process should make the base line determination. A relief system impact checklist is an effective tool in assisting responsible parties with such a determination. Furthermore, a relief system impact checklist serves as a very useful training tool.

7. Relief System Impact Evaluation

An evaluation should be made of the impact on the current relief system imposed by the projected change (MOC). Process engineers responsible for the relief system should determine the scope of the impact. Software tools can ease the analytical process and make the evaluation much more efficient. Once the analysis is completed, plant personnel impacted by the change would be notified.

Again, software tools will simplify and facilitate the evaluation process. Analytical software should have communication and integration capabilities with external software packages, e.g., Visual Flow, FlareSIM, HYSYS, Pro/II, Meridium, and other applications.

8. Risk Management and Mitigation Plan

After the impact evaluation is complete, recommendations for modifications to the relief system should be consolidated and presented to the plant's management team for consideration and approval. For complex changes, a relief system specialist should be engaged to verify the recommendations and mitigation methods.

A risk management and mitigation plan should be documented and include a timetable for resolution (both temporary and permanent resolution) of relief system impact concerns. The plan should be implemented and remain active until all relief system study findings are fully resolved.

At a minimum, the plan should include the following:

- An assessment of the risk (likelihood and consequence) for the particular concern. Risk matrices are commonly used as guidance in assessing risk of unmitigated concerns (in-place risk).
- Based on this risk assessment, a corrective action plan describing how the concern should be managed and mitigated both temporarily and permanently.
- An assessment of residual risk relative to in-place risk given the implementation of mitigative measures.
- Presentation of the plan to the plant's management team for consideration and approval.

The resultant risk management and mitigation plan should be documented and managed as PSI and be included as part of the final relief system study revalidation report. This plan should also be made available for PHA implementation.

9. Documentation and Data Management Control

A documentation process should be developed to preserve and track all actions taken. These include actions taken for the impact determination and analysis, actions taken to train or notify affected personnel, actions taken to ensure PSI is updated, and actions taken by the approval team. For good record keeping and document tracking, a work flow tool would be beneficial in supporting the interactions between the relief system change management system and the plant MOC system (see Figure 2).

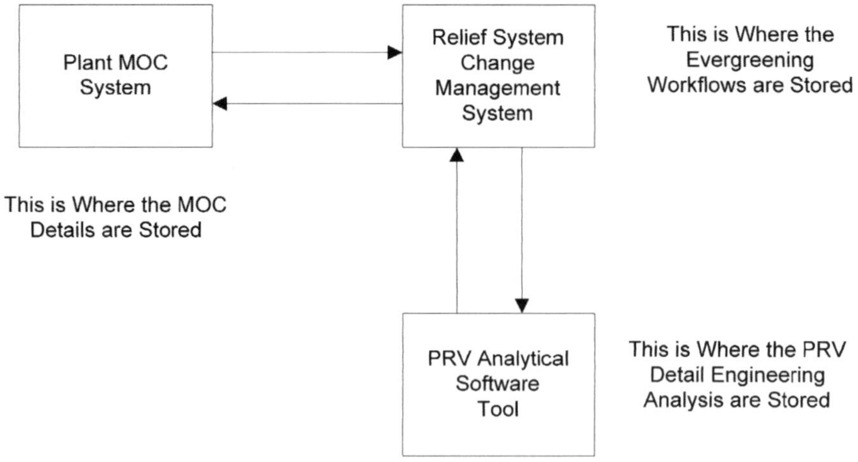

Figure 2: Interactions between plant MOC system, Relief System Change Management and PRV Tool

10. Relief System Review Closure and Documentation

The closure step is executed once the as-built information has been accepted and approved by the plant's management team. All action items associated with the relief system must be completed and documented, and process information must be updated and verified prior to closure of the relief system review. The relief system specialist should be engaged to verify that the performed actions meet all requirements.

It is essential that all action items and management decisions be documented, and that a system (preferably a software-based system) is utilized to track and manage final implementation of corrective actions.

11. Relief Study Revalidation Protocols

Based on specific criteria, the revalidation of a relief system analysis could take the form of a retrofit, update, or a complete redo. It is strongly suggested that revalidations of relief studies occur every 5 years in conjunction with PHA schedules. It is also suggested that a redo of a relief system study occur no later than the third cycle of the previous full relief system study. Regardless of how robust the management system associated with a particular PSM program and Management of Change process, a redo of a relief system study on the third revalidation cycle would appear appropriate to capture and address hazard scenarios or conditions which may have somehow emerged undetected. Over time, not even the most robust management systems can be expected to catch every variance or anomaly.

12. Key Performance Indicators

As previously mentioned, metrics (or key performance indicators [KPI's]) should be utilized because *"if it can't be measured, it can't be managed."* The continued success of any program is contingent on developing smart KPI's and acting appropriately to promote continuous improvement.

KPI metrics are measurable goals that stakeholders for an organization agree upon, usually mandated by management at the corporate level. KPI's provide feedback to the stakeholders and the responsible parties on the success of their programs. KPI's can be portrayed effectively by utilizing analytical software tools. An effective KPI system should include the following elements: success factors, targets, improvement actions, and performance measures.

The following are potential KPI metrics for a relief system evergreening program:

- Number of deficiencies in a given process unit.
- Breakdown deficiencies based on type.
- Compare type and number among other sites, both company and industry-wide.
- Track not-to-exceed (NTE) limits against relief limits (flaring events).
- Track number of pressure relief valve (PRV) activations.

- Track number of failed PRV activations.
- Track recommendation closure.
- Track types of scenarios and basis for each unit.
- Track throughput against design basis.
- Track number of MOC's that relate to relief systems as well as missed MOC's.

It is prudent to develop KPI's which can be incorporated into the plant's safety culture and strategy, such as the indicators listed in figure 3.

	MOC Initiated	PRV Study Conducted	MOC without PRV Study	Avg Lapsed Time for PRV Study	Open MOC and PRV Studies	Open PRV Action Items
FY2009						
FY2010						

Figure 3: Example of KPI's Associated with a Change Management System for PRV Studies

13. Pressure Relief System Management Program Audits

An evergreening auditing process cannot be successful without an unbiased auditing process. For continuous improvement and to help maintain the accuracy and availability of the pressure relief design basis, utilization of a *Three-Tier* auditing approach is recommended.

For *Tier I* audits, the plant could consider using the local site relief system specialist to conduct the audit. *Tier I* audits should be limited in scope and conducted on a frequent basis, e.g., monthly. These audits should be limited to simple reviews of the interaction of the site MOC workflow with the relief system management process and also accuracy reviews of the design basis for randomly selected relief systems.

For *Tier II* audits, the site could consider utilizing relief system specialists from other plants or engineering consultants with expertise in relief system review. A *Tier II* audit should be more comprehensive than a *Tier I* and should be performed at least annually.

As for *Tier III* audits, the site could consider using relief system specialists at the corporate level or engineering consultants with expertise in relief system design. *Tier III* audits should be a higher level review and conducted at least every three years in conjunction with the PSM compliance audit schedule.

The audits could include the following elements depending on the tier level of the audit:

- Interaction of the MOC workflow with the evergreening workflow for the relief system.

- Accuracy of the pressure relief design basis and data versus as-built for randomly selected systems of multiple process units.
- Review of KPI metrics as well as tracking and closure of action items.
- Knowledge and competency of the responsible parties about the relief system review workflow, checklists and evergreening process.
- A review of changes made following the last *Tier III* audit as well as the accuracy of information used and generated by the tiered relief system reviews.
- Throughput to the unit and latest pressure relief basis documentation.
- High PRV inlet/outlet pressure drop cases.

Audit reports should be generated and made available to affected personnel. The site management team should implement a plan of action to address the gaps identified. There should be follow up reviews to ensure corrective actions are properly conducted and on schedule for completion.

14. Training

Proper resources should be allocated to the evergreening process. Responsible parties should be properly trained on the relief system and its evaluation. The ultimate goal is to provide the requisite knowledge pertaining to the basis of plant relief and flare header system designs. Training should be conducted at the initial assignment and also periodically (annually) as refresher training. If an analytical tool is used to assist with PRV studies, responsible personnel should be trained on those tools. Subject matter experts also should be provided with training associated with applicable code and standards, such as API 520 and 521. This training is necessary to ensure the relief and flare header design always provides adequate protection for the process.

15. Conclusion

A pressure relief system is the last line of defense for overpressure protection, and it should be adequately maintained and fully capable of protecting the operating plant, its workers, and the environment. To maintain this capability, operating plants should implement an *Evergreen* Pressure Relief System Management Program. The evergreening program should include continuous revalidation of relief studies and continuous updates of process safety information. Refresher training, regular audits and key performance indicators should also be included as elements of the program. Operating plants should formulate a standardized evergreening process that can be replicated to fit the needs and achieve the goal of continuous understanding of the basis of the current plant relief and flare header design. To achieve these goals, the plan-do-check-act and continuous improvement aspects of a management system approach should be applied. Additionally, operating plants should utilize established tools such as Siemens' PS PPM™ to ease the analytical process and achieve the desired success of the program.

(79b) Emergency Relief Vent Sizing for Runaway Chemical Reactions Using Different Methods

Douglas Carson[1], Jean Pierre Bigot[2], Wassila Benaissa[1], Wilfried Minko[3] and Diogo Ferreira[2], (1) Direction des Risques Accidentels, INERIS, Verneuil en Halatte, France, (2)Centre des Sciences des Processus Industriels et Naturels (SPIN), …cole Nationale Supérieure des Mines de Saint-…tienne, Saint-Etienne, France, (3)IRSN

The sizing of emergency relief vents to mitigate runaway chemical reactions in batch or semi-batch reactors is important for the protection of the workers and the process unit. This has been a preoccupation in industry for many years and different design methods have been proposed depending on the type of system (gassy, vapor, or hybrid) and the flow characteristics of the reacting fluid. Industry needs simple, reliable and accurate sizing methods so as not to undersize nor oversize the vent area. In the present work, runaway reactions in two chemical systems were studied experimentally with various apparatuses in volumes from 30 µl up to 10 liters (DSC, PHI-TEC I, ARSST, VSP2, Vent Sizing Simulator, 800 ml reactor, and the standard 10 liter vessel used for the United Nations test for sizing rupture disks for vessels containing self-reactive materials). The two systems were the decomposition of a solution of DiButylPeroxide in butyrate and the esterification of acetic anhydride by methanol. The first system is virtually purely gassy, meaning that the overpressure in the vessel during runaway is produced by incondensable gas while the second is purely vapor, meaning that the overpressure is due to the vapor pressure of the reaction mass as the temperature rises. The data recorded were used to determine the A/V ratio for the emergency relief vent or rupture disk for a 1m3 vessel. Two categories of sizing methods were used. The similarity method was employed, whereby the A/V ratio is directly determined experimentally for a given set of conditions, with three reactors : 0.11, 0,8 and 10 liters. The calculation method was based on the DIERS approach and was employed using the data from bench-scale pseudo-adiabatic calorimeters. The results show that the similarity method predicts vent sizes which are much smaller than those predicted by the DIERS approach.

See more of this Session: New Relief System Solutions
See more of this Group/Topical: Topical 1: Global Congress on Process Safety

High-Fidelity Dynamic Modeling of Depressurizing Vessels and Flare Networks to Improve Safety and Reduce CAPEX

Dr James Marriott, Zbigniew Urban & Dr Stepan Spatenka
Process Systems Enterprise Ltd
6th Floor East, 26-28 Hammersmith Grove, London W6 7HA, United Kingdom
j.marriott@psenterprise.com

Keywords: flare, blowdown, dynamic simulation

Abstract

The use of dynamic modeling for relief system design can result in a considerable reduction in capital expenditure while simultaneously improving plant safety. This paper considers the application of dynamic analysis to two areas, vessel depressurization (or “blowdown”) and flare network design.

The paper first presents a next-generation model for vessel depressurization (or “blowdown”) calculations. The model incorporates an accurate description of the non-equilibrium interactions among the various phases and between these phases and the vessel wall. It also incorporates a 2-dimensional model of the metal walls taking account of the transfer of heat between regions of the wall in contact with different phases. This allows a more accurate estimation of the wall temperatures, and the direct computation of thermal stresses. The model is shown to be validated against the set of experimental data obtained from a full-scale vessel, as reported by Haque et al. (1992) and Szczepanski (1994).

This paper also describes an advanced system for model-based flare network design that addresses the above issues while being compatible with existing steady-state flare network technology. The system supports steady-state and dynamic analysis as well as wall temperature modeling within a single integrated framework.

1. Introduction

The Process Industries face significant challenges in driving down the capital cost of pressure relief systems while ensuring that they remain within the requirements of the relevant standards. Modeling is widely used to support the design of such systems. However, there is high degree of uncertainty in the results due to the simplified models that are generally applied. For example: steady-state tools are invariably used for flare network design; and lumped equilibrium models for vessel depressurization calculations.

In this paper, the use of next generation dynamic and rigorous modeling solutions is explored on both the process-side to model an emergency depressurization event and on the flare network side.

2. Depressurization of process vessels

The detailed dynamic modeling and simulation of the rapid depressurization (“blowdown”) of high-pressure vessels is a key element of the safety analysis of oil & gas production plants and other high-pressure installations. This depressurization not only determines the load imposed on the pressure relief system (e.g. flare network) but, more importantly, may result in significantly reduced temperatures of the vessel walls, which may lead to embrittlement and high thermal stresses.

Key questions include:

- **What are the relief loads entering the flare network**? For new designs, accurate information is needed in order to achieve an optimal design that minimizes the piping diameters required to meet Mach number and back-pressure constraints. Minimizing the piping sizes also provides benefits in terms of reduced support infrastructure, which can be particularly important in the case of offshore platforms where additional weight is heavily penalized. For revamps or expansions to existing process plant, accurate data can help determine whether the current flare system can handle the new loads acceptably. In either design scenario, CAPEX savings can be considerable.

- **Where is the temperature of the process and pipework metal walls too low?** This information is necessary in order to identify areas of potential embrittlement, and where (and when) unacceptable thermal stresses are likely to arise. Such information can be used to design out potential problems either by controlling the relief rates or by rerouting the relief flows.
- **What is the temperature of the relieving 'gas' streams?** (these streams may actually contain evaporating entrained liquids). This provides essential information for choosing the appropriate material of construction for the critical sections of pipework immediately downstream of the BDV.

The effects of low temperature can usually be addressed by using suitable materials of construction. Unfortunately such materials are expensive and it is highly desirable to minimize their use without compromising safety considerations. This requires accurate quantification of flowrate and temperature of the relieving stream, as well as the minimum temperatures reached in the metal walls.

Modeling requirements

The depressurization of a vessel involves a complex set of coupled physical phenomena which need to be characterized accurately in order to understand behavior and provide suitable design values.

Current depressurization modeling is often performed with off-the-shelf process flowsheeting simulators that use equilibrium thermodynamic methods. The latter provide some indication of the flow and temperature, but by no means adequately describe the complex thermodynamic phenomena occurring as a result of the rapid decrease in pressure. The fact that multiple phases can form within the vessel, and that these may not be in equilibrium with each other or with the vessel wall, can have a significant effect on both the relief flows and the metal temperatures of the vessel and relief system pipework.

The sudden decrease in pressure in a gas-filled vessel results in a rapid change in the thermodynamic state of the gas within the vessel. This can result in nucleation of liquids within the gas bulk to form a 'droplet phase'.

Some of the nucleated liquid leaves as entrained droplets in the high-velocity gas exit stream. Downstream of the vessel, and as the pressure further reduces, this exiting entrained liquid evaporates into the bulk gas stream, lowering the temperature of the cold exiting stream even further. This creates a risk of brittle fracture of the flare system pipework.

Some proportion of the liquid remaining in the vessel drops to the vessel floor. Initially this evaporates instantly because of the relatively warm temperature of the metal it encounters; the effect is similar to a drop of water falling on a hotplate. However, once the metal has cooled sufficiently (typically after a few tens of seconds), liquid begins to pool to form a continuous liquid phase.

The pool boils vigorously, cooling and reducing in size, and in turn reducing the temperature of the metal beneath it. This can lead to significant temperature differences between the metal immediately below the pool and its surrounds, presenting a very real threat of brittle fracture and rupture from the base of the vessel.

Of course this is only one possible scenario - other scenarios may develop depending on the initial inventory and state of the material in the vessel. For example, there may be 'bubblet' nucleation in liquid, rather than the droplet nucleation described here.

Model Validation

The model applied here describes all of the phenomena mentioned previously; it incorporates

- global mass and energy balance between the phases present and the vessel wall, at every stage of the calculation

- rigorous calculation of non-equilibrium interactions among the various phases
- rigorous thermodynamics through the use of equations of state common for oil and gas mixtures and able to predict phase behavior with good accuracy over a wide range of temperature, pressure and composition including the supercritical region
- a 2-dimensional description of the metal walls, including heat transfer between regions of the wall in contact with different phases
- axial, radial and tangential stress calculations based on a temperature map
- any number of vessels linked by pipework.

Implemented in Process Systems Enterprise Inc.'s gPROMS modeling and simulation environment, the model has been validated against the set of experimental data obtained from a full-scale vessel, as reported by Haque et al. (1992) and Szczepanski (1994) – Figures 1 & 2.

Figure 1 – Comparison of gPROMS depressurization model against experimental data reported by Haque et al. (1992) and Szczepanski (1994) – Experiment S9.

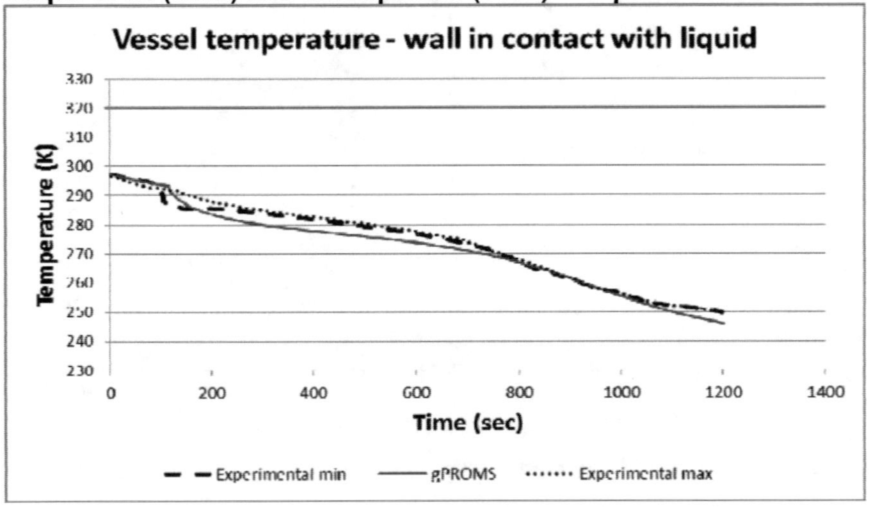

Figure 2 – Comparison of gPROMS depressurization model against experimental data reported by Haque et al. (1992) and Szczepanski (1994) – Experiment S9.

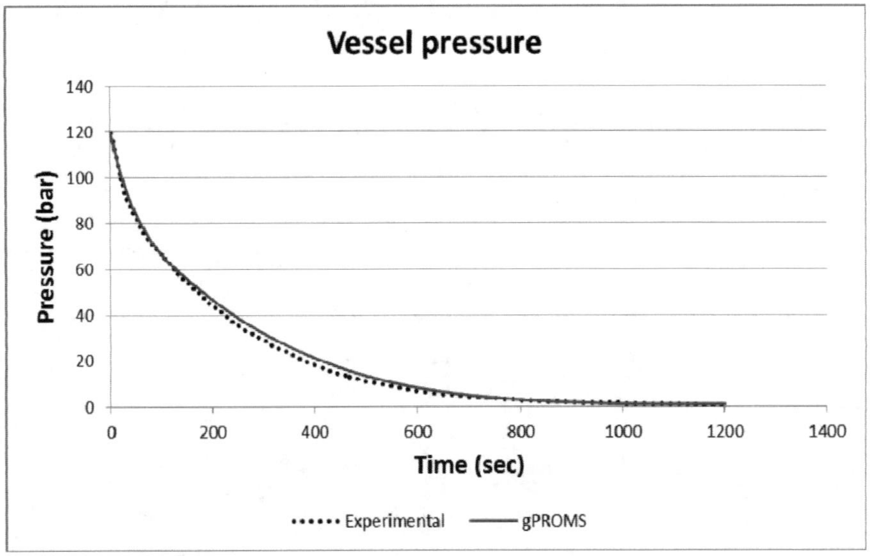

Case study - determining the lowest temperature

Results for the depressuring of a vessel filled with light hydrocarbon gas at 120 bar are shown in Figure 3. In the initial phase of depressurization, the gas temperature (black line) drops rapidly. The temperature of the metal wall in contact with the gas (green line) begins to drop too, but much more slowly because of resistance to heat transfer between wall and gas. After about 80 seconds, a droplet phase begins to form throughout the gas. Initially, droplets in contact with the metal heat up rapidly and vaporize (red spike). When relatively cool liquid droplets (at the temperature of the bulk gas) begin to pool on the vessel floor, the liquid temperature increases further above the bulk gas temperature as the liquid is heated by the metal wall and its composition changes. After a while, the gas bulk temperature begins to rise because of heat influx through the metal wall.

Figure. 3 –Temperature profiles over the duration of the blowdown

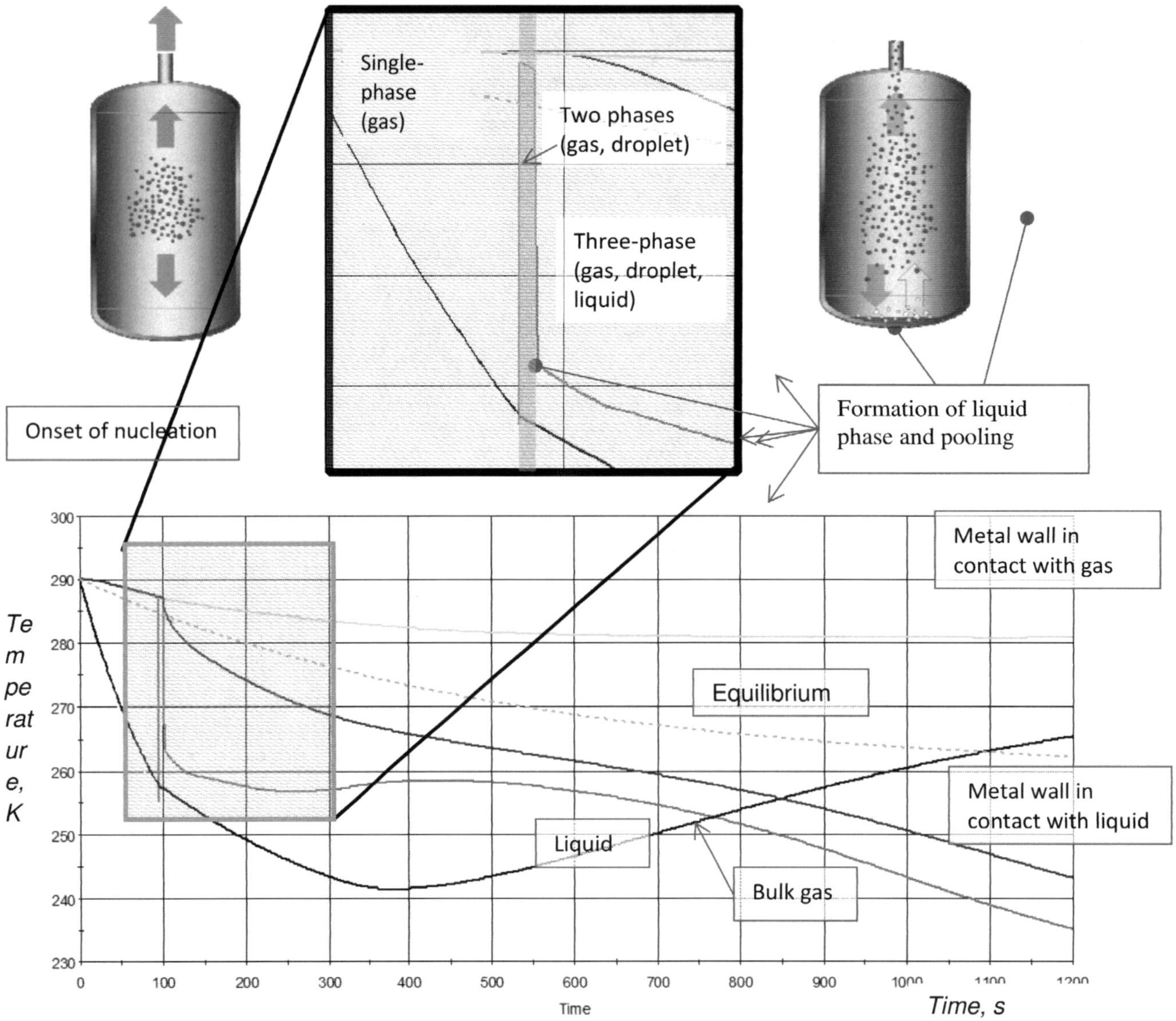

The items of most concern in Figure 3 are the rapidly decreasing temperature of the metal in contact with the liquid pool (blue line) and the difference between the temperature of this metal and the adjacent metal in contact with the gas (green line). The metal temperature can be seen to drop to nearly -30 C, introducing the risk of brittle fracture. The temperature difference between vessel floor and sides rapidly increases to over 20 degrees and is nearly 40 degrees by the end of the blowdown, which may give rise to unacceptable stress.

Because of the rapid change of conditions the three phases co-existing in the vessel (gas, droplet and the pool of liquid) and the vessel wall are not in equilibrium with each other throughout most of the blowdown event. In comparison, the dotted line shows the equivalent temperature curve obtained using an equilibrium model for the same blowdown, which predicts a much less severe drop in temperature and fails to identify the most significant safety-related aspect – the cooling effect of the liquid on the vessel bottom.

3. Dynamic analysis of flare network systems to reduce capital cost

The curves obtained from the type of study described above provide important input information for accurate flare network design.

The flare networks for major plants represent a non-negligible part of the overall capital investment. Current industrial practice for their design is primarily based on steady-state analysis, as described in standards such as API 521 and supported by widely-used software tools. However, it is a recognized fact that the application of steady-steady considerations to what is fundamentally a dynamic system inevitably requires the use of conservative assumptions which often result in significant oversizing of flare headers and other components of the network. Another major contributor to capital cost is the use of special materials for the parts of the system that may be exposed to low-temperature fluids – typically the tail pipes attached to high-pressure process vessels – and which are therefore at risk of embrittlement (as considered in the previous section). The key to being able to limit the additional capital expenditure is the accurate estimation of the length of piping that is subject to “abnormal” temperatures.

Conventional flare header design techniques use peak relief flows in steady-state simulation to assess system capacities and determine back-pressures downstream of blowdown valves (BDVs) and pressure safety valves (PSVs), Mach number in the headers and radiation at the flare tip.

This steady-state assumption is highly conservative. While conservative approaches may seem desirable in safety system design, they can nevertheless lead to gross overdesign throughout the system:

- **Oversized flare header**. Sizing the header for the sum of the maximum flows takes no account of effects such as:
    - system packing, where the gas pressurizes the available volume in the flare network
    - potential for sequencing of flare events. For example, depressurization initiated deliberately by an operator may be complete well before a fire causes PSVs to lift. Steady-state peak flow analysis, on the other hand, assumes that all events occur simultaneously.

  Reducing the peak flows used as the design basis by judicious analysis can significantly reduce the pipe sizes and hence materials and fabrication costs, which can be substantial for large-diameter headers. Reducing the size also creates knock-on savings related to the support structure and flare stack size.

- **Oversized flare stack**. The flare stack sizing depends on radiation emitted by the flame, which is a function of the volumetric flowrate of gas through the flare tip. Using unrealistically high flowrates determined from peak flows results in an over-long stack, creating weight problems in offshore facilities or adding stack support costs (or unnecessary additional header length) in onshore ones. Similarly a lack of accurate temperature information leads to a wide span between the minimum and maximum design temperatures used for gas arriving at the stack, resulting in unrealistic allowances for thermal expansion and contraction.

- **Over-use of expensive alloys for low-temperature service**. Though flare system pipework may be in contact with gas at extremely low temperatures, this typically occurs for a relatively short duration. The use of steady-state flows takes no account of the duration of such exposures to low temperature, which may result in very conservative (and hence expensive) requirements for use of alloys.

It can be argued that a good flare network design is one that minimizes capital expenditure while meeting all safety constraints. Hence overdesign should be avoided wherever possible.

Applying Dynamic Modeling to flare network design

An advanced model-based system for flare system network design that addresses the above issues while being compatible with existing steady-state flare network technology has been developed and implemented in gPROMS. The system supports steady-state and dynamic analysis of the flare network, wall temperature modeling and prediction of hydrate and ice formation within a single integrated framework.

By making relatively simple dynamic analyses using data that is mostly already available in some form, it is often possible to refine network designs to arrive at systems with a significantly lower capital cost but demonstrably meeting safety requirements. Similarly, it is often possible to find additional capacity during retrofit, thus removing the need for additional capital expenditure.

Typical examples of where dynamic analysis can bring significant new information that has an impact on capital cost are:

- **Peak flow rates.** The actual relief flow through any PSV is at the maximum only for a short period of time. Using steady-state methods based on peak flows is equivalent to making the assumption that all relief flows start at the same time and go on forever. In reality it is often possible to take credit for staged or staggered relief, in which different units are relieved at different times. Shifting depressurization of certain units by a few tens of seconds can make a significant difference to the peak flows through the system – an effect that cannot be represented at all by steady-state simulation.

- **Packing.** Steady-state approaches make the implicit assumption that the flare system has no volume – what goes in comes out, instantly. For larger systems, the impact of relief flows is partially 'absorbed' by pressurization of the flowing lines and the dead volumes in non-flowing parts of the system. This 'packing' effect can reduce both the calculated peak back-pressures or Mach numbers and the peak flows seen at the flare tip, allowing reduction in header and tailpipe diameters and flare stack length respectively. Dynamic simulation allows this important buffering effect to be taken into account in the design, as can be seen in the example below.

- **Duration**. Equally important, dynamic simulation can be used to determine the duration of peak flare loads. Engineering judgement can then be used to assess the risks of any infringements; for example a 5-second violation of back-pressure or radiation constraints may well be acceptable especially given the capital costs of oversizing the flare system to avoid such a contingency.

- **Temperature**. Relief system pipework that is likely to encounter low temperatures needs to be constructed from expensive alloys such as Inconel to avoid the possibility of embrittlement and consequent fracture. The true extent of pipework that needs to be constructed of such materials is impossible to gauge with steady-state simulation, as low-temperature flows are considered to continue 'forever', ensuring that calculated metal temperatures reach their minimum. In reality such flows may only last for a few minutes; the thermal inertia of the pipework metal and heat gain from the environment prevents the pipework from reaching the gas temperatures during this time (a similar effect can be seen in the bulk gas and metal temperature plots in Fig. 3 above). It is frequently possible to reduce the use of alloy significantly based on the more accurate information from the dynamic analysis

- **Flare stack temperatures.** Dynamics can also help provide a true picture of the temperature of gas arriving at the flare tip. Proper calculation of the effect of low temperature gas over a short time period taking into account metal thermal inertia and ambient heating provides much more accurate minimum and maximum design temperatures, allowing the designer to make sensible decisions on stack length and support mechanisms.

Case study – flare system capacity

Consider the flare system shown in Figure 4, where the header sizes are set primarily by a depressuring scenario from five units simultaneously, as highlighted in the figure. One of the key questions is the size of the long main header pipe leading to the flare stack. Typically such a system is designed by working back from the flare tip, sizing all the lines based on velocity constraints until reaching the relief valves, and then confirming that other constraints such as back-pressure constraints at the relief valves and limits on noise are not violated.

Figure 4 – Example flare network, showing active sources

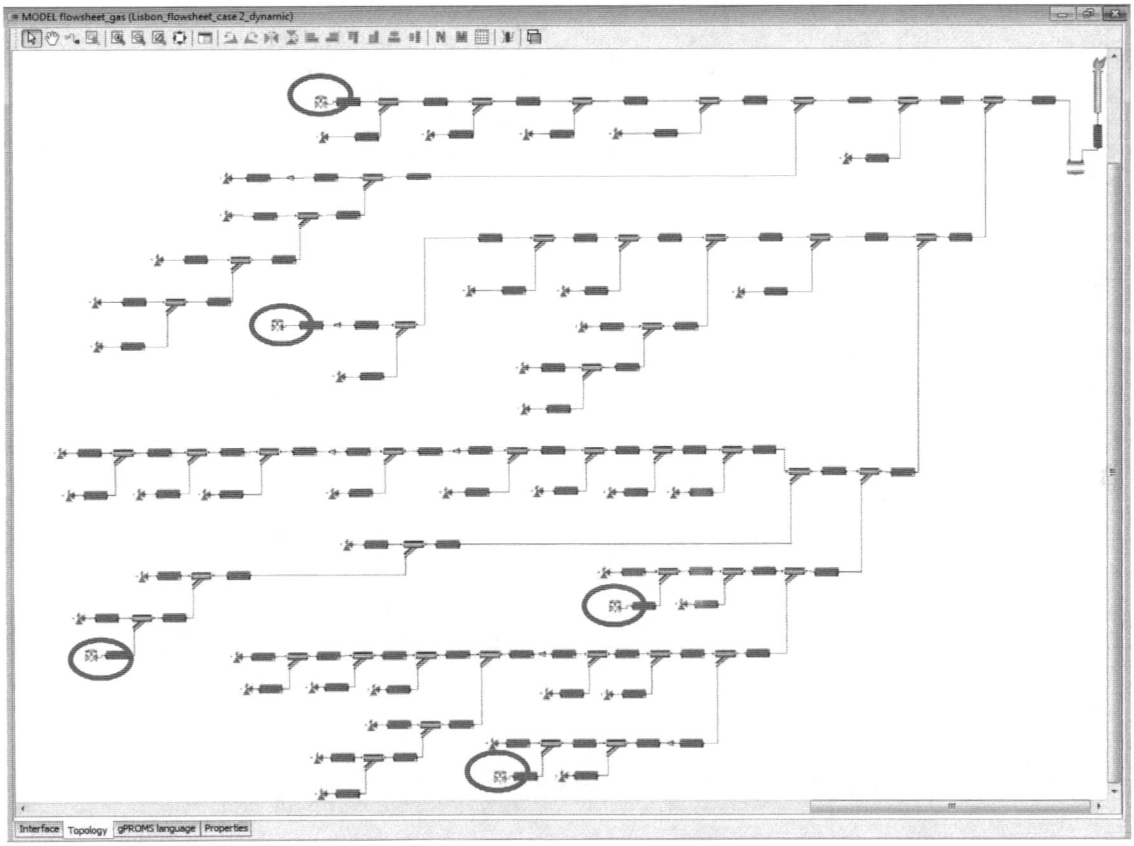

In this study, a system has been designed using steady-state techniques using a velocity heuristic requiring a Mach number between 0.25 and 0.35 in the main lines. A conventional steady-state approach calculates the Mach number using the sum of the peak flowrates. The maximum Mach number at the pipe outlet) is represented by the dotted line in Figure 5. At 0.29 this is well within the 0.25-0.35 range. The back-pressures at the five blowdown valve sources are well below the limit in this case.

For illustration it is useful to perform a 'pseudo-dynamic' run using relief flow curves but taking no account of the flare system volume. This shows a Mach number profile over time that has the characteristic sharp-peaked shape of relief flow curves (Figure 5, red line). As expected, the peak Mach number from this run is the same as for the steady-state case, at the sum of the individual peak flows (Figure 5, dotted line).

Though this case adds no new information to the design, it does provide some indication of the length of the blowdown event, allowing judgment to be applied in the case of constraint violation.

If a full dynamic simulation is performed, taking the volume of the system (both for active and inactive branches) into account, it can be seen that the effect of flare system packing significantly reduces the peak Mach number observed, to about 0.25 (Figure 5, black line). It is evident from these results that there may be potential to reduce the diameter of the 24" header, as the Mach number is nowhere near its limit.

A new series of calculations is performed with a 20" header diameter to see the effect of reducing the flare system line sizes. As expected, the Mach number obtained using steady-state peak flows (0.4) violates the system design constraints, indicating that the design is not viable (Figure 6, dotted line). The corresponding pseudo-dynamic case (Figure 6, red line) shows that the value is out of range for about three minutes, which is also unacceptable. However a full dynamic simulation taking into account line packing shows that the peak Mach number is within the 0.35 limit (Figure 6, black line); the back-pressures (Table 2) remain well within limit. The added information provided by the dynamic simulation thus indicates that the design is indeed viable. If further mitigation is required, it is possible to investigate the dynamic effects of staggering the depressurization, so that units depressure in sequence. The green line in Figure 6 shows the effect on Mach number of delaying the blowdown of unit 2 by several minutes.

This design study shows that the new and more accurate information provided by dynamic simulation indicates that a smaller piping system could be installed without violating the safe operation design constraints. Perhaps an even more important use of this approach is for plant revamps where an increased plant capacity or new equipment may require an expensive new flare network - unless by dynamic analysis it can be determined that the existing network has sufficient capacity.

Figure 5 – 24" header case

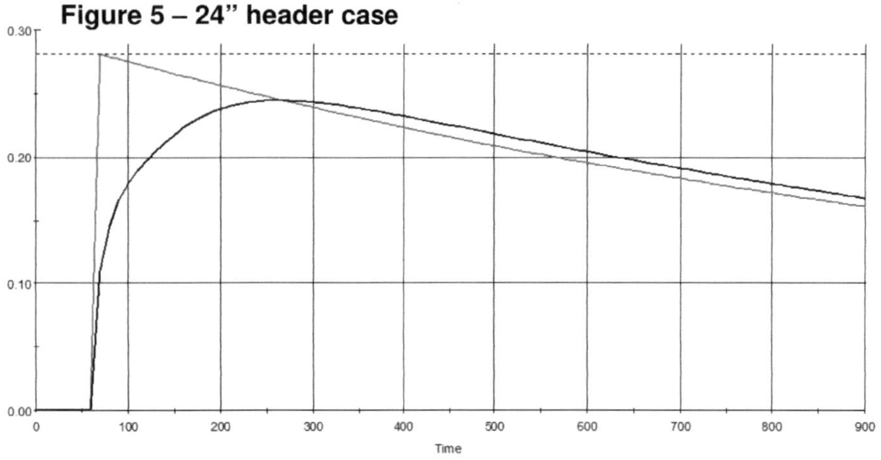

Figure 6 – 20" header case, also showing staggered blowdown

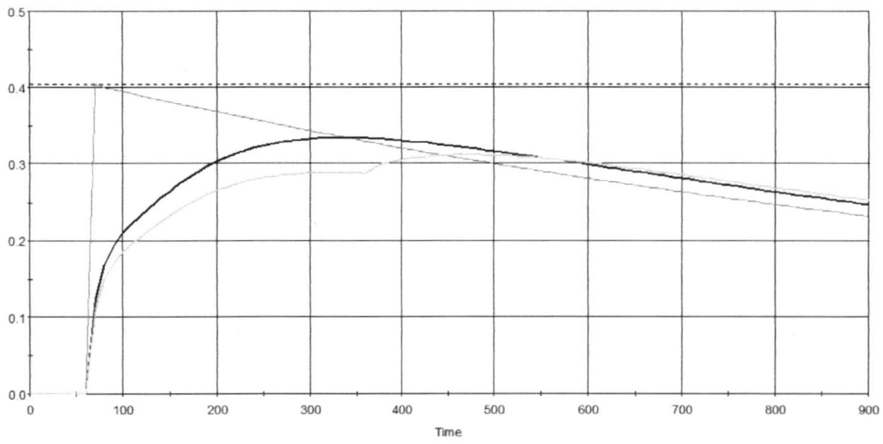

4. Conclusions

Rigorous dynamic modeling is critical for determining the safe design of both the process equipment and flare networks in the event of emergency depressurization, which is an inherently dynamic and in many cases thermodynamically complex process.

In the case of depressurization, rigorous dynamic simulation identifies potentially dangerous situations. For new flare system designs, it can lead – among other benefits – to a reduction in header size, resulting in significant capital savings. For existing headers, it provides a means to establish whether there is sufficient capacity to accommodate new sources, thereby avoiding the need for a new header and flare.

References

1. M.A. Haque, S.M. Richardson, G. Saville, *Blowdown of Pressure Vessels. I – Computer Model*, Transactions of the Institute of Chemical Engineers Part B: Process Safety Environmental Protection, 70(BI), 1 (1992)

2. H. Mafgerefteh, S.M.A. Wong, A numerical blowdown simulation incorporating cubic equations of state, Comput. chem. Engng., 23, 1309 (1999).

3. Gruber, D., D.-U. Leipnitz, P. Sethuraman, M.A. Alos, J.M. Nougues, M. Brodkorb, Are there alternatives to an expensive overhaul of a bottlenecked flare system?, Petroleum Technology Quarterly Q1 2010.

4. Chen, F.F.K., R.A. Jentz, D.G. Williams, Flare System Design: A Case for Dynamic Simulation, Offshore Technology Conference, 4 May-7 May 1992, Houston, Texas

5. Goyal, R.K., E.G. Al-Ansari, Emergency Shutdown devices and relief system sizing and design in oil refineries, Hydrocarbon World Vol 4 Issue 1, Touch Briefings (2009)

6. A Speranza, A. Terenzi, Blowdown of Hydrocarbons pressure vessel with partial phase separation, Series of Advances in Mathematics, available from http://www.i2t3.unifi.it/upload/file/Articoli/animp_2004.pdf (2005).

7. R. Szczepanski, Simulation programs for blowdown of pressure vessels, IChemE SONG Meeting (1994).

8. Kuo, A., R. Pitt, Flare line failure case, what have we learned? 16th Annual Ethylene Producers Conference, Session T8001 2004 Spring National Meeting, New Orleans, LA, April 25-29 2004 Paper 62f (unpublished)

Performance-Based Operator Training & Human Factor KPI

Don Nguyen
Siemens Energy, Inc., Houston, TX
don.nguyen@siemens.com

Keywords: operator training, key performance indicator, distribution control system

Abstract

Operator competency has been a major challenge for the petrochemical and refining industry. In recent days, human factors are believed to be the cause of many major process incidents. The system for determining whether an operator is competent to operate a process independently is a subjective process, often based solely on test scores and the opinion of the trainer. Tests are subjective because of the designer's bias, and the opinions of the trainers are limited to their knowledge and experience.

Since the promulgation of the OSHA PSM requirements in 1992, the training of process operators has somewhat evolved. Now, more companies use computer based training (CBT) rather than the conventional classroom training. Despite the standardization of training efforts and CBT systems already in place, PSM incidents continue to occur in refineries and chemical plants; human error continues to be one of the major contributing factors in such incidents. Human errors stem from the performance at the first respondent level.

The 2005 BP incident demonstrates that breakdown of human performance at the first respondent, i.e., an operator's level can impact safe operation of processes. Eliminating these incidents by improving human performance has been a big challenge for industry.

1. Industry Approaches to Training for Process Operators

Training is required for both new operators and seasoned operators. Training is typically conducted in classrooms, on computers, or hands-on via on-the-job training. In a refinery or a chemical plant, the primary objectives for training are:

- To assure operators are provided with the minimum performance knowledge for successful execution of job tasks
- To assure reliable human performance of critical, abnormal and emergency tasks
- To improve human performance
- To assure workers are aware of safety and health hazards associated with the process

Training for new operators assures that they are provided with the minimum knowledge to perform the job safely and successfully. Training for existing operators may be for closing the gaps between the operator's knowledge and the job, or it may be for refresher purposes. Subsequently, once training is provided, performance tests are administered to assure the operators in training understand the training materials and are capable of applying the training in a practical manner. Performance tests can be paper-based, computer-based, or conducted through verbal communication. Training is required to be completed prior to releasing the new operator to work independently. New operators must demonstrate that they understand the training and are capable of operating the process under all situations, including abnormal and emergency conditions.

Developing training materials is a complex process because operators in training have different educational backgrounds, technical experiences, knowledge and skills, and working cultures. In addition, people learn at different degrees. Training materials are usually compiled and developed by selected operators with many years of experience.

These operators typically use their knowledge, skills and experiences to develop the training materials; therefore, materials are usually subjective to the perspective of the trainers. The operational performances of the operators are not normally integrated into the training process.

2. Computer-Based Training

With today's global competitive market, refineries and chemical plants are relying more on software tools to become more efficient and competitive. Training is no different. Companies are utilizing computer-based training to be more efficient and save costs. With some initial capital investment, computer-based training will eventually yield the desired cost savings. As for proficiency, the quality of the training and comprehension may be sacrificed with computer-based training because it lacks the necessary human interface. Human interaction is a vital part of the training process, especially with today's diversified and multicultural workforce. Computer-based training lacks this vital human interface element.

3. Performance-Based Evaluation

Once an operator is released to work independently, his lack of knowledge or understanding of the process may not be obvious until his action or inaction results in costly mistakes. It is vital to implement some type of human performance assurance or follow-up program to ensure the new operator is performing at a safe and acceptable level. This requires multiple monitoring sessions for proper assessment. The normal practice has been to conduct periodic follow-up sessions that consist of informal conversations with the new operators. However, operating tendencies and behaviors are not evaluated because it is infeasible to shadow the new operator throughout the shift to make a reasonable assessment. If it was feasible, then the tendencies and behaviors of the operator can be analyzed. Any deviations from the norm could be corrected and thus, future mistakes and mishaps could be avoided.

4. Key Performance Indicator (KPI) for Training

One reason for developing Key Performing Indicators (KPI) for training is to measure the effectiveness of training programs so actions can be initiated for improvement. Conventional KPIs for training have been the number of trainings conducted annually compared to the number of overdue trainings. Test scores are also used to measure the success level of a training program. Historically, KPIs associated with operator performance have not been an effective proxy to measure operator performance and the lack of suitable KPIs remains a challenge for operating plants. The challenge is to develop a KPI system that can directly measure operating performance and tendencies so the information can be used to correct and reinforce good training strategies and improve performances.

According to the government investigators of the 2005 BP Texas Refinery accident where multiple workers were killed, the operating teams of the ISOM unit exceeded safe operating limits multiple times during prior startups of the unit. There were also evidences suggesting process operators have a tendency to operate on the high side of the tower to avoid possibility of damaging the unit furnace. Deviations associated with undesired tendencies should have been monitored, evaluated, and properly communicate to prevent reoccurrence.

5. Safe Operating Limits

Figure 1 illustrates an example scheme for safe operating limits. Processes and process equipments are required to have "safe operating limits." Safe operating limits as shown in Figure 1 should be determined using a qualitative evaluation approach among others. Once determined, they should be prioritized. Depending on the operating scheme of the process, some processes are equipped with "point-of-no-return" operating limits, where the process must be shut down (or placed in a "*safe-state*") and operator should not try to intervene for the process to remain in operation. A risk value should be assigned to each of the safe limits based on their associated consequential risk.

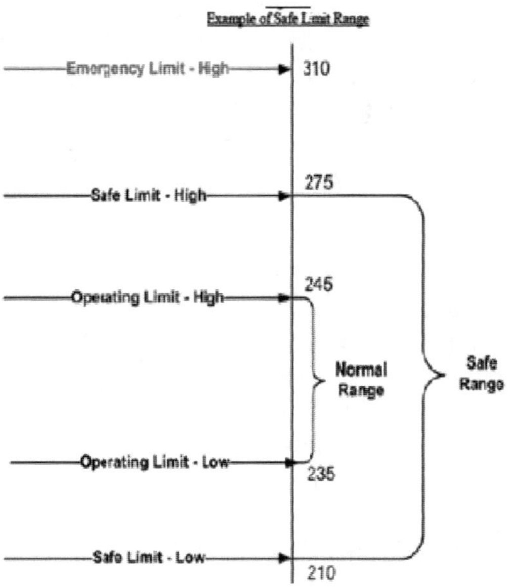

Figure 1: Illustration of different safe operating parameters.

As shown, within the ranges of the "*safe range*," operator intervention to maintain operation is allowed to return the process back to normal status (see Figure 1). The number of times an operating team exceeds "*safe range*" and how rapidly the process is returned to normal ranges depends on: experience, knowledge and skill, and training of the operating teams. Some console operators have undesired tendencies of operating the process too close to the "*point-of-no-return*" boundaries. This risk inadvertently shuts down the process and could create unnecessary risks to the process. Therefore, it is essential to evaluate these operating tendencies and behaviours and incorporate the results in the facilities' training and KPI system.

Solution

Most facilities use the Distributive Control System (DCS) to control and monitor process activities automatically. A typical process can have thousands of equipment pieces such as: heat exchangers, valves, pumps, pressure vessels, relief valves, etc., interconnected by piping. Each piece of equipment can have several safeguarding parameters and operating limits. Multiple instruments measuring variables like pressure, temperature, and flow are wired to the DCS for the purposes of controlling, monitoring, and archiving of the data. Typically the information is not evaluated unless there is an incident to determine "*what went wrong*" with the process. The same concept can be used to evaluate the performance of operating teams. Operator tendencies and performances can be evaluated using information available from the DCS). With careful consideration, operator performance of different conditions can be regularly assessed using DCS data. Critical deviations can be discovered revealing operator tendencies and process instabilities. After revealing all process deviations, further analysis can be performed to determine whether the deviations were the result of human error or process instabilities. The number of human caused deviations can be used for KPIs and for establishing refresher training material for process operators. Each deviation is evaluated qualitatively and assigned a risk value based on potential consequences. The risk value is subsequently compared to the acceptance risk matrix shown in figure 4.

The following charts and graphs show examples of information that can be extracted from the DCS to outline operating practices of different teams. Figure 2 shows pie charts indicating the percentage of: normal operation, operation outside normal ranges and operation outside safe operating limits. The charts show information for shift A, B, C, and D. With careful consideration, this information can be used to evaluate performance, competency, and effective of training programs. The information can also be utilized to develop goals and KPIs. For example, facilities can set improvement goals to operate their process within a safe range 90 percent of the time in the next 12 months.

The graphs from Figure 3 provide similar information with the pie charts. The graphs point out the specific weeks and shift of deviation occurrences. The graphs provide similar information for all 4 operating shifts, giving the reviewer an opportunity to assess the competency and performance of process operators and effectiveness of the facility's training programs.

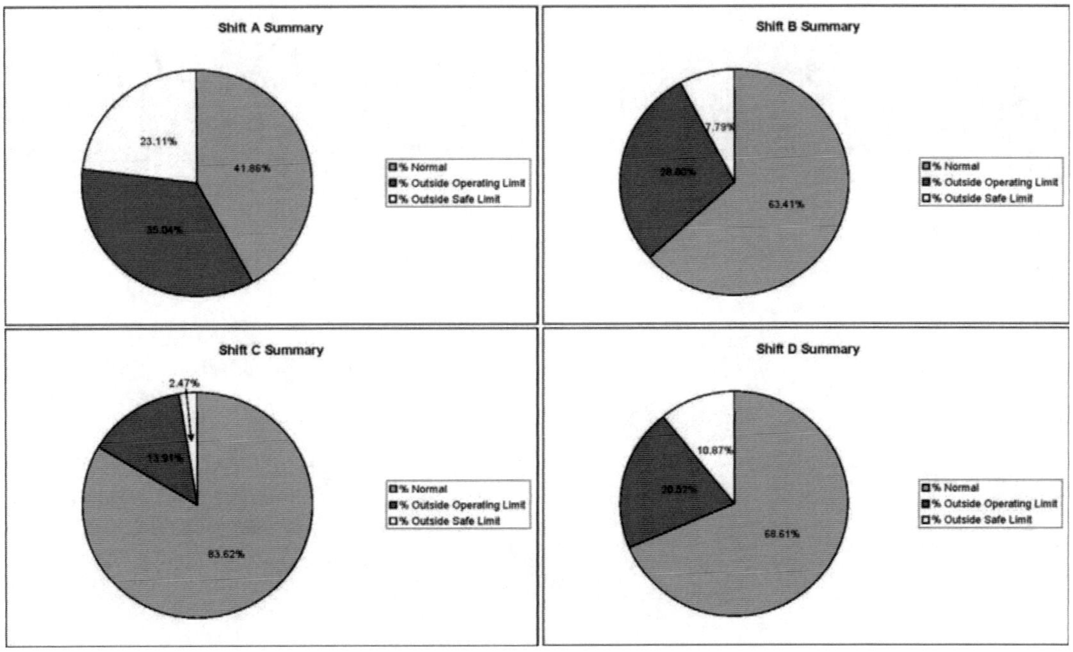

Figure 2: Pie charts showing deviations percentages of different operating shifts

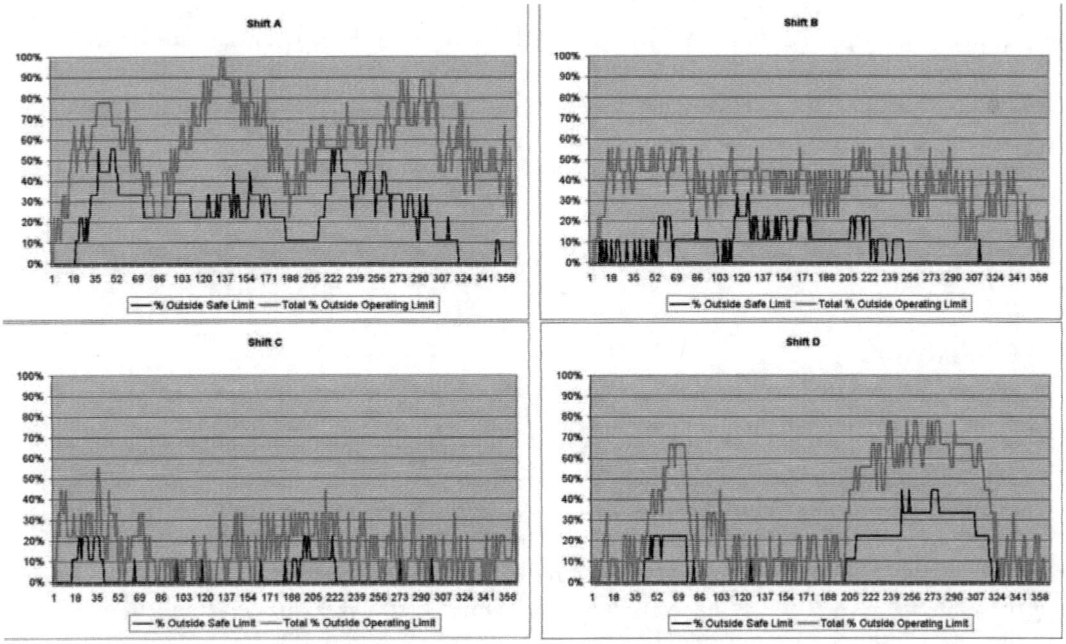

Figure 3: Graphs indicating deviations for each of the shifts. The outside limit includes outside of safe operating limits

A	**N**	**U**	**U**
A	C	N	U
A	A	C	N
A	A	A	C

Typical Risk Matrices		
A	Acceptable	No risk control measure are needed
C	Acceptable with Control	Risk control measures are in place
N	Not desirable	Risk control measures should be introduced within a specified time period
U	Unacceptable	Risk control measures should be introduced ASAP

Figure 4: Risk Acceptance Matrices

Performance-Based Training system utilizes objective process data stored in the DCS (*Distributive Control System*) to generate operating profiles of the operating teams. Significant deviations and trends discovered from DCS data are evaluated and a near miss investigation generated. Deviations associated with human performance are further evaluated. A qualitative approach is used to evaluate the information and a risk acceptance tolerance evaluation applied to determine high risk tendencies of the operating teams. Tendencies deemed high risk are further evaluated to determine the root cause. If operating proficiency is the issue, additional training or counseling is recommended to alleviate the issue. If the causal factor is related to deficiencies in the process, an additional safeguarding strategy is recommended to optimize the process. The information can also be used to develop Human Performance Indicators to measure operator performances.

6. Conclusion

There are major deficiencies with the current industry practices associated with operator training. Every year, operating companies spend hundreds of thousands of dollars to train and retrain process operators to establish adequate competency. In spite of this costly process, incidents continued to occur where human errors are contributing factors. Industry should consider a new training concept that integrates actual operational performance data with existing conventional training. Actual performance data archived in the DCS system is used to identify process deviations and undesired operating tendencies. After the deviations have been identified, they are further evaluated to identify undesirable operator behavior. The information is profiled and a risk acceptance concept is applied to identify the high risk tendencies of operating shifts so that training strategies can be developed using the information. Performance indicators based on this concept can be developed and used to improve process safety.

Jet Fire Protection -Where Risk Based CFD Analysis Pays Off

Eric Peterson
Scandpower Inc., Lloyd's Register Group
Houston, Texas
epe@scandpower.com

Chaitanya Sheth, Sverre Nodland and Julie White
Scandpower Inc., Lloyd's Register Group

Keywords: CFD, Process Safety, Process Engineering Design, Jet Fires.

Abstract

Chemical and refinery industry operators and engineering designers continuously search for means to improve plant design while maintaining or improving inherent safety. Computational Fluid Dynamics (CFD) use in the early engineering design stages of oil and gas process and plant facilities can produce counter-intuitive savings and improvements. Using detailed 3-dimensional CFD fire simulations with risk assessment models improve and optimize process plant design solutions in terms of fire risk. This is demonstrated with the jet fire scenario.

1. Introduction

In the quest to improve plant design while maintaining or improving inherent safety, chemical and refinery industry operators and engineering designers have begun to use CFD in the early design stages of oil and gas process plant design. Traditional 2-D modeling software tends to produce overly conservative results which are unable to account for changes in process plant structure, equipment and topography. CFD has been used extensively for fire and explosion analysis in the offshore oil and gas industry where space is at as premium[1]. Safety improvements and savings for these platforms are simultaneously generated with confidence from more realistic CFD results.

Several methods and tools can predict fire consequences as part of API RP 752:2009 facility siting studies to design mitigating measures. Mitigation measures for reducing hazards to building occupants from an external fire may include appropriate fire-resistant construction for buildings designated as shelter-in-place and adequate escape routes for evacuation. Such measures may be part of a site emergency response plan.

Engineering designers and operators realize that assessment uncertainty is a product of the methods and tools employed. Coarse consequence modeling of the type produced by 2-D products (e.g., TRACE, PHAST, etc.) generates greater uncertainty, necessitating larger design contingencies and attendant increases in capital cost.

Using detailed CFD fire simulations with risk assessment models are shown to improve and optimize design solutions in terms of fire risk. For example, optimization is achieved through CFD sensitivity studies for probabilistic jet fire scenarios leading to:

- increased process safety due to more accurate results; and
- greater understanding of risks and necessary mitigating measures.

CFD versus 2-D modeling results allow refinement of fireproofing requirements and adequate fire escape routing. Together these ensure greater safety in the personnel workplace. This improved safety is accomplished with better delineation of active and passive fire protection requirements, often in concert with reduced weight and cost of passive fire protection (PFP). Early knowledge of fire safety requirements and material savings lead to additional cost benefits for operators and design engineering and construction companies.

2. Fire Consequence Risk Analysis

2.1 Fire Scenario Type

Fire consequence modeling on process releases can determine the design thermal radiation flux levels for personnel consideration.

For a volume of air, a flammable region occurs between the lower and upper flammable limits (LFL and UFL). The LFL is the minimum volume fraction at which flammable gas - air mixture becomes ignitable. Gas plumes are directional when associated with high pressure releases but are also influenced by the prevailing wind, the obstacles that influence the volume they flow through and surrounding plant topography. It is expected that only a portion of the gas plume will be in the flammable range. If a release contains a liquid fraction, this will "rain out" on horizontal surfaces or gravitationally flow off those with a vertical component.

Three types of fire modeling are: flash; jet; and pool. A flash fire is an intense duration fire, typically less than 3 seconds with little impact on the structure. A jet fire can result from an ignited pressurized hydrocarbon gas release or hydrostatic impelled liquid from a vessel. Jet fires are directional, flame temperature typically approximately 650°C (1,200°F), producing high intensity thermal radiation. Local instantaneous heat fluxes within jet fires[2] approach 300kW/m^2. Coupling high temperature and thermal radiation flux pose an immediate threat to personnel and the facilities in terms of escalation. If a jet flame impinges upon a target such as a vessel, pipe or structural member, it can cause the item to fail in less than 5 minutes. Pool fires are typically: less well aerated than jet fires; tend to have lower flame temperatures; produce lower thermal radiation fluxes and have less potential for structural failure than jet fires. Jet fire is chosen for further examination and discussion. Analysis of jet fires following API RP 752:2009[3] may consider thermal dose (flux and duration) and flame impingement.

2.2 Computer Modeling-2D vs. 3D CFD

An example to pictorially reveal some limitations of 2D versus 3D CFD is presented in Figure 1. The top 2D (TRACE) chart shows a side view profile of a gas plume. The impact of obstructions is depicted for the same process conditions in the 3D (FLACS) view in the bottom of Figure 1.

From a dispersion vantage, the 2D case is seen to report overly conservative end distances. These would represent the radii or curves to account for downwind dispersions in all directions. For the 3D case, more realistic results are expected with the incorporation of 3D geometry. Adding probabilistic leak rates and ignition probabilities to the releases will provide an improved risk picture.

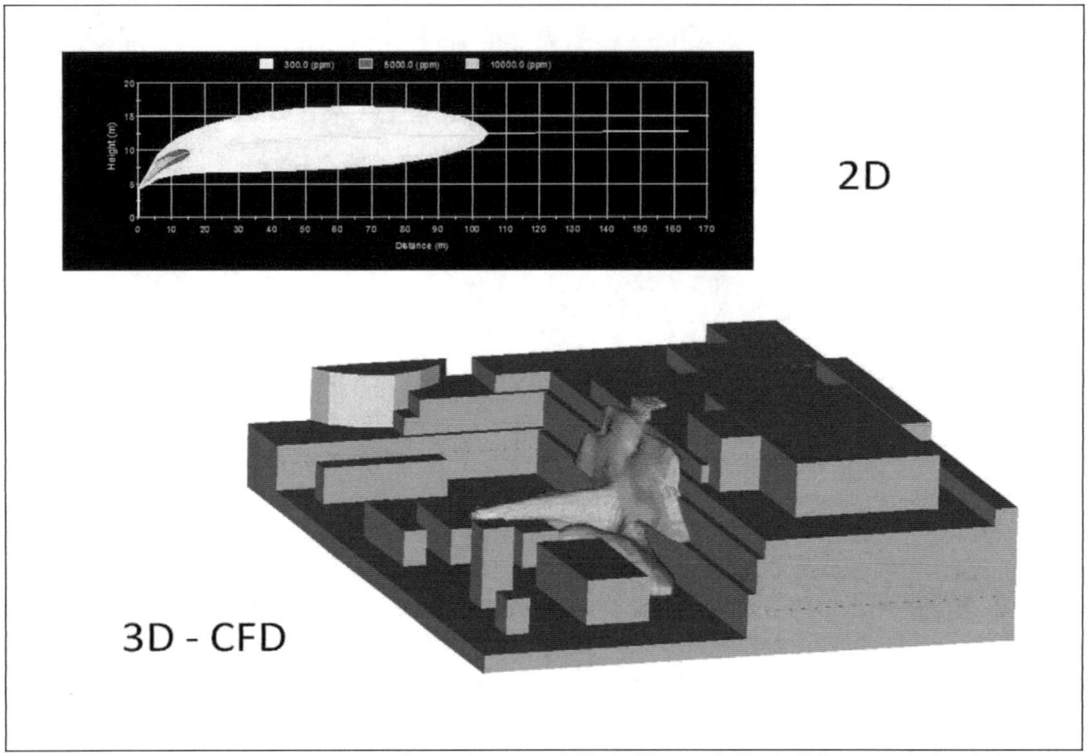

Figure 1: 2D and 3D CFD gas dispersion plumes (TRACE and FLACS respectively).

If a flammable plume were to be ignited, geometry can have a profound impact. Figure 2 shows the side view radiation of a jet fire from 2D modeling (PHAST). Here the ordinate is height and the abscissa distance downwind. The release and attendant jet fire is vertical. The isopleths of radiation flux are tilted with the wind. Another limitation in 2D software to geometry influences are that they do not generally allow for releases with component directions that are into the wind. This arises due to their choice in economy of equations allowing for faster computation.

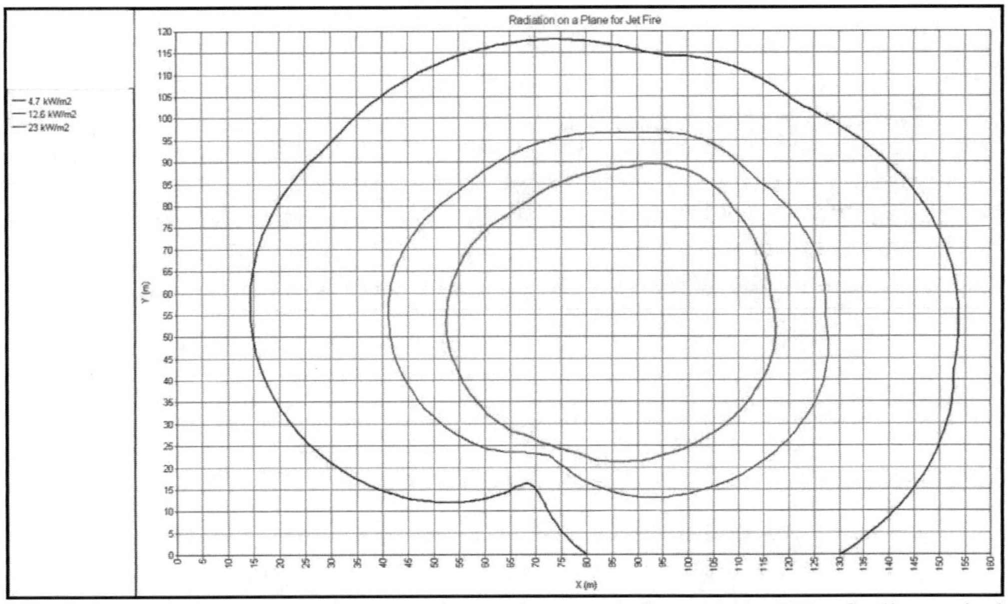

Figure 2: Side view radiation of a jet fire from 2D modeling (PHAST). Ordinate is height and abscissa distance downwind. The release and attendant jet fire is vertical.

2.3 Design Accidental Load

Consequence modeling for a fire Design Accidental Load (DAL) establishes: critical accidental loads; minimum separation distances between structures and their equipment; and personnel exposure to consequences. From API RP 521[2], the permissible DAL level for heat intensity on structures and in areas where operators are not likely to be performing duties, radiant shielding should be considered is 9.46 kW/m^2; where emergency actions lasting up to ½ minute may be required by personnel without shielding but with appropriate clothing is 6.31 kW/m^2 and; any location where personnel with appropriate clothing may be continuously exposed is 1.58 kW/m^2. With sufficient information, these may be determined and presented to the design engineer at a very early stage in the project.

Structures which eclipse a fire's radiant volume will impact the footprint of this radiation as they absorb, reflect, refract and re-radiate the radiation. The ability to predict the extent of these interactions is not within the general purview of 2-dimensional modeling capabilities. Where it may be of concern, to better assess how important secondary effects will influence the duration of structural integrity, radiation impact on fire escape routes, etc., it is recommended that CFD (Computational Fluid Dynamic) modeling be employed.

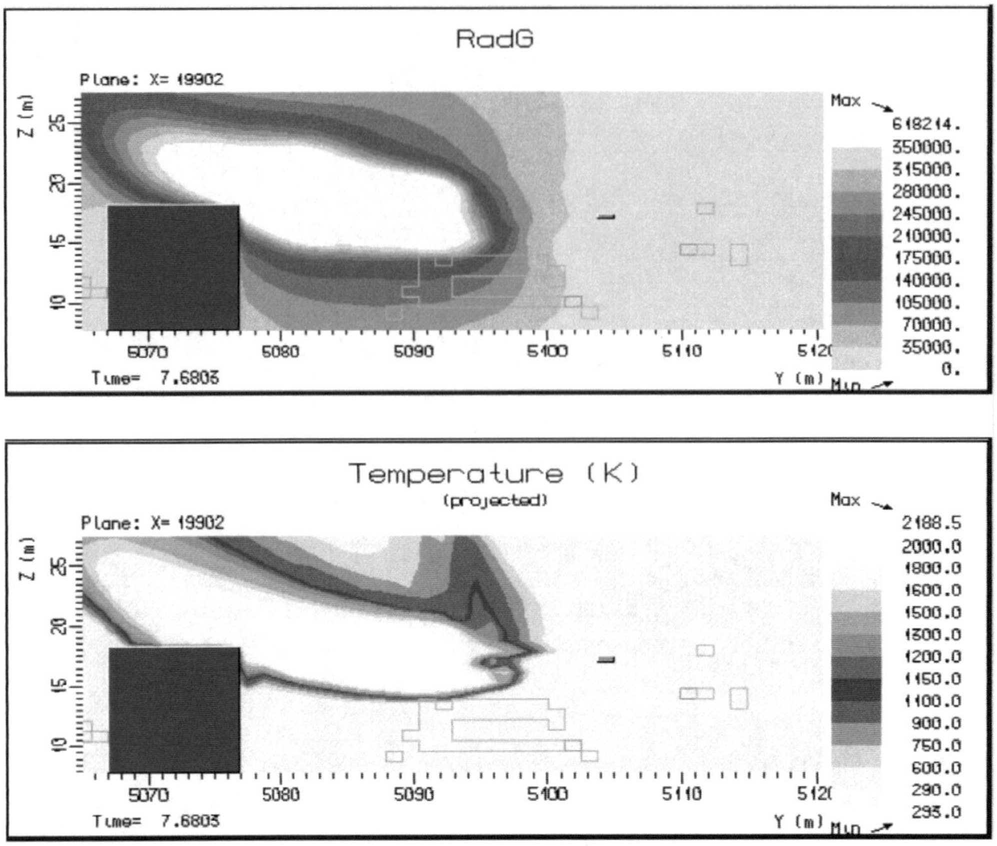

Figure 3: Side view profile of CFD jet fire loading on a structure without geometrical effects. Grey lines are where some geometry exists but have been excluded from the simulation (KFX).

Figure 3 shows the effect of fire loading on a structure using 3D CFD modeling. The top of the figure shows the radiation loading while the bottom displays the temperature profile. Light lines reveal a building outline which is not included in the modeling. Figure 4 shows a similar profile for temperature with the same process conditions on a similar cross section. Note that this yz-planar slice is displaced along x to incorporate the intervening building between a jet fire source and rectangular building.

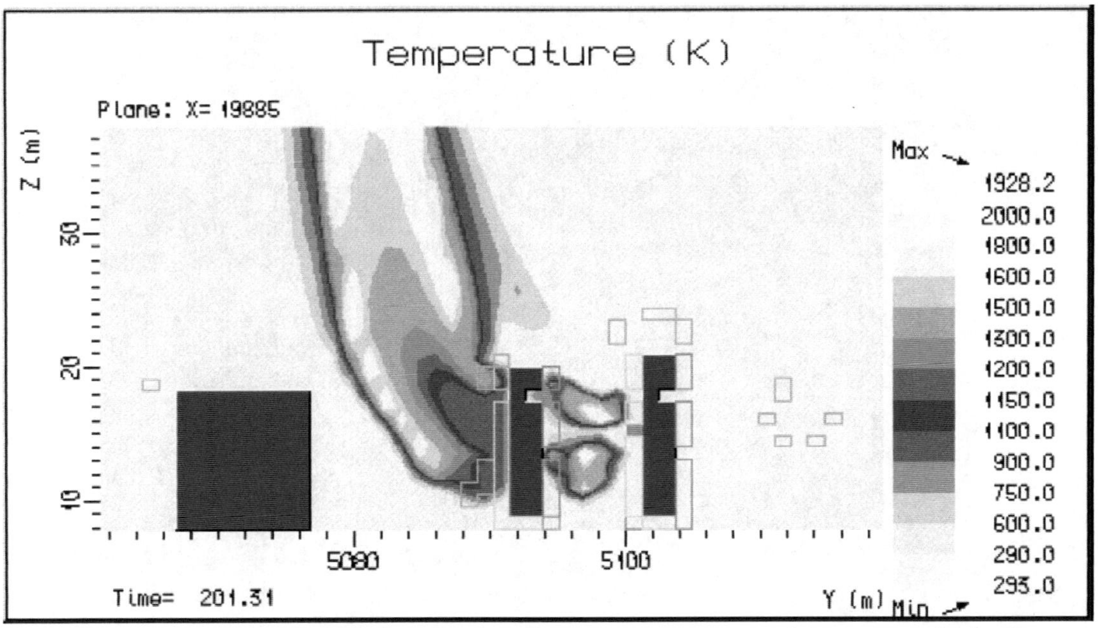

Figure 4: Side view profile of CFD jet fire loading on a structure with geometrical effects.

Thermal heat effects when geometry is included in the modeling are apparent when comparing Figures 3 and 4. If the square shaped structure to the left in both Figures represents a building to be designed for occupation, a thermal radiation flux picture will dictate the design fire proofing required. The 2D analysis of the building is shown to reach temperatures near 2000K; whereas, the 3D analysis shows no significant temperature elevation. If just Figure 3 is used as basis, a much higher rated passive fire protection may be design engineered for than necessary. This would require additional cost in material purchased and perhaps mass used.

In conjunction with Figures 3 and 4, probabilistic leak calculations and ignition sources reveal how DALs may be calculated to assist a client in the optimization of their facilities. Conclusions for this type of study include:

- Initial estimates were too conservative;
- Many buildings were outside large heat loads owing to intervening building interference with jet fire thermal radiation;
- Exposure of some buildings to higher heat loads than required standards;
- Large financial and ultimately safety advantage gains;
- Buildings in close proximity to potential jet fire concerns may be evaluated more fully for Shelter-In-Place designation using 3D analysis;
- Escape routes may be refined for consideration as part of an emergency response plan.

This is a great example of how a unit's central control room previously thought to be a jet fire concern via 2D could be evaluated further as a Shelter-in-Place designated building (assuming toxic, blast, etc. are at acceptable risks). Extremely fine resolution in results may be honed from detailed geometry like that in Figure 5. However, computational time is a limiting factor when extremely detailed geometry is used.

Figure 5: Detailed geometry may be imported for use at the expense of computational time.

Judicious selection of geometry based on careful consideration of the final desired outcome may bear excellent fruit. Figure 6 shows where reducing the size of a structure to elements and members of import allow them to be dynamically loaded with thermal radiation efficiently. Different materials and designs may be tested to see which will appropriately provide viable design solutions.

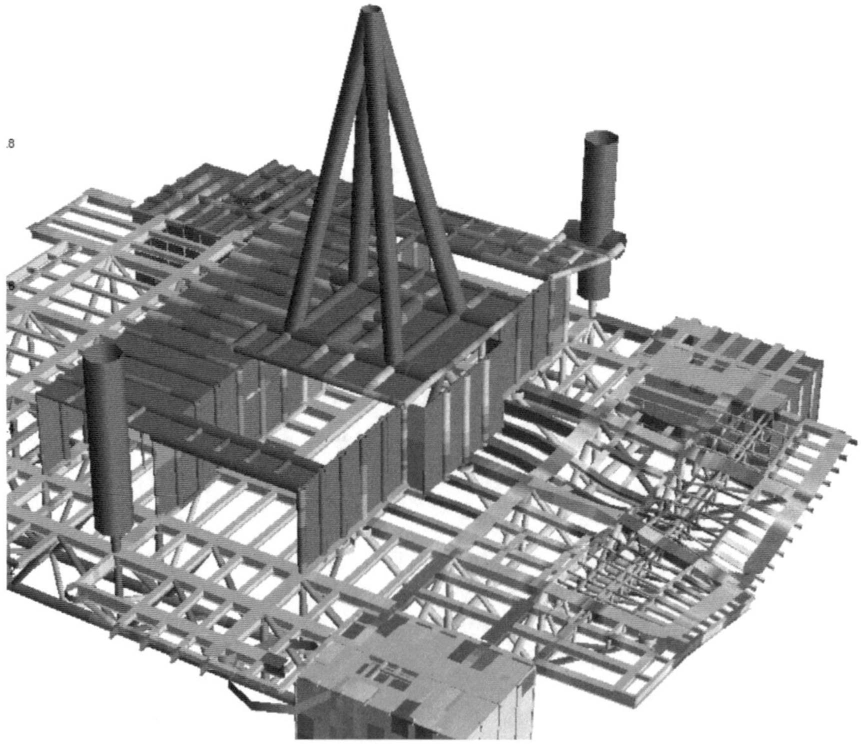

Figure 6: Load bearing and selected subset member geometry to investigate the response of a jet fire in determining design accident loads.

3. Conclusion

Employing 3D computational fluid dynamics (CFD) offers an opportunity to add value in honing economic process plant and facility designs which are inherently safer. Invoking appropriate probabilistic methods in conjunction with selective jet fire scenarios, (which may be culled from 2D modeling to filter), can prepare jet and other fire design accidental loads (DAL) for working environments which are better suited to meet operators risk tolerance then commonly utilized.

4. References

[1] Peterson, E., Nordstrand, H., Nodland, S., and Dalheim, J., "Using CFD in Platform Design", Mary Kay O'Connor Process Safety Center Symposium Proceedings 2010, Making Safety Second Nature, 2010.

[2] Guide for Pressure-relieving and Depressuring Systems, API Recommended Practice 521, Fifth Edition, January 2007, Addendum, May 2008.

[3] Management of Hazards Associated with Location of Process Plant Permanent Buildings, Downstream segment, API Recommended Practice 752, Third Edition, December 2009.

Additional references not cited:

[4] Lees, F.P., "Loss Prevention in the Process Industries", 3rd Ed, Elsevier, Butterworth-Heinemann, 2005.

[5] Spouge, J., "A Guide to Quantitative Risk Assessment for Offshore Installations", Centre for Marine and Petroleum Technology, 1999.

[6] "Fire Protection Handbook", 20th Ed., National Fire Protection Association, 2008.

Management of Risk for Large Flat Bottom Liquid Oxygen Storage Tanks

Martin A. Dennehy

Air Products plc.

Hersham Place Technology Park,

Molesey Road,

Walton on Thames,

England KT12 4RZ

dennehma@airproducts.com

Dennis P. Bernhard, Brian E. Farrell

Air Products & Chemicals Inc.

7201 Hamilton Blvd.

Allentown, PA 18051 USA

Abstract

Low pressure flat bottom tanks (FBTs) are widely used in the chemical and associated industries for storage of non - hazardous and hazardous liquids. The chemical industry has, for many years, considered the possibility of catastrophic failure of these tanks as a credible event and has designed hazardous material tank installations to reduce the risk from such events.

This paper summarizes the possible causes of failure of cryogenic Liquid Oxygen Flat Bottom Tanks (LOX FBTs) and describes incidents where failures of cryogenic FBT vessels have occurred. Fires can result from large spills of liquid oxygen inside and outside buildings and in vehicles on adjacent roads, causing fatalities or serious injuries. The likelihood of failure of LOX FBTs should be reduced by careful design and stringently controlled construction methods, and the consequences of failure should be mitigated by careful selection of the location of the tank or, if this is not possible, by measures such as secondary spill containment.

The paper discusses the risk reduction options that can be considered and recommends how the decision on selecting a location for a LOX FBT should be approached.

Figure 1 – A Typical Liquid Oxygen FBT installation

1. Introduction

Low pressure flat bottom tanks (FBTs) are widely used in the chemical and associated industries for storage of non - hazardous and hazardous liquids. One of these hazardous liquids, liquid oxygen (LOX), is stored in large quantities in cryogenic LOX tanks, usually on air separation plant sites. A typical installation is shown in Figure 1.Other hazardous liquids stored in FBTs include liquefied natural gas (LNG) and liquid ethylene.

The increase in recent years in the size and production capacity of air separation plants has led to a corresponding increase in the capacity of cryogenic liquid oxygen and liquid nitrogen (LIN) storages. It has therefore become more important to consider at the design stage the potential hazards associated with liquid products, the consequences and effects on the local environment of a major release of liquid and the preventive measures required. [1]

Generally, liquid nitrogen is not classified as a hazardous material.

Regulators and the chemical and energy industries have, for many years, considered the possibility of catastrophic failure of these flat bottom tanks as a credible event and the industry has designed tank installations to reduce the risk from such a failure. [2], [3]

2. Features of FBTs

The typical tank configuration for the storage of LOX in flat-bottomed tanks is a double wall, single containment type, where the liquid is contained in an inner tank and an outer tank serves mainly to contain the insulation (see Fig.2).

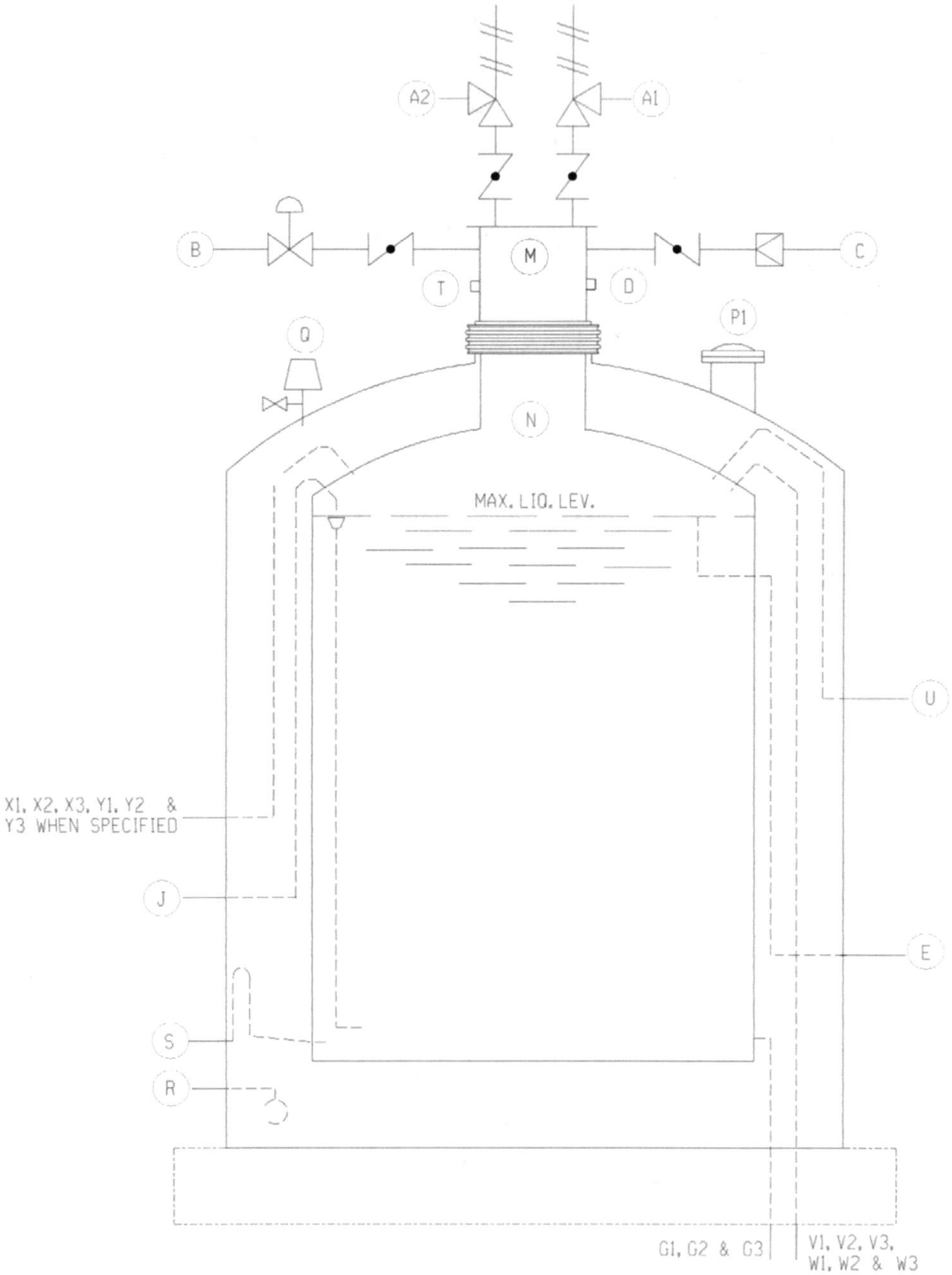

Fig 2 – Typical LOX FBT arrangement

NOZZLE IDENTIFICATION LETTER	CONNECTION
A1, A2	Primary and Secondary Relief Valve/Vacuum Breaker connections
B	Vent Valve connection
C	Rupture Disc connection
D	Pressure Build up Pressure controller connection
E	Overfill connection
G1,G2,G3	Liquid Outlet connections – typically feeding pumps
J	Liquid Fill connection
M	Outer Roof Manway
N	Inner Roof Manway
P1	Perlite Fill Ports/Emergency Vent connection
Q	Breather Valve connection
R	Annular Space Inert Gas Purge Inlet connection
S	Lower Liquid Level connection
T	Upper Liquid Level/Pressure Vent Controller connection
U	Vapour Return/Pressure Build Coil Return connection
V1, W1, V2, W2, V3, W3	Pump Recycle connections associated with G1,G2, G3
X1, Y1, X2, Y2 ,X3, Y3	Pump Casing Vent connection, where required. Associated with G1, G2,G3

Table 1 – Nozzle Connections for Figure 1

The FBT is designed so that only the inner tank is required to meet the low temperature ductility requirements for storage of the product.

Inner tanks are of all welded construction. Outer tanks may be of welded or bolted construction.

The annular space between the inner and outer tanks is purged with a dry, non flammable gas (usually nitrogen) that will not condense or freeze at the operating conditions of the inner tank. The annular space is not designed to contain the cryogenic liquid in the event of leakage from the inner tank. Sufficient pressurizing gas is needed to ensure a positive pressure is maintained throughout the annular space to maintain the insulation in a dry state; this provides the best insulation qualities. Wet or frozen insulation conducts heat better and will increase the boil-off rates of the stored material; ice formation can also cause damage to piping within the annular space.

The inner tank is fabricated of materials which are capable of withstanding the cryogenic temperature, the internal pressure of the stored fluid and which are compatible with oxygen when in LOX service.

Thermal stresses due to cool down and external pressure due to compaction of the insulation and gas purge maximum pressure need to be considered in the design of the inner tank for the worst case where the vessel is empty and at minimum pressure or partial vacuum.

The outer tank is designed to support the annular space insulation material and the gas purge pressure but it is not required to be fabricated with materials capable of withstanding the cryogenic temperatures of the stored fluid. As a minimum the outer tank is designed to resist the minimum wind loads required by national codes for the installation site. External loads such as snow and ice and mechanical loads such as stairs, platforms, and ladders need to be considered in the design of the system.

For FBTs located in earthquake zones, seismic design requirements must be considered for the complete tank system.

3. Codes

The design, fabrication, and testing of flat-bottom storage tanks must conform to all applicable national and local regulations. Since their design pressures are usually below 0.5 bar above atmospheric pressure, some pressure system legislation such as the EU Pressure Equipment Directive [4] will not apply to FBTs. However design, fabrication and testing of flat-bottom storage tanks can be guided by API 620[5] and API 625[6] or EN14620 [7]. Alternative codes that can be directly used include BS 7777, Part 4[8] or DIN 4119 [9] in its latest edition.

4. Safety Legislation

To minimize the off-site risk to the general public and the environment it is necessary for each of the likely risks and exposures to be considered individually. In particular, within the member states of the European Union, large bulk liquid oxygen storage tanks with capacity greater than 200 tonnes are subject to the EU Seveso Directive which, depending on the local details, will probably require significantly greater safety distances to be considered in the installation safety case than would smaller tanks. The assessment of the risks should take into account the various precautions provided within the site boundary, but should recognize that on-site emergency procedures and personnel training cannot normally be applied outside the site. [10]

5. Properties and Hazards of Oxygen

Gaseous oxygen is colorless, odorless and tasteless; it is not toxic; it is slightly denser than air at ambient conditions. It is not flammable but vigorously supports combustion. Breathing pure oxygen at atmospheric pressure is not dangerous although exposure for several hours may cause temporary functional disorder to the lungs. Some physical properties of oxygen are:

- Gas density at 1,013 bar and 15°C: 1.36 kg/m^3:
- Boiling temperature at 1,013 bar: -183 °C
- Liquid density at 1,013 bar and boiling temp: 1140 kg/ m^3

At ambient conditions, evaporation of one liter of liquid oxygen generates approximately 850 liters of gas.

Cold gaseous oxygen is heavier than air and may accumulate in pits and trenches.

The atmosphere normally contains 21% by volume of oxygen. Many materials, including some common metals, which are not flammable in air, may burn in an oxygen enriched atmosphere when ignited. In oxygen system design, the possible mechanisms for developing enough energy to ignite system components in oxygen service must be considered.
Most oils, greases and organic materials constitute a fire or explosion hazard in oxygen enriched atmospheres and on no account should these materials be used for equipment which is intended for oxygen service. Only materials acceptable for the particular oxygen service application should be used.

6. Hazard of Liquid Oxygen spills

Accidental releases of oxygen have caused fatalities. Examples of incidents which demonstrated the potentially fatal consequences of oxygen releases include:

- Germany 2001 - 1 fatality indoors - oxygen entered a control room and electrical equipment acted as ignition source
- Australia 1985 - 2 fatalities indoors - oxygen from liquid oxygen spill from a tanker entered a workshop and welding/brazing operations acted as ignition source
- USA 1978 - 5 fatalities in vehicle and outdoors - a van was engulfed by an oxygen cloud formed from a large LOX spill and caught fire

When oxygen is released into the atmosphere, without a source of ignition being present, the enrichment incident may pass by, unnoticed by many, without any harmful effect.
With a source of ignition close to a person, clothing may ignite, and smoking is the most obvious source of ignition. Cooking, uses of open flames and electrical equipment are other examples. This applies both indoors and outdoors. For vehicles, a running engine provides both the source of ignition and the fuel for the enhanced fire – as in the US 1978 incident.

Two of the incidents listed relate to relatively small releases of LOX. A large LOX FBT release can result in a cloud which can travel long distances, possibly reaching highly populated areas and heavily traveled roadways.

When an oxygen enriched cloud envelops a building, the persons indoors have some degree of protection, since for some period of time, the oxygen concentration indoors will be lower than that outdoors. This is because the oxygen has to permeate inside via door or window openings, or via the ventilation or air conditioning system. However for long duration oxygen releases, the concentration indoors will approach that outdoors. For oxygen enrichment indoors, as well as the risk of individual persons burning because they are close to ignition sources, there is some risk of fatalities from the internal furnishings and the fabric of the building burning, particularly at higher oxygen concentrations and under these conditions, persons not close to ignition sources still run the risk of becoming fatalities due to smoke inhalation or being trapped by the flames.

The closer the building is to the oxygen release source, the higher the concentration and hence the larger the percentage of occupants who could become burned and the higher the likelihood that they become fatalities.

Persons gathered together in public places may include persons with ignition sources close to them – smokers, portable electronic devices, etc. The likelihood of fatality depends on the oxygen concentration these persons are exposed to. Others not close to the persons with ignition sources close to them are not at risk.

Vehicles driving into enriched oxygen clouds are likely to catch fire and may crash, causing a traffic jam so that multiple vehicle fires result – the likelihood of fatality depends on the oxygen concentration at the road. Driving into the fog created by the cold oxygen vapour condensing the water vapour in the atmosphere can also cause fatalities in vehicle crashes because visibility is reduced.

7. FBT Failure mechanisms:

The potential causes of catastrophic failures of FBTs can be summarised as follows:

- Process causes
  - Overpressure, Over-filling, Vacuum, Hydrocarbon reaction
- External causes
  - Earthquake, Tornado, Flood, Lightning, External Fire, Aircraft impact, Vehicle impact, Projectile impact
- Causes related to design, construction, modification, and use (referred to as disruptive failures)
  - Defective design, Defective material, Defect in construction/welding, Failure of Non Destructive Testing during construction, Defective modification, Failure of inspection and testing (mechanical integrity) program in service

8. Examples of incidents involving failure of LOX or LIN FBTs

The 4 following incidents illustrate some failure modes which possible with cryogenic FBTs:

- 2 sites - France and Singapore - Road tankers were unloading product into the tank generating vapor from the liquid product in excess of the capacity of the tank pressure relief system. This pressure caused lifting of the inner vessel from its foundation, breaking liquid piping in the annulus.[11]

- Spain - A FBT was being filled, the first tank level transmitter was out of operation, the second level transmitter was wrongly calibrated, the tank overfilled by the transfer pump and the uplift forces caused a tearing of the weld between the roof and shell (a "pre-designed weak point") for about 75% of the circumference. Movement of the roof caused a pipe support welded to the inner vessel wall to tear off, and caused cracks in the wall of the inner tank. About 430 tonnes LOX flowed to bottom of annulus and cracked the outer vessel.[11]

- US - Over pressurization caused failure of the hold down straps of a 1000 Ton LIN FBT tank and the inner vessel lifted like a rocket and fell on the adjacent 1000 Ton LOX FBT liquid piping. The piping broke, and 800 tons of LOX were spilled – the oxygen vapor cloud drifted onto an adjacent chemical plant site, resulting in 5 fatalities.

The reported causes of these actual releases are process events, not mechanical defects or external impacts.

9. Reducing the risk of failure of LOX or LIN FBTs

The likelihood of significant damage to the FBT inner vessel from such process events can be reduced by providing the tank with:

- Multiple layers of overfill protection including level indication, high level alarm and independent high level safety instrumented systems to shut down liquid flow into the tank from any source.[12] These systems must be maintained and functionally tested at appropriate frequencies. It is recognized in the industry that filling a flat-bottomed tank to levels above the maximum design level can result in an uplifting, hydrostatic force on the inner tank dome that may cause the failure of a number of components including the inner tank anchor/hold down straps, the inner tank shell to floor joint, or the inner tank shell to roof joint. Therefore, the relief valves located as shown in Figure 2 may not protect the tank against the overfilling hazard.

- Multiple redundant relief devices (relief valves and bursting discs) including vacuum relief devices, to protect the inner vessel from overpressure due to heat leak or uncontrolled pressure build up or external fire. These systems must be maintained and the relief valves tested at appropriate frequencies.[12]

- Reaction of hydrocarbons at concentrations above the lower flammable limit with oxygen could result in fire or explosion. Analysis of hydrocarbon concentration in the LOX flowing into the tank or in the tank itself reduces the likelihood of this hazard occurring . Analyzer(s) must be calibrated and alarms tested at appropriate frequencies.

The likelihood of failure of a FBT due to a tornado can be reduced by designing the tank for appropriate maximum wind loading. Similarly likelihood of failure of a FBT due to earthquake can be reduced by designing the tank system for appropriate seismic loadings. Flood impact can be addressed by elevating the FBT above predicted floodwater levels on suitably designed supports, and lightning protection can be provided for the FBT. The likelihood of aircraft impact is increased by proximity to airport flight paths and this can be taken into account in selecting the location for the FBT. Vehicle impact risk can be reduced by the availability of protective structures between site roadways and the FBT. External fire risk can be reduced by limiting or eliminating the storage of combustible materials or fluids on the air separation plant site and missile impact risk can be reduced by appropriate security measures.

With appropriate independent safety instrumented systems and relief devices, it should be possible to reduce the likelihood of LOX FBT tank catastrophic failure due to process causes to the order of $1x10^{-7}$/tank year. It should be possible to reduce the likelihood of LOX FBT tank catastrophic failure due to external causes to the order of $1x10^{-7}$/tank. year to $1x10^{-8}$/tank. year by adopting the measures listed above.

Disruptive or mechanical failures are currently addressed by

- design and fabrication to recognized industrial consensus standards
- performance based company purchase standards
- use of qualified suppliers
- site construction oversight

However site construction of these FBTs involves a large amount of on-site welding and code requirements for inspection during construction are not as rigorous as for pressure vessels. The tanks are constructed at locations all round the world, using localized labor and so these factors all provide challenges for maintaining equipment quality.

10. Prediction of failure frequency of LOX FBTs

In 1997, the European Industrial Gases Association, EIGA, updated its collected data from its members for liquid oxygen, nitrogen and argon storage tanks in Europe (the previous survey had been completed in 1988):

- Production sites 11,760 tank years
- Customer sites 712,000 tank years

EIGA reported [10] that there had been no catastrophic failures due to mechanical design, manufacturing or construction defects and that this data resulted in a predicted median frequency of 1.4×10^{-6} /yr (50 % confidence) for a catastrophic tank failure from these causes. Confidence limits for this frequency could also be predicted as $0.06x10^{-6}$ /yr (5% confidence) to $4.1x10^{-6}$ /yr (95% confidence). However nearly all the customer site tanks and a proportion of the production site tanks will have been shop fabricated pressure vessels, where the control of fabrication quality and inspection requirements would be expected to be easier to implement and hence the figure of $1.4x10^{-6}$ /yr (50 % confidence) can be regarded as a under-estimate for site constructed FBTs.

An AEA SRD/HSE paper by Edmondson and Michell, dated Oct 1988, [13] surveying worldwide data on failures of cryogenic tanks estimated an average catastrophic failure rate of $5.8x10^{-5}$/tank. year and an average frequency for significant liquid release of $2.3x10^{-4}$/tank. year. These are averages based on reported failures across the globe, and some of the failures were related to hydrocarbon/air explosions, which would not be applicable to LOX tanks. The main failure causes reported were overpressure and mechanical failures, e.g. cracks in welded joints.

Based on this data, a potential cryogenic FBT catastrophic event frequency range of $1x10^{-5}$ to $1x10^{-6}$ per tank. year for failures due to mechanical design, manufacturing or construction defects would appear to be more appropriate.

This suggests that a catastrophic failure at this frequency should be considered in risk analyses of new and existing LOX FBTs, where the failure could result in potential of fatalities to the public. Examples of such situations could include:

- Office or apartment blocks, shopping centers, sports stadiums, public squares, railway stations (or other areas where populations congregate) in the vicinity of the LOX FBT installation
- Busy roads in the vicinity of the LOX FBT installation, particularly where at peak traffic periods the number of vehicles in traffic jams could be large.

11. Risk reduction options

It can be seen from the above sections that causes related to design, construction, modification, and use (referred to as disruptive failures) which are expected to be in the order of $1x10^{-5}$ to $1x10^{-6}$ per tank. year, dominate the overall FBT catastrophic failure frequency, when compared with failures due to process causes (in the order of $1x10\text{-}^{7}$/tank. year) and failures due to external causes (in the order of $1x10^{-7}$/tank. year to $1x10^{-8}$/tank. year).

Reducing the disruptive failure frequency by reducing the applied stresses in specific tank components (i.e. reducing the structural challenge to these components) is possible to some extent. Examples include:

- Reducing the pressure plus liquid head hoop stresses in the cylindrical shell to 2/3 of the design stress defined in API-620 620 by using thicker material.

- Designing the hold down strap anchorage for at least 1.5 times the uplift due to the design pressure. This is intended to reduce the applied stress in the straps to 2/3 of the design stress defined in API-620 and to increase the strength of the anchor to shell attachment accordingly.

However it is expected that such measures would reduce the disruptive failure frequency by one order of magnitude, and bring the disruptive failure frequency closer to the frequency of failures due to process causes.

Hence where there is risk of public exposure, the need for other precautions should be considered.

The kind of moats and sumps typically used to contain LOX releases from LOX tank pipe work leakage may not mitigate a LOX FBT catastrophic release – the LOX will have sufficient momentum to be projected over a moat or sump under the tank and to be largely unaffected by normal site grading. (14)

A risk mitigation option adopted in the LNG industry is to design the tank with a secondary containment outer wall and this approach can also be used with LOX FBTs. [3] The secondary containment outer wall would be constructed of concrete or a metal such as stainless steel, fixed to the tank foundation and would be of sufficient height and strength to contain the full LOX contents of the inner vessel, in the event of failure of the inner vessel. The liquid pool contained by this secondary containment would be of limited area and hence LOX evaporation would be small when compared with a tank failure spreading a liquid pool across the air separation plant yard. In this way the potential impact on the public off-site would be significantly reduced. However the secondary containment outer wall creates a confined space into which personnel inspecting the tank jacket and its piping connections could only enter with rigorous safety procedures in place; furthermore the cost of the wall would be considerable.

12. Recommendations for selecting a location for a LOX FBT

It is recommended that a location for the LOX FBT is selected that minimizes the risk to the public, by having adequate separation distances to buildings, places of public assembly and busy roadways.

To establish these distances it is recommended to use a recognized computer dispersion model (or computational fluid dynamics model) which can characterize cryogenic spills and model the dispersion of oxygen from two cases:

- A catastrophic LOX FBT failure
- The failure of the largest liquid piping line which connects to the LOX FBT.

In the modeling, appropriate wind speed, atmospheric stability, surface roughness, pool evaporation surface characteristics and receptor height should be selected which are appropriate for the proposed tank location.

It is recommended that the dispersion isopleths out to 30% total oxygen concentration (11.45% oxygen in excess of that in the air) should be superimposed on a suitably scaled map of the area around the tank and rotated through 360 degrees to help identify buildings, places of public assembly and busy roadways which might be covered by the isopleths.

Once the populations within the isopleth radius have been identified, the computer dispersion model (or computational fluid dynamics model) can be used to predict the peak oxygen concentration at each population area. For indoor populations, knowledge of building ventilation rate and the duration of presence of the oxygen cloud can be used to estimate the peak indoor oxygen concentration.

In predicting the number of fatalities at each population area, an approach based on BCGA Technical Report TR2 [15] is recommended. Prediction of the frequency of fatalities at each population location should take into account the frequency of the release case, the probabilities of atmospheric stability and wind speed and wind direction and the probability of presence of people in the area.

Judgment on the acceptability of the risk to people in the identified populated areas can be made against corporate risk elevation criteria, or national or regional risk guidelines and this comparison will aid the decision on acceptability of the proposed location.

13. Conclusions

The likelihood of a LOX FBT catastrophic failure is very low, but not so low that it can be ignored. Failures of flat bottom tanks have occurred. [11], [16]. It is recommended that an approach should be taken to select a location for the LOX FBT which minimizes the risk to the public, by having adequate separation distances to buildings, places of public assembly and busy roadways.

14. References

[1] EIGA Document IGC 127/04 – "Bulk Liquid Oxygen, Nitrogen and Argon Storage Systems at Production Sites", European Industrial Gases Association, Avenue des Arts, Brussels, B-1210, Belgium.

[2] "Guidelines for Quantified Risk Assessment", CPR18E, published by the Committee for Prevention of Disasters", RIVM, Netherlands, first edition 1999.

[3] "Developments in the safe design of LNG tanks," N.J. Fulford, M.D. Slatter, British Gas plc, 326 High Holborn, London WCIV 7PT, UK.

[4] EU Pressure Equipment Directive (PED) 97/23/EC.

[5] API 620 – "Design and Construction of Large, Welded, Low-Pressure Storage Tanks", (with Addendum 2) American Petroleum Institute, 1220 L Street, NW Washington, DC 20005-4070, USA. Aug 2010.

[6] API 625 - "Tank Systems for Refrigerated Liquefied Gas Storage", 1220 L Street, NW Washington, DC 20005-4070, USA. Aug 2010.

[7] EN 14620, Design and manufacture of site built, vertical, cylindrical, flat-bottomed steel tanks for the storage of refrigerated, liquefied gases with operating temperatures between 0 °C and -165 °C. 2006.

[8] BSI BS 7777-4 "Flat-Bottomed, Vertical, Cylindrical Storage Tanks for Low Temperature Service Part 4: Specification for the Design and Construction of Single Containment Tanks for the Storage of Liquid Oxygen, Liquid Nitrogen or Liquid Argon (Q)". British Standards Institution, Publication, Jan 1, 1993.

[9] DIN 4119 "Above-ground cylindrical flat-bottom tank structures of metallic materials". DIN, Deutsches Institut für Normung e. V., Burggrafenstraße 6, 10787 Berlin, Germany.

[10] EIGA Document IGC 60/04 – "Prevention of Major Accidents", European Industrial Gases Association, Avenue des Arts, Brussels, B-1210, Belgium.

[11] Safe Design of Flat Bottomed Storage Tanks", J A Somavarapu, M-J Masson, CGA Safety and Reliability of Industrial Gases, Equipment and Facilities Seminar, Oct 21-23, 2008, Orlando, Florida, USA.

[12] Alternatives for improved safety in storage and handling of LOX and LIN", Cryogenics, 1993, Vol 33, No 8 , G. E. Schmauch and G.A. Peters, Air Products and Chemicals, USA.

[13] "An Estimate of Operating Experience over the period 1954-1984 with Low Pressure, Flat Bottomed Metal Tanks storing Refrigerated and Cryogenic Liquids and of the Associated Historical Incident Frequencies", J N Edmondson, P D Michell, October 1988, Health and Safety Executive and AEA Technology Safety and Reliability Directorate, HSE/SRD/41068/WP1, UK.

[14] "Bund Overtopping – the Consequence of Catastrophic Tank Failure, A M Thyer, I L Hirst, Journal of Loss Prevention in the Process Industries, 15 (2002) 357-363, (Elsevier).

[15] "The Probability of Fatality in Oxygen Enriched Atmospheres due to Spillage of Liquid Oxygen" – R2 (LOX), British Compressed Gases Association.

[16] "Review of Failures, Causes and Consequences, in the Bulk Storage Industry", W Atherton, J W Ash, Liverpool John Moores University, Faculty of Technology and Environment, School of the Built Environment, The Cherie Booth Building, Byrom Street, Liverpool L3 3AF, UK.

The Path Forward from Facility Siting

J. Kollin Kenady
ABS Consulting
14607 San Pedro Ave. Suite 215
San Antonio, TX 78232
KKenady@absconsulting.com

Michael D. Weinberg, Matthew J. Kraemer, Eric S. Becker
ABS Consulting

Keywords: Facility Siting, Hazard Analysis

Abstract

Facility siting is the vehicle for quantifying hazards from accidental fires, toxic releases, and explosions to building occupants. Due to the large number of processes and buildings involved in a typical facility siting study, cost effective and conservative approaches are used in the engineering analysis. Hence, it is not unusual that many buildings at a site are identified as "at risk." One approach to risk reduction is to make structural upgrades to withstand the predicted hazardous loads (fire, toxic, and explosion hazards). Building upgrades can be expensive and may interrupt production. Prior to implementing building upgrades, a combination of scenario reduction and the use of advanced analysis should be considered to reduce conservatism in predicting hazard loading.

Hence, the initial facility siting study should not be seen as the final answer. The initial study should instead be viewed as a conservative screening tool that will identify critical hazard areas in the facility arrangement, processes, and building construction. The path forward from facility siting could include verifying assumptions, performing more advanced analysis, utilizing more advanced hazard criteria, and implementing mitigation measures, including building retrofits. This perspective focuses on accurately determining the hazard and minimizing it before designing structural upgrades possibly resulting in fewer buildings requiring retrofits or less extensive building retrofits for those in need. The facility siting process becomes a series of studies and actions, all with the same goal to identify and reduce the vulnerability to building occupants.

1 Introduction

The facility siting assessment will identify buildings that offer an unacceptable level of protection to its occupants. It is the intent of this paper to offer guidance on possible paths forward to decrease the predicted vulnerability to occupants. There are three paths for consideration to modify the results of a facility siting assessment, they are: 1) to verify assumptions, 2) perform more advanced analyses, and 3) utilize a more advanced hazard criterion. These paths should be used to refine the analysis to achieve a more accurate, less conservative answer. It should be noted that a more accurate answer may not be a less hazardous answer. These approaches are not intended to be used to inappropriately "game" the analysis for a more desirable result. Paths can be used individually, or in combination, to reassess the hazards to building occupants.

2 Background

Facility siting is a requirement in federal regulation 29 CFR 1910.119 (OSHA, 1992) as a part of the PSM process hazard analysis (PHA). This regulation contains requirements for preventing or minimizing the consequences of a catastrophic release of flammable, toxic, or explosive chemicals. Part of minimizing consequences is assessing how a release can affect personnel within buildings. The assessment of these hazards is the goal of facility siting.

The American Petroleum Institute has published recommended practice documents that provide guidance on the siting evaluation for new and existing permanent and portable buildings. The API RP 752 (API, 2009) and API RP 753 (API, 2007) provide an overview of some technical aspects for the facility siting analysis. Guidance is presented on which buildings to include in the analysis, what assessment approaches can be used, and what hazard effects need to be considered.

In the API RP 752 and RP 753, there are three approaches to hazard assessment, i.e., "consequence-based", "risk-based", and "spacing tables" (fire hazards only). A "consequence-based" assessment will be assumed in this paper because it is the typical screening-level analysis and most often used in facility siting. If further study shows any of the maximum credible events (MCEs) to no longer be the controlling scenario, other scenarios in the area should be reconsidered because they may have been screened out as a lesser event.

3 Verify Assumptions

All hazard analysis methods apply some assumptions or idealizations to determine what might happen if a release occurred. When buildings are identified as having an unacceptable level of protection, these assumptions should be revisited to see if they can be refined to increase the accuracy of the analysis. Assumptions are necessary because the analysis methods cannot take into account all existing variables and sometimes the input data is insufficient or unavailable. This data could be about the process, the specifics of where the release could occur, or could be a lack of scientific data about the chemicals and their associated hazards. It is for these reasons that assumptions are a necessity when performing a screening-level hazards evaluation.

Assumptions to the input of a method are largely the assumptions that define the release scenario. They define the conditions of the unreleased chemicals, the release conditions, and the environment that they are released into.

The conditions of the unreleased chemicals are usually assumed to be at typical conditions without regard for the possibility of deviation or contamination. Accurately describing the conditions of the release scenario increases the fidelity of any calculation method.

Due to the lack of scientific data on many chemicals and to the limited ability of many analysis methods to handle mixtures, there are often assumptions made in the form of substituting chemicals for other chemicals or substituting few chemicals for process streams composed of many chemicals. These assumptions can further affect the results by changing the phase of the chemical released. If the process material is a vapor, but because of the material assumption it is analyzed as a liquid, the resulting consequences can be orders of magnitude more severe than they should be. The opposite also holds true.

Process parameters like connected inventory and process flow rate are used to determine the amount of material that can be released and how fast it is released. Often these parameters are not specified and assumed to be non-limiting as a way to add conservatism to the analysis. But they can have a large effect on the outcome of the analysis.

The release size is arguably the most subjective assumption that has a significant effect on the outcome of the analysis and there are different approaches in industry of how to define the release size. Some approaches include:

- Assuming a maximum leak size,
- Assuming a leak size determined by a fraction of the pipe size,
- Assuming a range of leak sizes,
- Releasing the entire inventory at over a specified time period or instantaneously, or
- Calculating a frequency-based leak size.

Caution should be exercised in re-evaluating a release size. Any change in the release size must be justifiable and not subjective; it should not be changed arbitrarily because of undesirable results.

4 Analysis Methods

It is often advisable to perform the facility siting assessment using simple methods to screen out scenarios that do not require the investment of a higher accuracy, advanced analysis. This section serves as an overview of the different analysis methods identified as "Simple," "Moderate," and "Advanced" for each step in a hazard analysis and gives some guidance of the merits and weaknesses of each method, as well as an example comparison case.

4.1 Cloud Size Estimation

These analysis methods are used to define the size of the flammable or toxic vapor cloud that would result if a release occurred.

4.1.1 Simple: Assume a Cloud Size

Dispersion is not a requirement in facility siting in that the analyst can assume a process area is filled with flammable vapors or all buildings are surrounded by toxic vapors. This is the most conservative method to assess toxic impact to buildings. The method of filling the congestion is the simplest way to determine vapor cloud explosion volumes.

For explosion volume calculations, this method is applied by simply estimating the unit area and using an average congestion height to calculate a unit volume. Unfortunately, results from this method tend to be very conservative but may be very unconservative in some situations. It would be conservative to assume the entire unit volume is available for an explosion if the release were small and the unit relatively large. But, if the release is large and the unit is relatively small, then this method does not recognize the potential for the vapor cloud to cover adjacent units as well and, assuming only one unit filled would under estimate the size of the release.

For toxic releases, API RP 752 allows the analyst to assume all buildings will be within the toxic cloud. The response to this is to either evacuate or shelter in place. However, a safe evacuation route is a requirement in API RP 752 when evacuation is the selected response. It should not be assumed people will be safe unless dispersion is performed. Performing a toxic dispersion will help determine if an evacuation can be performed safely without PPE or what level of protection is needed for evacuation or if people should shelter in place. If the response selected is to shelter in place, occupants may be at risk if they are inside buildings that do not have proper ventilation systems or a building envelope that would prevent ingestion of toxic vapors since infiltration of small amounts of some chemicals can be hazardous. Thus, knowing the toxic environment surrounding a building is critical regardless of the response selected.

4.1.2 Moderate: Empirical Dispersion

Empirically based dispersion modeling calculates the shape and position of the flammable cloud based on the released material and release conditions. This method is a significant step up in terms of accuracy over methods involving an assumption of cloud size.

Empirically based dispersion models have several distinctive advantages over simpler methods. These models take into account the material's properties (boiling point, molecular weight, etc.) during the release. The models also calculate the release conditions (velocity, temperature, etc.). The inclusion of these properties allows the models to predict liquid flashing, pool formation, rising/sinking vapors, etc.

The primary disadvantage of this model is that it calculates a free-field vapor cloud using surface roughness parameters or other methods to try to simulate the actual dispersion. While predicting the flammable cloud path and shape, it cannot accurately account for the influence of site geometry. So clouds are predicted to pass through solid geometry such as buildings and vessels instead of flowing around or over them.

4.1.3 Advanced: Numerical Modeling

Numerical modeling analysis of a flammable dispersion is a detailed method for predicting vapor cloud formation that includes the effects of flow around geometry (e.g. buildings, equipment, and piping) and air mixing. This method is commonly used to model releases within process buildings.

Advantages of using the numerical modeling method for a flammable dispersion is the detailed wind fields generated for specific atmospheric conditions. The dispersion of a flammable vapor is greatly influenced by the mixing rate with air and wind speed. Specific wind directions can be used to see effects on the flammable dispersion and vapor concentrations. Additionally, these advantages allow for areas or pockets of varying concentrations to form.

Disadvantages for the numerical modeling method is that it is more labor intensive than empirical methods, although increases in computing power have made this a viable option for many companies. The model must have a detailed three dimensional representation of the objects in the area where the dispersion is predicted to take place. CFD models require a significantly longer simulation time compared to empirical methods. Finally, due to the numerical nature of CFD several parameters (e.g., geometry, gridding, and boundary conditions) can have a significant impact on CFD results. Therefore it is essential that these methods be performed by a trained and experienced analyst.

4.1.4 Method Comparison Example

The following example is a plant where a large C2 hydrocarbon release, from the north most area is analyzed. Utilizing the simple unit filling method results in a long east-west cloud (Figure 1-top). Using empirical methods, a large tear drop cloud is predicted (Figure 1-left) along the west side of the plant. When directed south, this cloud partially fills the three southern units, crosses four roads, and extends beyond the plant boundary. When the same release is modeled numerically, the cloud is influenced by equipment and pipe blocking its flow and forms an amorphous cloud (Figure 1-right) filling both east and west sides of the plant. This cloud mostly fills the three northern units, having about twice the breadth and half the extent of the empirical cloud.

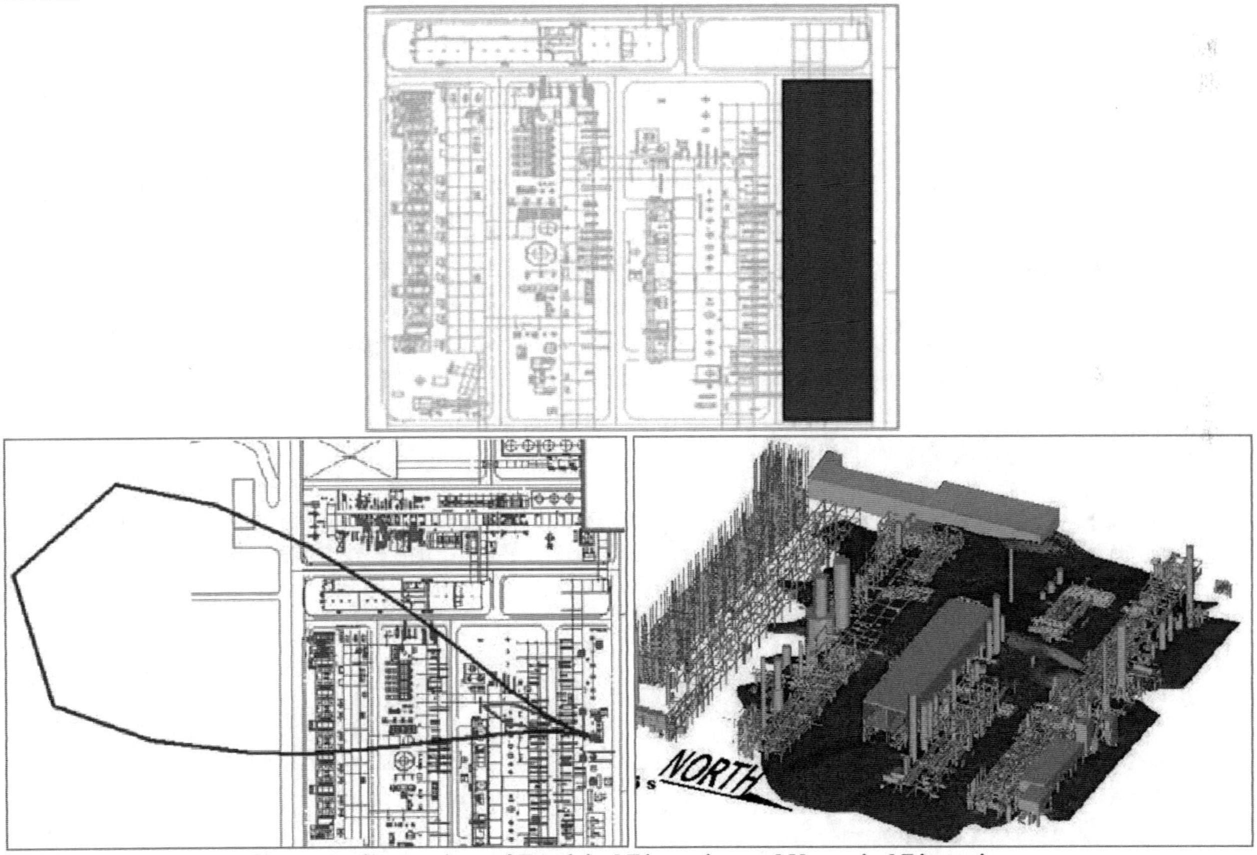

Figure 1. Comparison of Empirical Dispersion and Numerical Dispersion

4.2 Fire Analysis

4.2.1 Simple: Spacing Tables

Utilizing spacing tables is the simplest form of fire analysis. This method is used by matching the equipment and process areas with those listed on the table. The standoff distance can then be read off the table. The primary advantage of this method is its ease of use. A limitation of this method is that only process equipment locations have an effect on the spacing distance, regardless of the fuel properties and release parameters. Thus, it can easily be unconservative for some applications and overly conservative in others.

4.2.2 Moderate: Empirical Model

Empirical models use model parameters such as fuel characteristics, release conditions, and weather conditions. Typically, these models calculate a simple geometric flame shape (or a point) and flame surface emissive power based on the release conditions and fuel type. Together with the ambient conditions a worst case view factor is determined. The calculated flame characteristics and worst case view factor allow for the calculation of flame length and radiant heat flux versus distance data.

The primary advantage of this method is that it takes into account specific fuel properties and release conditions. This method can be quickly performed using existing commercial software.

One disadvantage to this method is the conservatism introduced by calculating view factors based on a simple geometric flame shape (or a point). Additionally, these methods do not typically account for potential shielding from radiant heat offered by obstructions.

4.2.3 Advanced: Numerical Modeling

The use of numerical modeling for fire analysis allows for a more detailed assessment of the burning of fuel, flame shape, and the effect on surrounding geometry. Numerical modeling programs require the specification of fuel location and properties as well as a detailed representation of the site geometry (piping, vessels, buildings, etc.). This method uses a mixture of theoretical and empirical data to predict how a flame will form and change over time based on the geometry and ambient conditions. A more accurate representation of the flame allows for better assessment of the radiant heat flux. This method is most commonly used as a refinement following an empirical model analysis.

The primary advantage is the increased accuracy in determining the flame shape and how it changes over time. Unlike previous methods, numerical modeling is not limited to fixed flame geometry (tilted cylinders or truncated cones). This method can also model smoke or soot generation which can have an impact on radiated heat. In addition, various complex geometries can be evaluated for their heat exposure as well as their ability to shield other geometry. The result is less conservative and is uniquely linked to the modeled event.

The most significant disadvantage of numerical modeling methods is the level of effort required over empirical based methods. The site models require a higher level of detail, thus requiring more time to assess the site and create the model. In addition, numerical modeling requires a longer simulation time compared to empirical methods. Finally, due to the nature of numerical modeling, several parameters (geometry, gridding, boundary conditions, etc.) have a significant impact on the results. Therefore, it is essential that these methods be performed by a trained and experienced analyst.

4.2.4 Method Comparison Example

The following example is a unit in which a C7 hydrocarbon release is analyzed as a MCE. Utilizing the spacing tables from *Guidelines for Facility Siting and Layout* (CCPS, 2003) unit filling method results in the spacing contours in Figure 2-left. Using the release calculated pool fire, the radiant heat flux contours are shown, for multiple release points, in Figure 2-right. Figure 3 and Figure 4 show comparisons of empirical and numerically modeled pool fires. The numerical fire is influenced by the solid deck above a portion of it (Figure 4-right), but the empirical model is not.

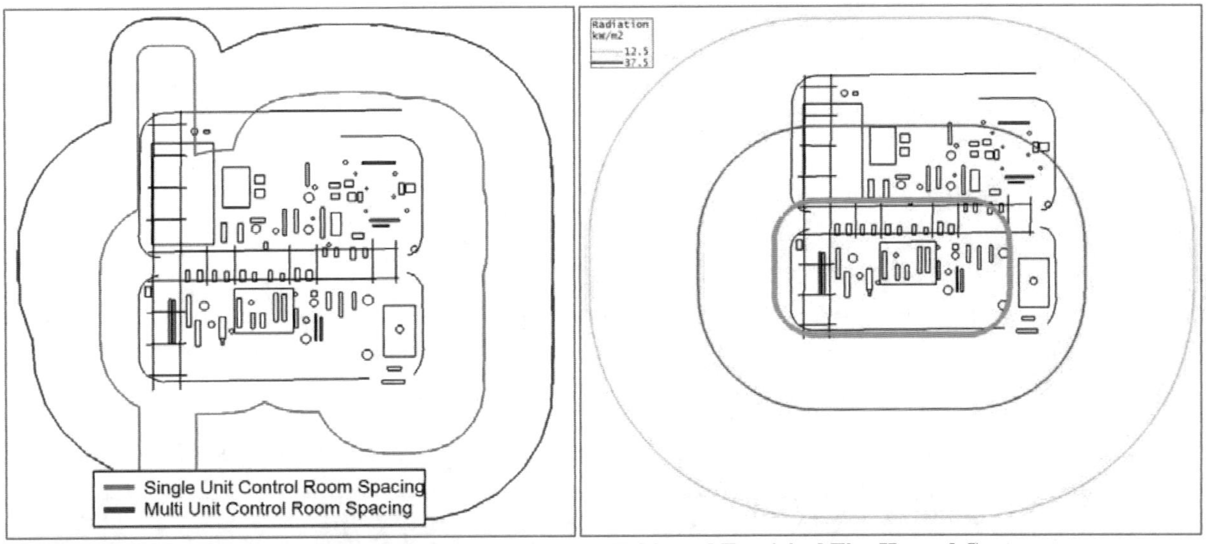

Figure 2. Comparison of Fire Spacing Table and Empirical Fire Hazard Contours

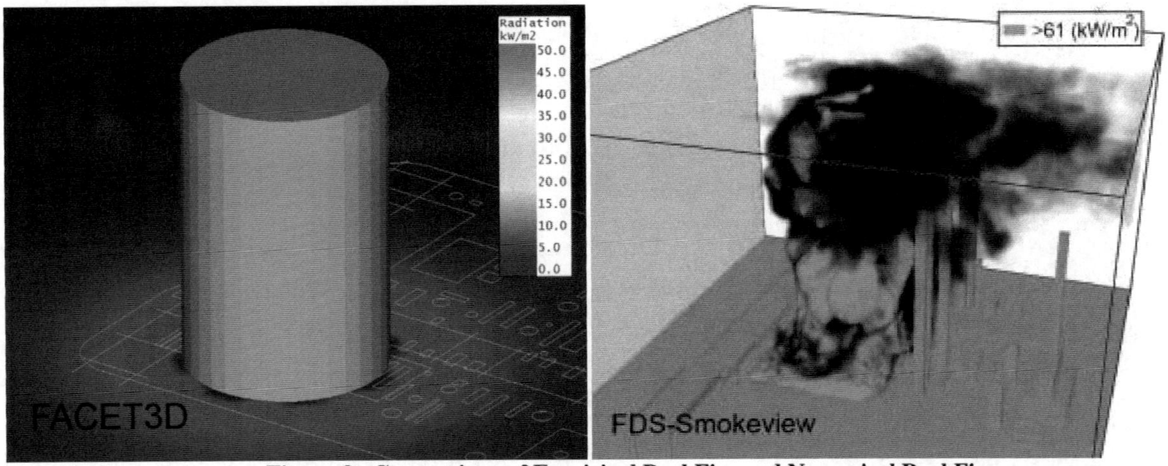

Figure 3. Comparison of Empirical Pool Fire and Numerical Pool Fire

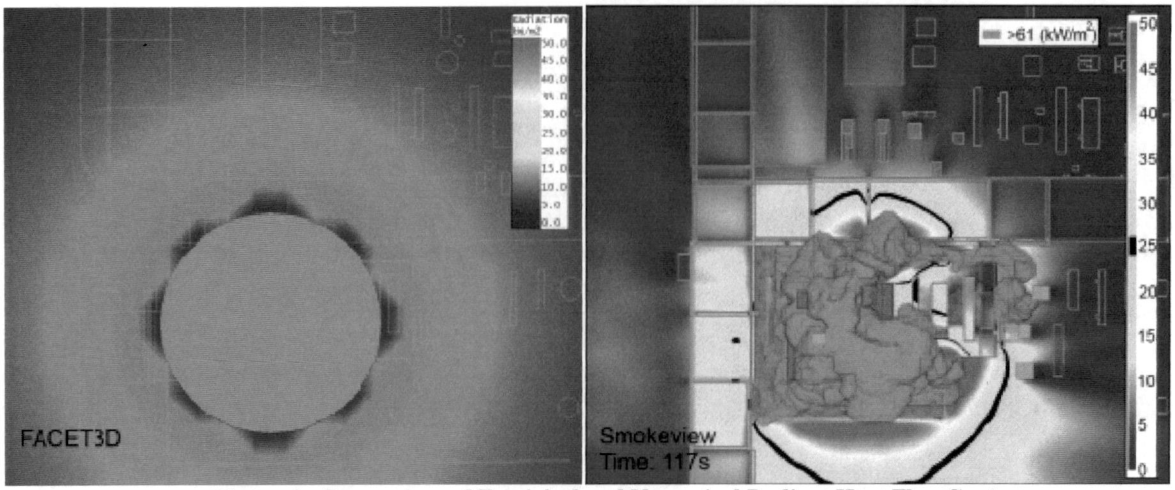

Figure 4. Comparison of Empirical and Numerical Radiant Heat Flux Contours

4.3 Explosion Analysis

4.3.1 Simple: Empirical Curves

Empirical curves are based on testing and computer analysis of explosions. There are curves for high explosives, pressure vessel bursts, and vapor cloud explosions. They are scalable and simple to use.

The TNT equivalence method equates an explosion event to an amount of TNT and uses TNT curves to determine loads at target distances. API RP 752 explicitly states that "TNT equivalency method shall not be used to assess VCE blast loads" (API, 2009, p. 14). But, this method may be appropriate for other explosion events.

Empirical vapor cloud explosion curves can be used by determining vapor and geometric parameters and then applying an empirical curve to calculate pressure and impulse vs. distance. There are several specific VCE curve methods available (e.g., TNO Multi-Energy Method, Baker-Strehlow-Tang, Congestion Assessment Method, etc.) that all attempt to account for event specific factors such as fuel, congestion, and confinement.

The primary advantage of this method is the introduction of the site specific conditions for congestion and confinement (e.g., piping, equipment, walls and decks) that influence the severity of a vapor cloud explosion(Herrmann, 1999).

One disadvantage of this method is the requirement of assigning a single congestion/confinement to represent an entire effected area. Another is that the determination of the congestion and confinement parameters are fairly subjective in practice, but some methods can be less subjective than others. Additionally, the resulting blast wave expands outward spherically, uninfluenced by plant and building geometry. Although these methods can be quite accurate in the far field, they can vary significantly from the real physics at close range. A limitation for these curves is that they are not validated for application to interior releases with significant confinement (more confinement than just a roof or deck). Application of blast curves to interior releases is often performed but must be done with a good measure of conservatism and knowledge of its limitations (Fitzgerald, Knight, & Gutierrez, 2007).

4.3.2 Advanced: Numerical Modeling

The use of numerical modeling for VCE analysis allows for a more detailed assessment of the burning of fuel, a more accurate use of the site geometry, and more realistic calculation of blast wave generation and propagation. It is the only validated method for accurately evaluating an explosion inside a building. Numerical modeling software requires the specification of vapor cloud location as well as a detailed representation of the process unit geometry (piping, vessels, buildings, etc.). This method uses a mixture of theoretical and empirical data to predict how a flame front will move through the process unit. The flame front is then used to calculate the blast wave propagation.

The primary advantage of numerical modeling is the increased accuracy of blast wave development and propagation. Through the use of fuel burning algorithms and detailed process unit models, burning through various vapor mixture (lean or rich) areas with changing congestion and confinement can all be taken into account. This method can simulate multiple points of explosion from single cloud by allowing the flame front to accelerate and decelerate depending on process unit geometry. This combined with the ability to account for shock reflections off surfaces, numerical modeling methods generate irregularly shaped (directional) blast waves. The result is less conservative and is uniquely linked to the modeled event.

A limitation of the numerical models is that they are currently unable to accurately assess the phenomenon of deflagration to detonation transition (DDT) which occurs when the shock front becomes the burning zone for the fuel and explosion strength is no longer dependent on congestion or confinement. Parameters may be tracked to determine if this condition becomes more likely which would indicate the results may not be accurate. A DDT evaluation requires the use of the VCE detonation blast curves. However, determining the onset of this phenomenon with numerical modeling is helpful in reducing the explosion energy to use with blast curves.

The most significant disadvantage of numerical modeling methods is the level of effort required over curve based methods. However, as previously stated, advances in computing power have made numerical modeling a much more attractive option than in recent past. The site models require a high level of detail, thus requiring more time to assess the site and create the model. In addition, numerical modeling models require a longer simulation time compared to curve based methods. Finally, due to the nature of numerical modeling, parameters like geometry, gridding, boundary conditions, etc. have a significant impact on the results. Therefore it is essential that these methods be performed by a trained and experienced analyst.

4.3.3 Method Comparison Example

The following example is a unit filled by a C2 hydrocarbon release that explodes. Figure 5 (top) shows the calculated loads from an empirical VCE curve model and Figure 5 (bottom) shows the loads calculated by numerical methods. It can be seen in the numerical model that the load is not uniform throughout the unit, but rather decreases in severity in the middle and then increases again towards the opposite end. The influence of the building just south of the unit can be seen as there is an increase in loads on the north face of the structure and a decrease on the south and west faces, from what was calculated empirically.

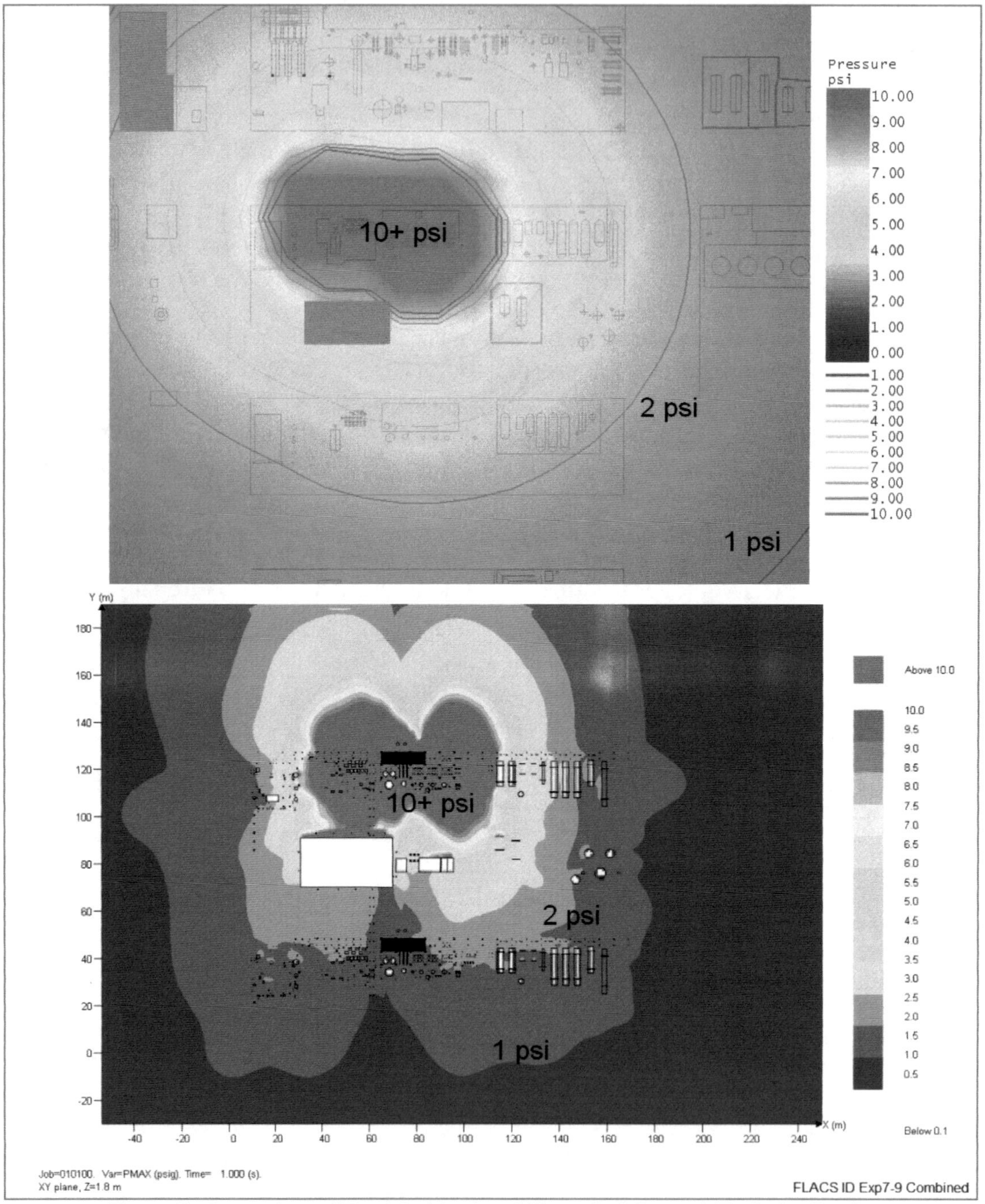

Figure 5. Comparison of Empirical (top) and Numerical (bottom) Explosion Loads

5 Hazard Criterion

For flammable and toxic releases, personnel often have time to react and either find shelter, shelter-in-place, or evacuate. If an occupied building is designated as an emergency shelter, then the building needs to be able to protect personnel for the duration of the event or until it is safe to evacuate. If an occupied building is not designated as a shelter in place, then personnel are expected to evacuate. This may expose them to new hazards during their evacuation that should be considered. Conversely, buildings are also intended to protect the occupants in the event of an explosion because the release and explosion happens so fast that personnel have little or no warning in which to react.

5.1 Flammable Dispersion Hazard

Personnel within a flammable vapor cloud are very vulnerable because of the possibility of being engulfed by the flame if the cloud is ignited. The hazard effect measurement for flammable dispersions is flammable gas concentration or some fraction thereof. Buildings can offer shelter from this hazard until the flammable vapor gets inside the building at a concentration within the flammable limits. Buildings with adequate positive pressure systems or without significant opening can prevent the buildup of flammable vapors within the building. Outside muster areas should be located outside the flammable dispersion zone.

5.1.1 Simple: Flammable Concentration Outside Building

Criterion for a flammable dispersion hazard to a building is often set as the flammable concentration of the material around the building. This is a conservative approach since the ingress rate of the vapors into the building is often unknown.

5.1.2 Advanced: Building Flammable Vapor Ingress

A supplementary analysis could be performed to determine the concentration of vapors inside the building, given a concentration-time function of flammable vapors outside of the building and the ingress rate of the building envelope. The concentration inside the building could then be used as the hazard criteria.

5.1.3 Evacuation

If the building is not designated as an emergency shelter and personnel are expected to evacuate in the event of a hazardous release, then in the event that the building exits are within a flammable cloud, evacuees will be exposed to a high vulnerability until they are outside of the cloud. Whether the occupants are more vulnerable sheltering in place or evacuating should be evaluated.

5.2 Toxic Dispersion Hazard

The hazard criteria for measuring toxic inhalation affects are toxic concentration, toxic load, or vulnerability.

5.2.1 Simple: Toxic Concentration

Hazard criteria based on a concentration are often used because they are usually conservative due to a basis of relatively long exposure times. Hazardous concentration thresholds are specified by associations and government organizations for many chemicals. ERPG, IDHL, or TEEL are examples of these concentration thresholds. But the time of exposure is an important variable to determining toxic hazards and could be considered in conjunction with concentration. Each of the example thresholds mentioned above have a time of exposure and hazard consequence associated with it. Using concentration alone to determine hazards typically makes for conservative criteria unless being used where long duration exposures are expected, like outdoor muster points or inside emergency shelters.

5.2.2 Advanced: Toxic Load

An alternative hazard criterion is the toxic load, which is a combination of the toxic concentration and exposure time of a chemical. The toxic load is more specific in its approach because it can be correlated with vulnerability. For comparison, Figure 6 shows endpoint contours for a toxic dispersion using the ERPG 3 toxic concentration and the toxic load given a low vulnerability for evacuees and assumed short exposure duration.

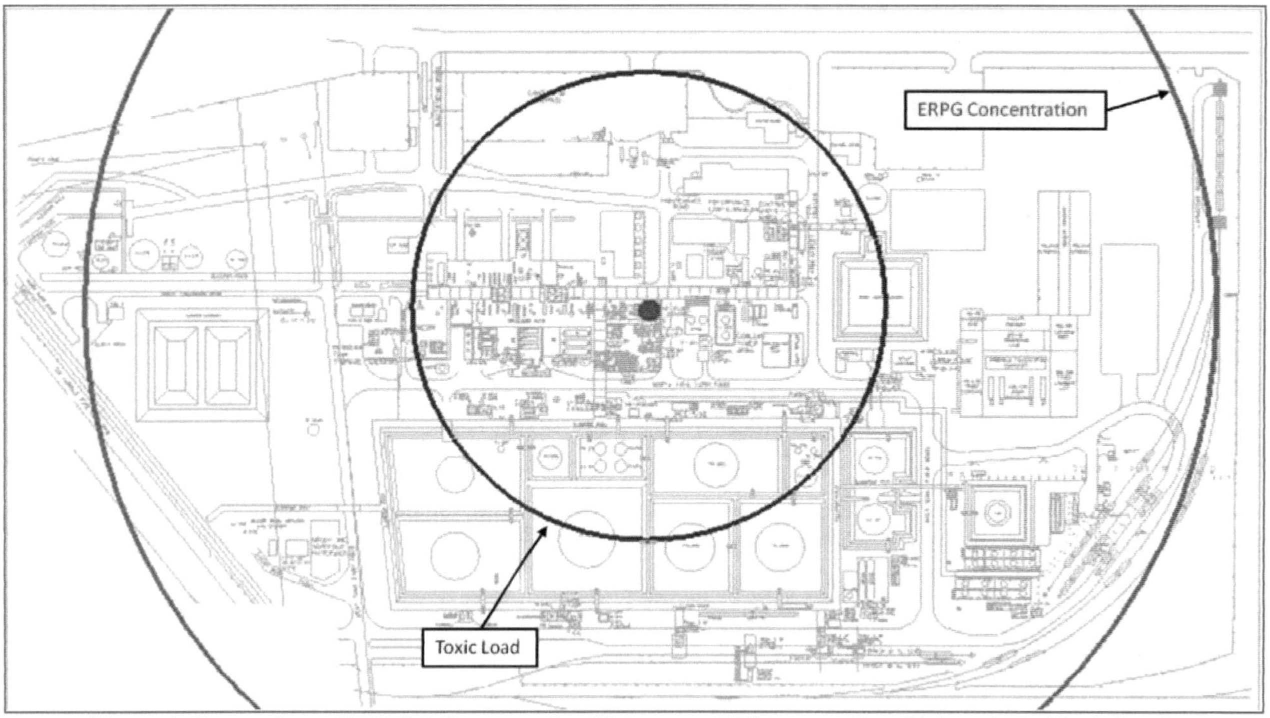

Figure 6. Comparison of ERPG-3 Concentration Distance and Distance to Low Vulnerability Toxic Load

5.2.3 Advanced: Building Toxic Vapor Ingress

Buildings can offer temporary shelter from toxic hazards until the vapors get inside the building above the toxic threshold criterion. Sufficient positive pressure systems prevent the buildup of toxic vapors within the building as long as their ventilation intakes are elevated sufficiently above the cloud and intakes are monitored for vapor ingestion. Therefore, the criterion for inhalation hazards from toxic dispersion is often set to toxic concentration/load around the building. This is a conservative approach since the ingress rate of the vapors into the building is often unknown. As with flammable ingress, a supplementary analysis could be performed to determine the concentration and time history of vapors inside the building.

5.3 Fire Hazard

Personnel within the flame of a pool fire, jet fire, or fireball are assumed to sustain lethal burns, even with flame resistant clothing (FRC). Outside of the flame, the hazard effect measurement is radiant heat flux or thermal load.

Buildings can offer shelter from this hazard until it catches fire or its shielding capability diminishes. If occupants are expected to evacuate, then the vulnerability of evacuees should be established. This can be done by determining the evacuation path, rate and the evacuee's theoretical thermal load over that path. This can be used to determine if the occupants are more vulnerable sheltering or evacuating the building.

5.3.1 Simple: Radiant Heat Flux

Hazard criteria based on radiant heat flux is often used because it is conservative and not a time dependant calculation. Known relationships between radiant heat exposure for long durations and different types of building materials are available. However, if personnel are evacuating from a fire, they will not be exposed to a constant heat flux and use of injury relationships to constant heat flux will produce very conservative results.

5.3.2 Simple: Thermal Load

Thermal load can also be used to take into account the changing heat flux over time. This is best applied to evaluating evacuating personnel vulnerability. Integrating a variable heat flux over the time for someone to evacuate a building will provide a heat load that can be related to potential injury. Prevention of second degree burns is paramount when evaluating safe evacuation routes.

5.4 Building Response to Explosion Hazard

5.4.1 Simple: Free Field Pressure Threshold Tables

Pressure threshold tables contain lookup charts consisting of several simple building construction types related to free field overpressure failure limits. They are simple to use and require limited knowledge of blast effects. There are many issues with using these threshold tables that limit the effectiveness of their use.

Much of the data from which the pressure limits are based are high explosives testing and field measurements dating back to World War II. This is a major problem because petrochemical facilities are primarily subjected to vapor cloud explosions that have greater durations than high explosives. This difference in impulse will create a different structural response at the reference pressure limit.

Another problem is that no impulse threshold exists in these tables. All structures have two limiting load conditions that can cause failure of structural components, the pressure limit and the impulse limit. The impulse limit is related to the duration of the applied the blast load. As the duration of the blast load increases then the impulse applied to the structure grows as well.

Also there are limited construction types in the pressure threshold tables that do not fit many of the buildings found in process facilities.

Overall these tables are not capable of determining accurate structural response for blast loading and are explicitly prohibited by API RP 752 unless their duration or impulse is known.

5.4.2 Simple: Library Building

The library building screening-level analysis uses pressure-impulse (PI) limits of typical buildings developed using a database of typical buildings and components.

The screening-level analysis evaluates the predicted blast loads against a set of PI curves for each building. A PI curve is an iso-response curve (each pressure-impulse pair results in the same response) for a structural member. The advantage of this approach is the simplified approach of having predefined buildings, components, and sets of pressure-impulse threshold combinations. This allows for a quick analysis of structures with limited knowledge of blast effects.

Although this analysis is more accurate than the pressure threshold tables it is still not applicable to all buildings that can be encountered in a petrochemical facility. Variations in building construction are fit to the closest resembling building or component so there is no way to precisely account for the difference in spans, member spacing, components, and material properties. This can result in inaccurate results.

5.4.3 Moderate: SDOF Analysis

The next highest fidelity method for structural damage predictions is the single degree of freedom (SDOF) method. The strength and geometric properties of each component are used to determine the SDOF systems including transformation factors for mass, stiffness, and force. Transformation factors for mass and load are based upon the deflected shape of the system being modeled. Refer to Biggs (Biggs, 1964) for a more detailed discussion of the equivalent SDOF technique. SDOF models are then run repeatedly to develop PI curves. This allows PI curves to be compared to several different blast loads quickly and determine the expected damage to a building component without performing the additional analysis.

This method of analysis can be conducted for each component of every building, and takes into account any difference in construction, while still maintaining an efficient calculation time.

The disadvantage of this approach is that knowledge of structural response to blast effects is required to provide accurate damage predictions.

5.4.4 Advanced: Numerical Modeling

Numerical modeling of structural response is conducted in special cases where an SDOF model is determined to be insufficient because of the need to capture localized failures and the transfer of a load path through a connection.

Accurate results from numerical modeling generate a more realistic representation of the actual structural response of the components of interest and include localized failure modes.

Nonlinear dynamic numerical modeling requires advanced knowledge of structural mechanics and blast effects on structures. The results of numerical modeling are very sensitive to the skills and abilities of the analyst creating the model. Due to the inherent requirements of this method, an accurate model must be generated and the amount of analysis time is greater when compared to the other calculation methods described previously.

5.4.5 Method Comparison Example

The following example demonstrates the wide range of results each of the different methods of structural response result in for a reinforced concrete framed structure subjected to blast loading. The pressure lookup tables, which are the simplest method, dramatically over predict the response (Table 1), as does the nearest FACEDAP(USACE PDC, 1991) library building. The SDOF method also over predicts the response (Figure 7), but the resulting damage is within the typically accepted range of damage. Results for the finite-element analysis were the lowest of the three different methods resulting in approximately half the deflection predicted by the SDOF method.

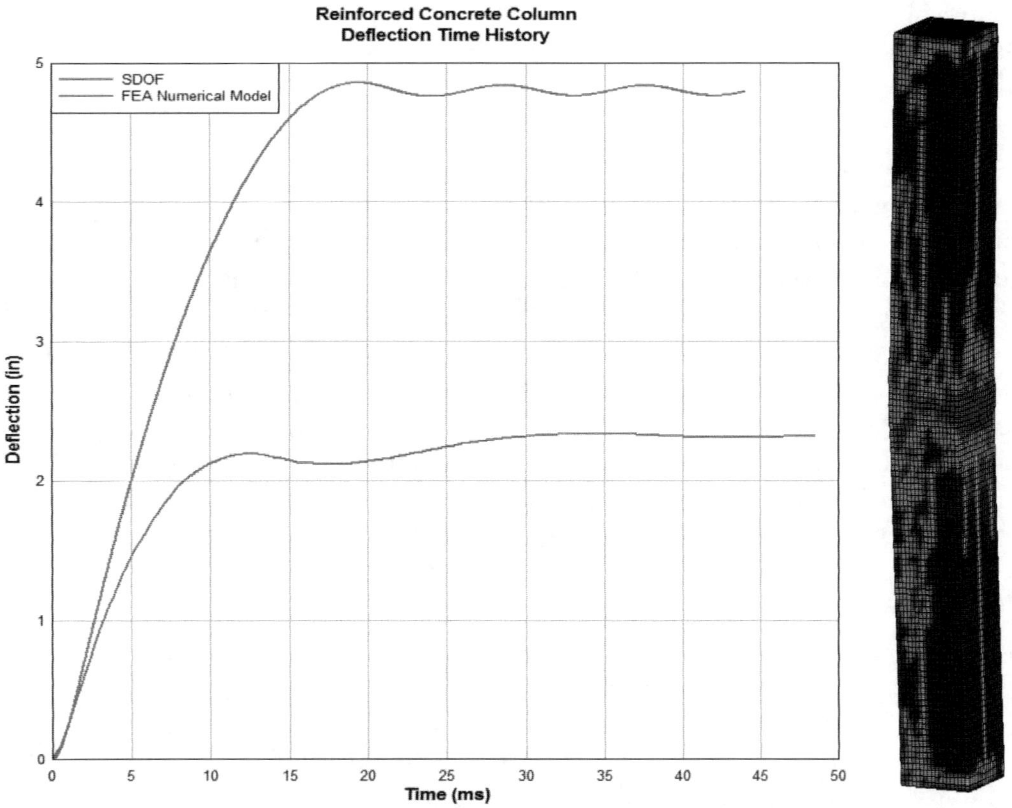

Figure 7. Comparison of SDOF Analysis to FEA Numerical Model

Table 1. Comparison of Predicted Structural Response of Reinforced Concrete Column

Building Response Method	Deflection	Component Response
Pressure Look Table	N/A	Collapse
Library Building	N/A	High Damage ($\theta > 4^o$)
SDOF Model	4.8 inches	Medium Damage ($1^o < \theta < 4^o$)
FEA Model	2.3 inches	Medium Damage ($1^o < \theta < 4^o$)

6 Mitigation Measures

When buildings are identified as providing an insufficient level of protection to its occupants, mitigation measures should be considered to remediate the hazards to occupants. API RP 752 discusses the mitigation approach and provides an excellent table outlining mitigation approaches and ranking them by reliability (API, 2009, p. 10). The most reliable mitigation measures are identified as "Passive Measures" where the measure just being present helps to reduce the risk. Less reliable are "Active Measures" which automatically react to an event in order to reduce the risk. The least reliable are "Procedural Measures" which require personnel to perform tasks to reduce the risk.

Often supplementary analysis is necessary to assess the effectiveness of the measure to remediate the hazard.

7 Conclusion

The facility siting assessment often identifies buildings that are predicted to offer an unacceptable level of protection to its occupants. Guidance has been given on possible paths forward to increase the accuracy of the analysis and decrease the reliance on assumptions and subjective assessments built into many simple approaches. Three paths were discussed that work to establish different levels of analysis accuracy. Paths can be used individually or in combination to reassess the hazards to building occupants. It should be realized that if a "consequence-based" approach was used in facility siting, the mitigation of MCEs leads to new MCEs that may have been screened out as a lesser event, but should now be evaluated. These paths should not be utilized to "game" the analysis to obtain a more desirable answer, but to obtain a more accurate answer with greater confidence that remedial actions are necessary.

8 Bibliography

API. (2009). *Management of Hazards Associated with Location of Process Plant Permanent Buildings* (3rd ed.). Washington DC: API Publishing Services.

API. (2009). *Management of Hazards Associated with Location of Process Plant Permenant Buildings* (3rd ed.). Washington DC: API Publishing Services.

API. (2007). *Management of Hazards Associated with Process Plant Portable Buildings.* Washington DC: API Publishing Services.

Biggs, J. M. (1964). *Introduction to Structural Dynamics.* USA: McGraw Hill, Inc.

CCPS. (2003). *Guidelines for Facility Siting and Layout.* New York: AIChE.

CCPS. (2010). *Guidelines for Vapor Cloud Explosion, Pressure Vessel Burst, BLEVE and Flash Fire Hazards* (2nd ed.). Hoboken, NJ: John Wiley & Sons, Inc.

CPD. (2005). *Methods for the Calculation of Physical Effects* (3rd ed.). The Netherlands.

CPD. (1992). *Methods for the Determination of Possible Damage* (1st ed.). The Netherlands.

Fitzgerald, G. A., Knight, G., & Gutierrez, A. (2007). Application of CFD Models for the Onshore Process Industry. *Loss Prevention Symposium.* Houston, Tx: AIChE.

Herrmann, D. D. (1999). Improved Vapor Cloud Explosion Prediction by Combining CFD Modeling with the Baker-Strehlow Method. *International Conference and Workshop on Modeling the Consequences of Accidental Releases of Hazardous Materials* (pp. 479-494). San Francisco, California: CCPS of AIChE.

Lees, F. P. (1996). *Loss Prevention in the Process Industries* (2nd ed.). Oxford, Great Britain: Reed Educational and Professional Publishing Ltd.

OSHA. (1992). Final Rule on Process Safety Management of Highly Hazardous Chemicals; Explosives and Blasting Agents. In *29 CFR 1910.119.*

USACE PDC. (1991). *PDC-TR-91-6 Blast Analysis Manual Part I - Level of Protection Assessment Guide*. Omaha, Nebraska: U.S. Army Corps of Engineers.

When Management Systems Collide....

Jack Lee, Ed Bailey, Richard Jones
Talisman Energy

Keywords: Management Systems, Standards, Implementation

Abstract

This paper will:

- Outline the recent evolution of the health, safety (occupational and process) and environmental management system at Talisman Energy.
- Describe the objectives and development plan for an integrated HSE management system.
- Document the challenges experienced and the implementation strategy from the initial steps in implementing a fully integrated HSE management system.

Talisman began a number of broader corporate HSE development initiatives in 2007, including a global review of process safety management systems, development of a new corporate HSE policy, and a comprehensive review of the Company's global HSE structure and management systems. The overall goal was to deliver a more resilient corporate HSE function, to improve standardization of the HSE management system and processes worldwide, and to ensure safe profitable growth.

To achieve these objectives in a structured way, Talisman is developing a HSE management system, which integrates the disciplines of health, occupational and process safety, and environment into one system. The integrated management system is derived from a basic quality model driven by key management processes, and contains industry best practices (i.e. CCPS's risk based process safety, ISO 140001, ISO 180001) as well as Talisman's own experience.

The challenges and lessons learned from the initial implementation stages have highlighted the complexity and effort required to combine the different management systems together. Some of the challenges discussed are:

- Breaking down regional biases and functional silos;
- Different organizational maturity in different regions;
- Integrating competing objectives from a multi-disciplinary approach; and
- Chronology of management system.

1.0 Introduction: Scene Setting

As the title of this paper suggests, the integration of different management systems is similar to a nuclear fusion reaction - large amounts of energy and expertise are required to make the process work and be sustainable. Traditionally, organizations have had isolated and separate management systems for health, safety, and environment. Individually, each can stand alone and function well, but in reality there are a lot of common elements within the different disciplines. Using the synergies between them can result in a more robust and effective management system.

Talisman Energy has nearly 20 years of experience in the upstream oil and gas exploration and operation. The company was born from the sale of BP Canada's assets in 1992. Initially, operations were solely rooted in Western Canada; however the desire to grow the company was strong. Within two years, the company expanded their Western Canada holdings through exploration and began international operations with entry into the UK North Sea in 1994, followed by entry into Indonesia in 1997; both through acquisitions.

By 2006, the company was truly established as a global oil and gas company. Three core geographic areas were well established; North America, North Sea and South East Asia. In addition, there are planned expansions into Latin American, Africa, and Middle East. [1]

Despite the rapid growth of the company, each regional Health, Safety and Environment (HSE) management system was largely unchanged and left to the regional management to organize their HSE programs and systems. This philosophy satisfied the desire to keep the programs and systems locally meaningful and address their unique risks and priorities. Unfortunately with a global company, this philosophy resulted in functional silos, little in the way of shared knowledge and knowledge transfer, different priorities across the company from region to region, and lack of consistency in HSE processes.

Talisman started 2008 with a focus on expanding process safety within their existing HSE management system. Additionally, the whole management system was reviewed to determine whether it would meet the future requirements and the challenges of managing the ever increasing risks in our business.

An early decision was made not to develop process safety as another silo management system but rather to integrate it with health, occupational safety and environmental management systems. This required a more global vision than what the current fragmented management system could offer. This desire for a global integrated HSE management system would allow for more consistently applied and clearly communicated expectations, better support for our global operational activities, and the ability to define and set company-wide performance indicators.

This paper documents the development process of the global integrated HSE management system at Talisman, the implementation challenges encountered, and some observations on the next stages that have not been completed.

2.0 Integrated HSE Management System Development

The expansion of the company in the late 1990's through acquisitions meant a number of inherited systems, in which both system processes and biases were inherited. Each acquisition was integrated within the regional operations resulting in multiple independent HSE management systems, where the region had its own priorities and developed its own initiatives. This model served Talisman well in the past, allowing flexibility and freedom in the region to focus their attention on specific HSE needs that were meaningful to them.

Talisman has matured into a global company now and it was determined that the time was right for a new HSE management system. The vision for the new system consists of 5 building blocks:

- *Manage Risk* is the cornerstone of the five blocks. The objective is to improve the hazard identification and risk management in all aspects of our business.
- Embed the *accountability* with the right person, in the right place to ensure HSE is delivered into the operating line.
- Work to *one global management system* that drives continuous improvements across all regions.
- Measure, manage and motivate *performance.* The belief is that one can only improve what one can measure.
- Build *capability* in all staff through education, experience, and exposure.

The focus of this paper is on the development of the global management system.

Figure 1 The Vision

2.1 Structure

With the objective to ensure clearly communicated and consistently applied expectations, it was decided that the management system be developed in a structured way. This hierarchy is shown in Figure 2.

There are three tiers that set the accountability, responsibility and roles. At the top tier, a centralized global HSE function will set the policy, principles and standards. Expertise and guidance will be available to the regions from this centralized global function. At the top level of tier one, a common HSE policy is established for the whole company. Next, guiding principles are set that support the policy. And lastly, standards are developed that sets forth the common vocabulary, expectations, processes, and tools (such as software applications). Essentially, the standard translates the principles into global expectations.

Figure 2 The Management System

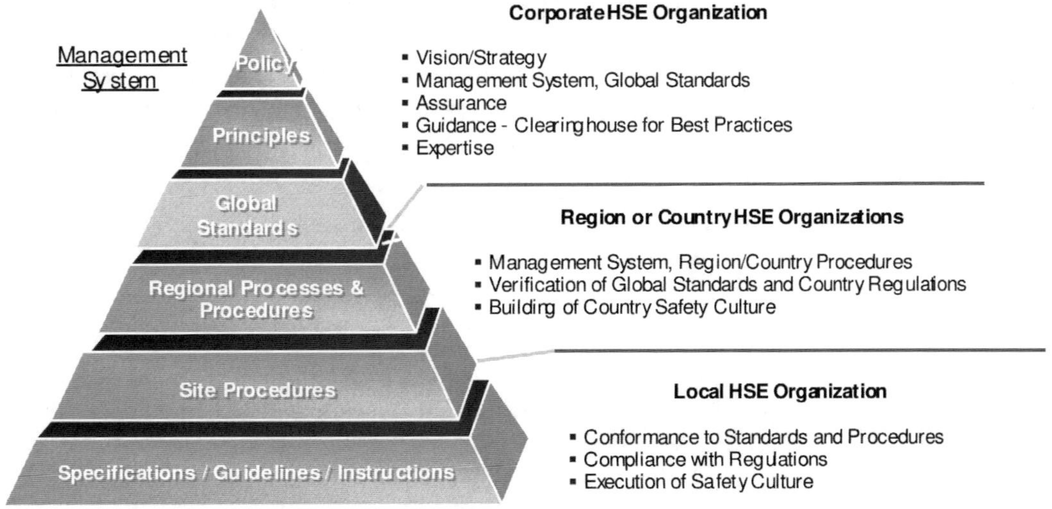

The next tier is the regional or country HSE organization. At this level, regional practices and procedures are established that take into account the global standard's expectations and any unique regional operating requirements including regulations and culture. The regional or country organization is responsible for compliance to the global standards and also the verification of compliance. A very high degree of interaction and communication between the regional or country HSE organization and the regional operational groups is critical at this level.

The last tier is the local site HSE organization. Here the site-specific procedures, guidelines, specifications and instructions are developed and executed by the operational group. If the accountability and responsibility are appropriately embedded into the operational group, it becomes a part of their routine activities and culture.

2.2 Policy

A new HSE global policy was rolled out in January 2009. Similar to the previous policy, the objectives were to create a working environment such that we cause no harm to people and where we minimize our negative impact on the environment. To achieve this, we will:

- Always comply with the law, or Talisman standards, whichever are higher;

- Operate our business to ensure proactive risk management and continuous improvement;
- Set goals and targets, and measure our performance against them;
- Hold ourselves and our contractors accountable to meeting Talisman standards;
- Communicate openly with those who may be affected by our activities.

In addition, safety was adopted as a core company value. The philosophy was that the processes and systems that are required for compliance with business standards, such as Sarbanes Oxley, are the equivalent processes and systems that govern HSE management systems. Improvement in safety performance was not perceived as a cost to the company, but rather an opportunity for better business performance. From this belief, the mantra "**safe, profitable growth**" was born for better managing our risks.

2.3 Principles

Supporting the policy were 12 principles. These 12 principles are comparable to those held by our peer companies and are in the areas of:

- Leadership and Commitment
- Internal and External Communication
- Information and Documentation
- Personnel and Training
- Risk Management
- Management of Change
- Contractor Management
- Facilities, Design and Construction
- Emergency Preparedness
- Operations and Maintenance
- Incident Reporting, Investigation, and Analysis
- Assessment and Improvement

2.4 Standards

To implement the principles, 22 global standards are being developed and embedded throughout the company to drive continuous improvement [2]. The objective of the global standards is to establish a common vocabulary, expectations, processes, and tools for how the company will manage HSE and Operational Assurance around the world to support our global operations. The full development of all the global standards and their implementation is estimated to take 5 years.

Talisman could have decided to procure and implement, or modify, an existing system such as the DuPont STOP program or others. Although we recognized that the adoption of proven best practices may have enabled faster training and implementation, the decision was made to develop the system in-house because each region was at a different HSE maturity level and had strong cultural biases. In-house development allows for a more collaborative approach, greater regional buy-in, and more robust implementation.

In addition, recognition was made that each discipline, such as Occupational Health, Process Safety, Environment, within the HSE management system had similar components to the other disciplines. For example, all disciplines had some form or need for a management of change process. From this realization, it appeared a natural decision to integrate all the disciplines together within one management system.

A consequence of this decision is that global standards were better organized into two types: Foundational and Framework standards. Foundational Standards are standards that have common expectations across multiple disciplines whilst Framework Standards are standards that, for the most part, apply only to one technical discipline. Furthermore, it became clear at the early stages that for this type of management system there was a need for strong collaboration between the different disciplines..

Figure 3 Foundational and Framework Standards

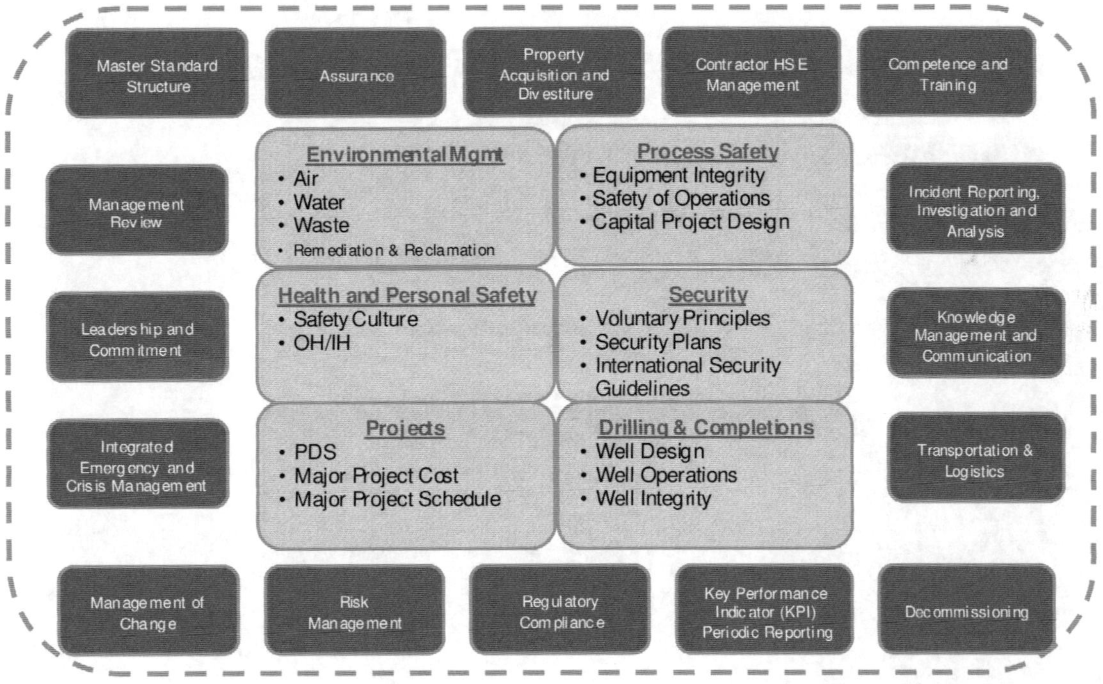

2.5 Regional Practices and Procedures

Underlining the global standards are the regional practices and procedures. Here, the regions develop the detailed practices and procedures whilst taking into account any local regulations and culture, and the requirements for compliance to the standards. A high degree of interaction and communication between the regional HSE organization and the regional operational groups is critical for establishing the right culture and embedding the process into the operation's line activities.

2.6 Standard Development Process

Due to the collaboration required between the different disciplines and between the global and regional functions, a rigorous process was established for the development of the global standards. The standards were developed in a ten-step process which ensured consistent engagement with all the stakeholders both across functions and between the various management levels. Standards were written internally by one principal author and then reviewed by all the stakeholders. This proved more efficient than writing the standards by committee, and also helped break down some of the regional biases and functional silos.

No process is immune from criticism. Having different principal authors caused some confusion because of different writing styles and terminology. The standards were written at a management level rather than for the local user. Lastly, there were some disjointed standards as they were developed singularly and lacked the overall cohesiveness of a set of documents that are all developed concurrently.

Figure 4 Process for Global Standard Development

	What	Who	Key Actions
1	Prioritize & Identify Resources	Global HSSE/OI Leadership Team	▪ Prioritize standards for implementation ▪ Establish PMO role and implementation schedule and qualitative risk assessment ▪ Identify technical authorities to be consulted for Global Standard from corporate and regional staffs ▪ Assign authorship for each Global Standard (Corporate, Regional, or Third Party)
2	Draft Corporate Standard	Designated Author	▪ Draft initial Global Standard based on industry best practices ▪ Circulate to technical authority for comment
3	Provide Feedback	Technical Authority	▪ Provide quantitative and data driven feedback ▪ Feedback process may take several iterations
4	Finalize Standard	Designated Author	▪ Determine when feedback process is complete ▪ Finalize draft Global Standard, noting key areas of dissent if consensus is not reached ▪ Send draft standard to approval authority
5	Approve or Disapprove Standard	Corporate HSSE/OI Leadership Team	▪ Approve or Disapprove submitted Global Standard ▪ If Standard is approved, it is sent forward to VP HSSE/OI for stakeholder review process ▪ If Standard is disapproved it is sent back to author with deficient areas noted for revision
6	Stakeholder Review	Regional VPs and SVPs	▪ SVP HSSE/OI reviews approved standard with Regional VPs/SVPs ▪ Regional VPs/SVPs provide final comment on local implications of approved standard ▪ SVP HSSE/OI determines if approved Global Standard requires revision to address local concerns
7	Ratify Standard	OpCom	▪ If Global Standard is ratified, it is sent to Corporate HSSE/OI for implementation ▪ If Global Standard is not ratified, it is sent back to Corporate HSSE/OI with deficient areas noted for revision
8	Issue Standard	Corporate HSSE/OI Leadership Team	▪ Communicate Global Standards to regions ▪ Identify expected timeline for implementation and compliance with qualitative risk assessment for implementation
9	Implement Standard	Regional HSSE/OI Teams	▪ Identify actions steps needed to comply with Global Standard (i.e., resource loading) ▪ Implement changes as needed ▪ Provide corporate HSSE/OI periodic feedback on implementation plan
10	Confirm Implementation	Corporate HSSE/OI Audit	▪ Follow up within 180 days of full implementation with staff assistance and courtesy audit ▪ Report back to Corporate HSSE/OI team following assistance visit if regions are in conformance, and if a region is deemed not in conformance submit a plan for remediation and follow up assessment

3.0 Implementation Challenges

The following are a number of observations made over the last year regarding the challenges that were encountered whilst implementing the new integrated management system. Of those challenges, there were four main ones:

3.1 Past Biases and Functional Silos

Given Talisman's history, it was not a surprise that there was resistance by established disciplines within a region to a perceived excess of "corporate oversight" on their activities. In fact, there were very good existing practices within most regions. The perception that management was using a top down approach was mitigated largely through the standard development process. Engagement and communication are essential to building buy-in and support from the region. Also, the collaboration with the regions allowed for the sharing of best practices and knowledge that previously did not exist.

However, at times the depth of the established functional silos and past biases, made implementing any rigorous corporate oversight challenging. We found that any weakness in the corporate center allowed for regional interests to drive the agenda. The need for strong corporate governance is necessary to fulfill the mantra of "safe profitable growth", but having the right balance between prescription and guidance is required. It is anticipated that the degree of governance will evolve as the management system develops over time and becomes more embedded.

In a traditional HSE management system ownership is held by the HSE function, often creating competing interests between operations group and safety function. Pushing the ownership of the management system to the line (operations) rather than the function (HSE), removes the competing nature that traditionally existed between them.

3.2 Organizational Maturity

Since Talisman was not building a new management system from a clean sheet of paper, implementation needed to consider the different maturity of the regions. Clearly, there are different levels of operational maturity across the company - from new country entries such as Latin America to well-established areas like Western Canada and the North Sea. This varying degree of knowledge and maturity necessitated a different implementation time scale, different guidance and different coaching for each region. No single implementation approach would fit all. New country entries had nothing to start with so implementation was akin to a new start up with no previous history. For established regions this was more of a continuous improvement and evolution model. Working with the established biases and reshaping them, required a longer time scale and a more collaborative mindset.

3.3 Integrated Systems

Integrating the elements of process safety into this type of management system appears relatively simple and it allowed for the adoption of any of the elements of a process safety management (example CCPS's risk based process safety, DuPont Stop, OSHA, API etc.). However, not having an explicit management system specifically for each discipline within Talisman can make external audits of the management system against any specific criteria (such as OSHA or ISO) very difficult because of the intended integration. Thus, the need for clear documentation, complete with identification of the connections between industry best practice and our own standards, is essential.

Integrating the different disciplines together into one management system is not without its own issues. Should each discipline develop their own management system in isolation and then integrate on implementation, or integrate the disciplines at the start and implement a single system? Talisman has chosen the later. Developing each discipline's own management system has many advantages:

- industry-established systems have been used and applied;
- training material is readily available;
- resources, both people and software; and
- implementation strategies are available to develop such a system.

Integrating all the disciplines into one management system will perhaps not be accomplished by a common approach. There are pit falls and challenges to getting it right, and there is no obvious strategy to follow. The path is unique to the company. In addition, these systems alone have their own objectives that may compete against each other when integrated together. For example, minimizing discharge to the atmosphere and the need to blow down the plant.

As the title of this paper suggests, integration of different management systems is similar to a nuclear fusion reaction - large amounts of energy and expertise are required to make the process work and be sustainable. When the reaction goes awry, there is chaos. When it works, the integrated management system will be more diverse, independent, and robust by design. To achieve this, good communication and collaboration between all stakeholders are required to get it right.

3.4 Chronology

The time scale adopted to develop and implement the management system is over 4-5 years. Over this period all 22 standards will be developed and implemented. The first few standards implemented will not be fully supported by other standards as they have not been fully developed. As a result, the system is weakened by the absence of not having all standards available at the same time. This choice was made so as to not overwhelm the operations groups.

As standards will be developed on different time scales, there will be some disjointedness inherited in the implementation. It is important for the system architecture to be aware of how everything interacts, and understand there will be periods when standards are revised in order to smooth out any discontinuities between them.

3.5 Other Challenges

During the development and implementation of the management system, some other challenges arose.

3.5.1 Lifecycle Challenges

The integrated management system was developed not just for HSE but to include the full lifecycle of the operation. The full lifecycle requires input from exploration, drilling and completions, engineering design, construction and commissioning, operations and maintenance, and finally decommissioning to be integrated into the HSE management system that traditionally functioned in isolation. Including these functions into the management system fostered increased communication across the various disciplines and different functional groups.

The ideal system would have the HSE management system woven into the existing business processes and systems such that activities would be performed in the normal course of business without competition between operation and safety, and between different functions. This weaving of the system throughout the lifecycle enhances the system and can bring about other business benefits, such as inherently safer design.

3.5.2 Regulation Challenges

Regulation regimes in each region differ - ranging from very little regulation, through highly prescriptive, to goal seeking. For a global company, this variation in regulation allows the company to operate to the minimum legal standards for a region in the absence of any corporate standards. In lightly regulated or un-regulated countries, corporate standards are often more stringent and can infer that the region is marginal or uneconomical. An integrated HSE management system allows for consistent and clear communication of expectations, better supports for our global operational activities, and defines and sets company-wide performance indicators regardless of the regional regulatory regime.

3.5.3 Cultural Challenges

Within different countries, or even regions, come many different cultures. The culture in an operating area can have a significant impact on the management system. If all regions are to work as one, implementing a single management system requires clear understanding of the different values and norms that the workforce has. A lack of cultural sensitivity can result in implementation stagnation, lack of engagement, and resistance to change.

Culture within a company is the first, middle and last element that must be constantly worked on. Developing a strong culture takes time and strong leadership to get right and to sustain it. Even the best implemented management system will fail in the absence of a strong workforce culture.

4.0 Conclusions and Future Challenges

Integration of the various management systems into one cohesive system is challenging but potentially rewarding from both a safety and business perspective. The development of the management system should be flexible yet sufficiently prescriptive to resist regional fragmentation.

Taking a collaborative approach in the development and implementation of the management system, rather than a competing one, has been successful in removing some existing regional biases and functional silos. Engagement and communication are key to building buy-in and support from the region. This collaboration has also fostered a knowledge sharing approach between regions.

The varying degree of knowledge and organizational maturity necessitated a different implementation time scale, different guidance and different coaching for each region. No single implementation approach would fit all. A structured development and implementation process ensures proper engagement and buy-in from the various regions.

As Talisman progresses with the journey of our new management system, some future challenges include:

- implementation of the standards,
- consistency across the company,
- cohesion in the regions,
- equalization of operational maturity,
- evolution and continuous improvement of the standards, and
- evaluation of the effectiveness of the integration.

The performance of the management system, once implemented, should be directly evaluated against these criteria, and will be the subject of a future paper.

References (examples given to show format)

[1] Talisman Energy Website, www.talisman-energy.com.

[2] Center for Chemical Process Safety, "Guidelines for Risk Based Process Safety," Center for Chemical Process Safety/AIChE, 2007

Additional references not cited:

[3] Cadigan, Michael F., Stough, Joe., "Implementation of a Management System Execution Framework to Establish a Culture of Accountability and Improve Safety Performance," 2007 SPE E&P Environmental and Safety Conference, SPE 106825, March 2007.

[4] McBride, Marc., Chen, Harry., "Process Safety Management in the Energy Sector," 6th Global Congress on Process Safety Conference, March 2010.

[5] Ramaswamy, Ravi., "Journey towards PSM Excellence," 6th Global Congress on Process Safety Conference, March 2010.

[6] Davis, Karen., Rasmussen, Elmer., "Implementation of Process Safety Management for Non-Petrochemical Industries," Session 505-2, ASSE Professional Development Conference and Exposition, June 9 - 12, 2002, Nashville, Tennessee

[7] Hart, Katherine A., "Safety and Health Integration: Understanding Systems Thinking to more Effectively Implement ANSI Z10," Session 636, ASSE Professional Development Conference, June 24 - 27, 2007, Orlando, Florida

PSM Metrics Impact on Process Safety Culture

Dawn Wurst
Safety Manager
Flint Hills Resources, LP
P.O. Box 64596 St. Paul, MN 55164
dawn.wurst@fhr.com

Bill Patton
Asst. Safety Manager
Flint Hills Resources, LP
P.O. Box 64596 St. Paul, MN 55164
bill.patton@fhr.com

Keywords: process safety, PSM, culture, refinery, union, USW, metrics, improvement

Abstract

Five years ago our facility recognized the need to expand from a PSM Program managed by a small group to a PSM Culture owned by the full facility. Development and implementation of effective and consistent PSM metrics were utilized to achieve the following:

(1) Set the vision for a PSM Culture
(2) Gauge basic awareness
(3) Educate on basic PSM elements
(4) Identify gaps via near miss incidents and
(5) Build momentum for improvement- an effective 'Case for Change' short of a major incident

This paper covers the path of this facility through these five points and provides example tools, methods, and results of each step along the way towards PSM culture improvement.

1. Background

The Flint Hills Resources (FHR) Pine Bend Refinery (Pine Bend) is a 320,000 barrel per day capacity refinery built in 1955 which processes sour crude oil primarily from Canada, and is the 11th largest refinery in the Unites States. Pine Bend is situated on 1,000 acres southeast of the Minneapolis/St. Paul, Minnesota metro area, supplying 40% of the petroleum products used in the upper Midwest. The refinery employs 970 employees and approximately 500 contractors on a daily basis, ramping up to 2000 contractors for large Turnarounds or unit maintenance outages. The facility front line workers are members of the United Steel Workers (USW) Union-662.

Figure 1: Flint Hills Resources Pine Bend Refinery, Rosemount, MN

The refinery is staffed with a Process Safety Management (PSM) Group within a 29-member Safety Department, and includes a PSM Manager, MOC Engineer, PHA Leader, and PSM Technician.

2. The Challenge

Like many refineries its age and size, Pine Bend had implemented its PSM Program back in the early 90s. The compliance aspects of the PSM Program were implemented, and Process Hazard Analyses (PHAs) were occurring and improving with every cycle. After several successful 3-year compliance audits, the PSM Program was 'under the radar' from a full organization perspective.

The leadership vision for PSM, which had existed for a number of years, was "No one gets hurt from process safety incidents."

In early 2007, the refinery Safety Department developed a comprehensive measures sheet to include key metrics from each area of the safety business. The metrics included traditional personal safety measures such as near miss and recordable injury counts, counts of employees with restrictions, number of employee safety suggestions made, and number of field permit and housekeeping/VPP audits. The metrics list, though was lacking in PSM-focused outcomes to analyze performance.

During an oversight meeting led by FHR Leadership who was visiting from FHR headquarters in Wichita, Kansas, Pine Bends' Safety Managers were challenged to create a more complete measures set which would reflect the refinery's performance in PSM specifically.

Our refinery, like many others, is a very metrics driven organization. "What gets measured gets done" is especially applicable in this case. Although it wasn't specifically stated, the challenge of creating more PSM measures was setting the vision for the entire organization to become engaged in understanding and improving process safety.

2.1 *Industry PSM Metrics*

Starting the effort to meet the challenge, Pine Bend analyzed CCPS metrics guidance, as well as those of the American Petroleum Institute (API) and the National Petrochemical and Refiners Association (NPRA). Reviewing all of those sources, Pine Bend supplemented the existing measures set to include several new line items such as count of late action items, number of missed Management of Change (MOC) jobs, late PHAs, late completion of PHA action items, as well as a number of mechanical integrity metrics.

Many of these measures were perceived as compliance metrics rather than measures of process risk. Exceptions to this perception were mechanical integrity (MI) metrics, since this change started to clearly show MI as a process safety indicator, instead of purely a reliability or unplanned event driver.

2.2 *The Missing Link*

Though all of these metrics are useful indicators for PSM system health and started showing things in a different light, there was still an element missing. None of these metrics were the type to be reviewed on a daily basis, analyzing the last 24-hours of refinery operation. As previously mentioned, Pine Bend was already a very metrics-centric facility. The daily 'Morning Summary', which includes hundreds of metrics used to gauge all aspects of the operation, is used as a foundation of daily meetings throughout the plant.

Without PSM as a part of daily conversation, Pine Bend as a facility had a low awareness of PSM. In fact, if you would have polled the majority of employees, they would not have been able to tell you what the acronym PSM *stood* for.

Safety Department Leadership decided to attempt to fill this gap by determining criteria for an incident in the refinery to be flagged as a PSM risk. The intent was to keep a pulse on the day-to-day PSM performance of the facility, and addressing gaps found via minor incidents, while also answering the question "How are we doing today?" from the perspective of PSM.

2.2.1 PSM Event

Similar to a spill being categorized as an 'environmental event', six categories of event types were created and labeled 'PSM Events', meaning incidents that occurred within the refinery that could be considered a leading indicator to a larger event impacting PSM, such as a release of 10,000 lbs or more of a hazardous substance per OSHA 1910.119.

The first generation of the categories included the following:

1 Fire or explosion affecting or with potential to affect process equipment
2 Mechanical failure resulting in a pressure breach significant enough to require emergency response
3 Process PSV/Rupture Disc overpressure (excluding thermal relief)
4 Safe Limit of Operation (SLO) exceeded
5 Critical Safety Instrumented System triggered or should have triggered
6 Exceeded equipment design limit (temperature/pressure; Excludes those with established SLOs)
7 Operation of Emergency Isolation Device to gain control of an event

3. Analysis of Current Performance

With the categories determined, Pine Bend analyzed all 2007 daily reported incident data against the criteria to determine how many "PSM Events" had occurred. Even within the first few months, the results were surprising. Pine Bend as a whole was having a Safety System triggered every 3-4 days, with Safety Systems defined as a safety instrumented system which serves as a layer of protection in a Process Hazard Analysis. This was much more frequent than either the PSM Group or the Refinery engineering and leadership teams had estimated.

Pine Bend also used existing data from Safety Instrumented System (SIS) tracking. The Instrumentation Group had been tracking spurious trips for a few years, but that data had not been viewed by other departments.

3.1 Sharing the Results

Analysis of the first 4 months of performance, along with the descriptions of the categories were communicated by Safety Management to the refinery production and engineering leaders.

3.1.1 Change in PSM Perspective

The first time the categories and facility results were shared, several managers expressed confusion about the incidents that were being flagged as a PSM Event. For example, if a pressure safety valve (PSV) relieved to flare, it was now called a PSM Event, requiring an investigation. General reaction was, "Why? It did what it was designed to do and there was a safe result by routing the excess material to the controlled flare system."

The missed point was seeing the PSV overpressure event as a near miss PSM Event. These same managers were asked to answer the question of what would have happened if the PSV did not function as designed? It was explained that if we determined the cause of the overpressure, we could prevent a potential explosion in the future, similar to investigating a fall where no one was hurt because they were wearing their fall protection.

3.1.2 Limits Focus

Concurrent with the PSM event growing focus, engineers were encouraged to 'Know Your Limits'. Close analysis of some refinery unit constraints found that they were based on a safety system limit (i.e. PSV Lift pressure). Several risk profile discussions occurred, questioning if the unit constraint should be shifted down to an Operating limit rather than leaving it at the Safe Limit.

3.1.3 PSM Events go Mainstream

After the PSM Event categories and initial data had been vetted with refinery Managers and corporate leaders, Pine Bend added a PSM column to the daily Morning Summary. This column allowed every reported event to be evaluated for its potential to be considered as a PSM Event. Safety Leadership at the same time worked to search for ways to discover that incidents that met the criteria were occurring, such as electronic records that safety systems had triggered. The Organization was then required to report these events through the refinery's normal incident tracking system and perform an investigation.

Refinery shift managers were trained on the categories and criteria so that they could classify the events as they came in as well.

The addition of PSM to the Morning Summary proved to be the most impactful change made to improve PSM Awareness at this stage, because it started the refinery personnel asking questions about PSM. "Why is this incident PSM? How does this impact PSM? What is PSM again?"

4. "It CAN Happen Here"

Eight months into the PSM Event genesis at Pine Bend, an alarming event occurred which sent shock waves through the refinery.

4.1 The Explosion

On August 14, 2007 an employee in one of Pine Bend's Polymerization Units was adjusting a chemical addition pump located near a caustic treatment vessel and an atmospheric tank when suddenly the tank exploded, launched off its pad, and came crashing down in an open area nearby. A ground fire followed which was extinguished within one hour. Pine Bend was extremely fortunate that no one was injured due to the explosion, including the employee who stood within 40 feet of the tank location. Significant equipment damage resulted.

The explosion served as a major culture wake-up call the entire refinery. 2 years prior to Pine Bend's explosion, when the BP Texas City incident occurred, there was an attitude by many assuming "that couldn't happen here." The August explosion changed that attitude dramatically.

The explosion was also unfortunate proof that a PSM vision that drew the line at an injury occurring as the result of an incident ("No one gets from a process safety incident") was not aggressive enough. Over the next year, Refinery Leadership revised the vision to state, "Significant process safety incidents do not happen while the facility stays in full PSM compliance."

Figure 2: Atmospheric Tank after explosion

4.2 The Investigation

A detailed investigation immediately followed, and ultimately determined that a combination of contributing factors was involved.

Contributing Factors:

- Air introduction into the tank during normal atmospheric vacuum breaker operation.
- Vaporizing LPG was present in the upstream vessel
- No vortex breaker was present in the upstream vessel
- 2004 PHA on Tank did not note explosion risk represented by air/hydrocarbon mixtures
- 2004 PHA on upstream vessel did not note a potential explosion scenario or vapor entrainment risk inside the atmospheric tank
- Flare gas components could remain in headspace of upstream vessel
- Contents of upstream vessel were able to siphon into atmospheric tank

- Procedures did not address risk of simultaneously draining an LPG Contactor into the upstream vessel while dumping the upstream vessel into the atmospheric tank

4.3 Impact on the Organization

As Pine Bend learned more about the factors that caused the explosion, their hunger for PSM-related information grew quickly.

4.3.1 Sense of Unease

Concern grew rapidly in the process units, wondering what was in place to prevent a similar occurrence in other process areas. Suddenly the PSM Group was receiving frequent questions and requests for unit hazard information. Many employees were unfamiliar with the PSM regulation, specific elements, and requirements, despite past training efforts by the PSM Group.

Similar to conclusions from the BP Texas City explosion in 2005, results of the Pine Bend tank explosion indicated that better adherence to the existing fundamentals of the PSM rule could have prevented the explosion.

4.3.2 Changes to PSM Event Categorization

Safety Department Management realized after the tank explosion occurred that there was a fundamental flaw in the existing categories defined for PSM Events. The measurement system did not distinguish between (1) a low impact pressure relief device activation and (2) a major tank explosion. Severity had not been considered in the categories.

To address this gap, the PSM Group added another dimension to the PSM Categories, converting the categories into a matrix which incorporated severity at 3 levels. 3 levels were selected because that division was consistent with other severity levels already in place for other refinery metrics. Levels were labeled A, B or C, with C being the most severe.

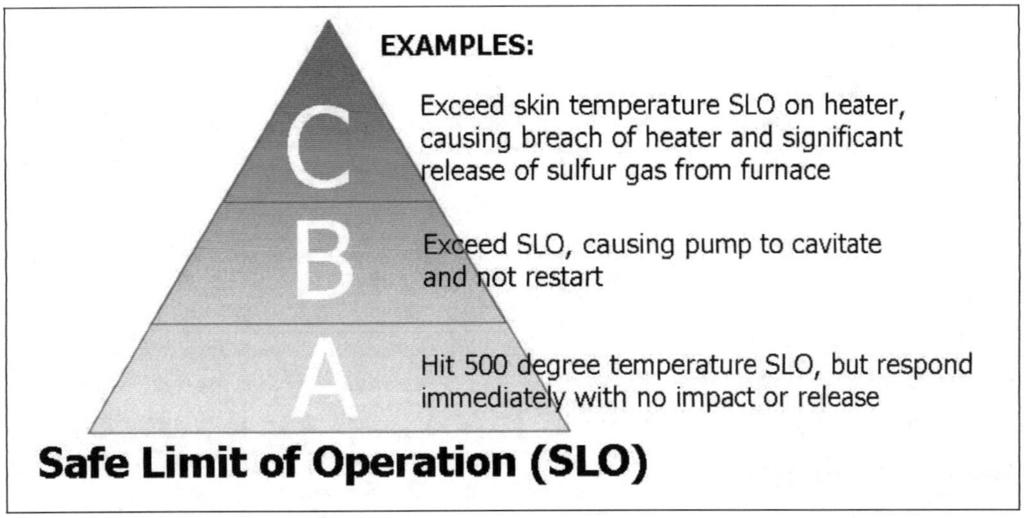

Figure 3: PSM Event Examples

The full matrix was further developed with detailed criteria for each category and severity level. The matrix, shown in Figure 3, was reviewed by Refinery Leadership and mid-level managers to validate that the criteria were clear that that the risk profile for the most severe (C) level was agreed upon.

Criteria for the C PSM Events was designed to align with existing CCPS and API criteria, such as $25,000 direct cost resulting from a fire or explosion event and release threshold quantities referenced by the CCPS Process Safety Leading and Lagging Metrics document[i]. Use of these benchmarks allowed Pine Bend to have an objective view of the most severe events relative to the rest of industry who were tracking Process Safety Incidents.

The differentiation of Pine Bend's efforts from CCPS and API metrics was the fact that <u>two</u> lower levels of severity were also tracked and investigated as leading indicators to larger events.

PSM A Events, the least severe of the 3 levels, have carried the most significant and numerous learnings for the Pine Bend Organization.

FLINT HILLS resources

Process Safety Event (PSE) Classification

Type	PSE A	PSE B	PSE C
General Description	*Leading Process Safety Indicator : Event which serves as a warning signal for a potential Process Safety Issue*	*Lagging PSM Indicator : Process Safety Event*	*Lagging PSM Indicator : Process Safety Event with 2nd or 3rd Order consequences*
Fire	1 Class A, B, or C fire in a process area that does not have the potential to cause a loss of containment 2 Fire resulting from a release of flammable or combustible process material in an process/storage/utility area which results in less than $5,000 direct cost to FHR 3 Minor release of flammable material that flashes and immediately extinguishes itself due to depletion of fuel	1 Fire resulting from a release of flammable or combustible process material in an process/storage/utility area which: - results in between $5,000 and $25,000 direct cost or potential direct cost to FHR; <u>or</u> - had the potential to cause serious injury (i.e. Lost Time Injury)	1 Fire resulting from a release of flammable or combustible process material in an process/storage/utility area which: - results in greater than $25,000 direct cost to FHR; <u>or</u> - caused serious injury (i.e. Lost Time Injury)
Release [Meets criteria in ≤ 1 hour-duration]	1 **Flammable or Combustible Liquid:** Minor release above flash point and greater than **5%** but less than **10%** of chemical release threshold quantity (i.e. 110 lbs or 18 gallons of gasoline) 2 **Flammable Gas:** Minor release. Volume greater than **5%** but less than **10%** of chemical release threshold quantity <u>OR</u> release is not visible (can be detected more than 1 foot away from leak point using LDAR or VOC detection equipment) 3 **Toxic Gas:** Minor release. Volume greater than **5%** but less than **10%** of chemical release threshold quantity <u>OR</u> potential for worker exposure 4 *Combustible Dust: [Pending]*	1 **Flammable or Combustible Liquid:** Release above flash point and equal to or greater than **10%** but less than **100%** of chemical release threshold quantity (i.e. 220 lbs or 37 gallons of gasoline) 2 **Flammable Gas:** Release greater than **10%** but less than **100%** of chemical release threshold quantity <u>OR</u> potential to cause serious injury with visible vapor cloud 3 **Toxic Gas:** Release greater than **10%** but less than **100%** of chemical release threshold quantity <u>OR</u> potential to cause serious injury with visible vapor cloud 4 *Combustible Dust: [Pending]*	1 **Flammable or Combustible Liquid:** Release in process, storage, or utility area <u>equal to or greater than</u> the chemical release threshold quantity (i.e. 2,200 lbs or 367 gallons of gasoline) 2 **Flammable Gas**: Release in process, storage, or utility area equal to or greater than the chemical release threshold quantity <u>OR</u> caused serious injury, with visible vapor cloud beyond immediate area or near an ignition source. 3 **Toxic Gas**: Release in process, storage, or utility area equal to or greater than the chemical release threshold quantity OR caused serious injury 4 *Combustible Dust: [Pending]*
PSV Relief [Excludes Thermal Reliefs unless Release is created]	1 PSV relief to flare system 2 PSV Relief to atmosphere where no visible vapor cloud is formed;	1 PSV relief to atmosphere resulting in a B release (see above) <u>or</u> where a vapor cloud is formed;	1 PSV relief to atmosphere resulting in a C release (see above) <u>or</u> ground-level vapor cloud;
Safe Limit of Operation / Hard Limit	1 Safe Limit of Operation/Hard Limit exceeded during normal operation with no impact or release	1 Safe Limit of Operation/Hard limit exceeded during normal operation causing an A or B release (see above) <u>or</u> equipment damage.	1 Safe Limit of Operation/Hard limit exceeded during normal operation causing a C Release (see above).
Safety Systems	1 Activation of a Safety System during normal operation 2 Safety System found to be ineffective during routine test/ inspection.	1 Unsuccessful activation or absense of Safety System during a process upset 2 Safety System failure to activate or restriction present in a safety system	1 Safety System failure to activate or safety system restriction that results in a C Release (see above)

Figure 4: PSM Event Matrix

In addition to daily tracking of PSM Events on the Refinery Morning Summary, Safety Department Management created PSM 'Smart Cards' for use by the entire refinery, explaining the basics of PSM on a reference card. The reference card, which included the PSM Event category explanation, was rolled out to all refinery Supervisors, and used in a series of safety talks.

SMART CARD LEADER COMMUNICATION BRIEFING 1

TOPIC: Process Safety Management
Focus: What is PSM? / What is a PSM Event?

RLT/PLT talking points for discussion with Production Area Teams.

What is PSM?

- OSHA Standard put in place in 1992 as result of petrochemical disasters impacting communities and employees, including the Phillips Petrochemical disaster in 1989 which killed 23 and injured 300 in Pasadena, TX, and the 1990 Arco Channelview chemical waste tank explosion that killed 17.
- PSM applies to facilities which have processes involving (A) a chemical on the PSM hazard list above the specified threshold quantity or (B) 10,000 pounds of a flammable gas or liquid in one location.
- The purpose of PSM is to prevent catastrophic releases of hazardous chemicals in areas where employees could be exposed to serious harm

What is Pine Bend required to do?

- OSHA 1910.119 requires us to have a Process Safety Management Program, which includes 14 different elements:
  - Employee Participation
  - Operating Procedures
  - Process Safety Information
  - Process Hazard Analysis
  - Training
  - Mechanical Integrity
  - Management of Change
  - Pre-Startup Safety Review
  - Contractor Program
  - Safe Work Practices
  - Incident Investigation
  - Emergency Planning
  - Compliance Audits
  - Trade Secrets

2

TOPIC: Process Safety Management
Focus: What is PSM? / What is a PSM Event? SC-08-07

What is a PSM Event?

A PSM Event is a situation/incident which caused or could have potentially caused a *catastrophic release*, defined as a major uncontrolled emission, fire, or explosion involving one or more highly hazardous chemicals, that presents serious danger to employees in the workplace.

FHR is looking at PSM Events in these categories:

Category	Example
Fire/Explosion	Coker feed line fire - valve body holed through due to manufacturing defect
Toxic Release	Chlorine gas release (chlorine is on PSM Chem. List)
Significant Mech. Failure	Leak on 1200# H2 header at manifold - fatigue cracking near welded joint
PSV relief	Compressor spillback system overpressure - lifted H2 PSV
Safe Limit of Operation	While trying to max feed drum temp, hit 500 degrees SLO on TI
Safety Instrumented Systems	Compressor trip due to high level in knockout vessel; Heater Trip due to high Fuel Gas pressure

Pine Bend has started to track PSM Events, along with all other KII facilities. You will soon see 'PSM Event' as an incident type on the Morning Summary. The purpose is to track where these incidents are occurring and why, in order to improve the health of our Process Safety Management Program.

Figure 5: PSM Smart Card

4.3.4 Realizing the PSM Awareness Gap

Judging by the interest in the PSM Smart Card and the impression by the many in the organization that PSM was a 'new' concept, Safety Management moved to start a large PSM Awareness campaign in 2008.

5. PSM Awareness

Though many aspects of PSM are complex and technical, the most significant PSM gap inside Pine Bend was cultural. Employees did not understand the requirements and systems that constitute the PSM Program, therefore they could not tie PSM to their day to day business. Pine Bend's PSM Awareness strategy included graphics and posters, routine safety talks, training presentations, and new web pages to make PSM a very visible part of daily work.

5.1 PSM Elements

Initial training efforts found that the 14 elements of the PSM rule could appear complex to employees not already familiar with the requirements. Safety Management sought to define the elements in terms employees could relate to. To graphically show the elements, one piece of equipment was chosen that was recognizable by all employees.

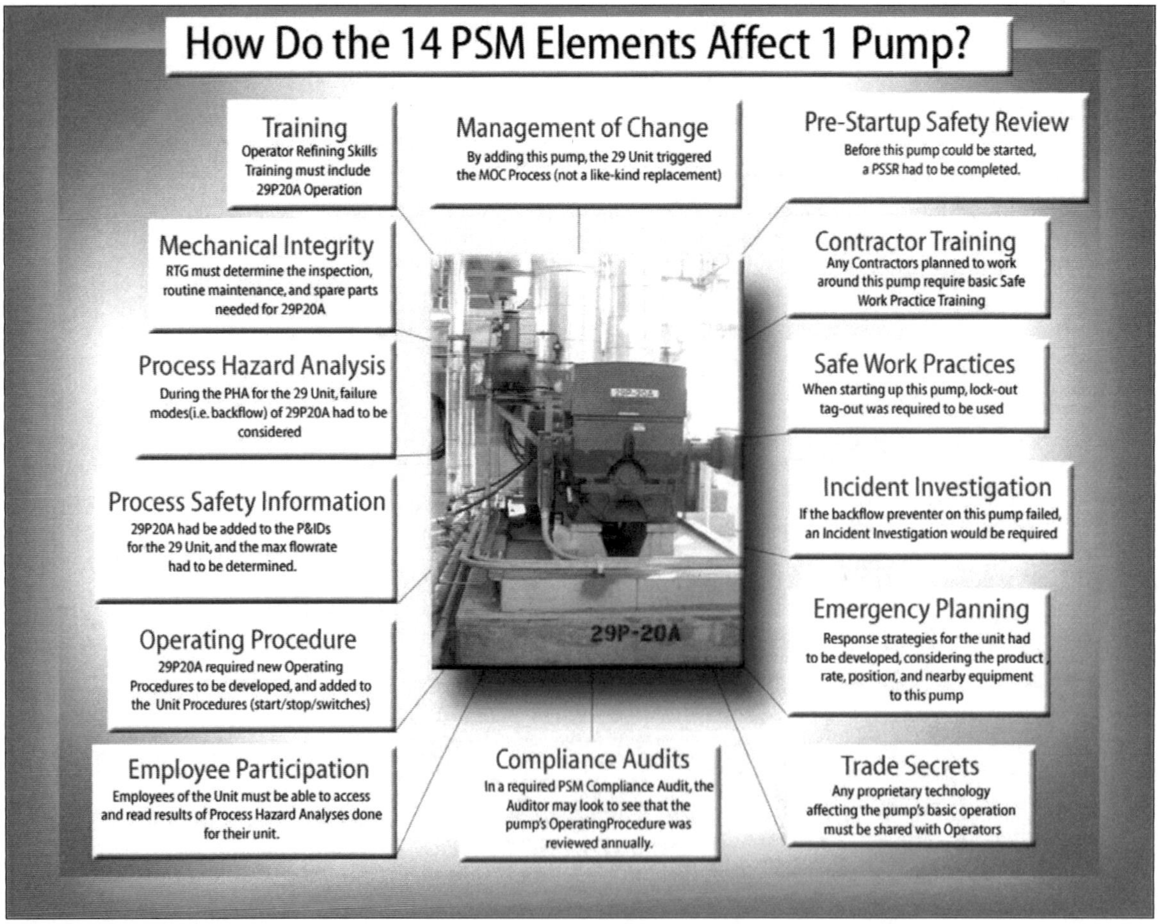

Figure 6: PSM Element Pump Poster

This poster was mass-produced, and used as a training tool and reminder in control rooms and shops throughout the refinery how PSM applies to a simple piece of equipment. Later versions of the poster were customized for Maintenance workers and Equipment Operators, in their own words.

To show how the elements fit together, another graphic was developed, showing each element as a literal layer of protection such as a bullet-proof vest.

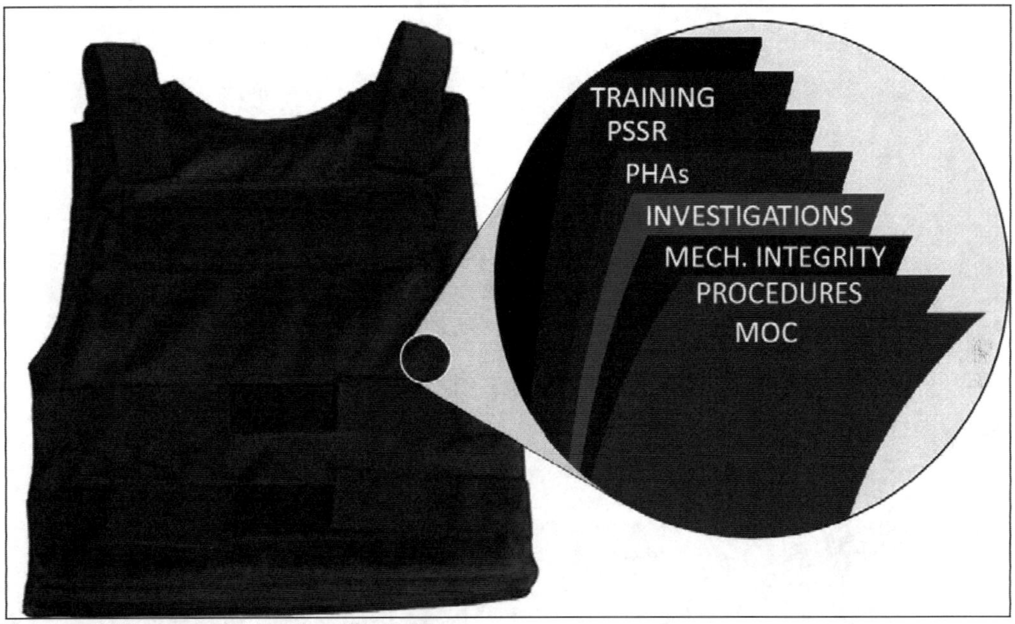

Figure 7: PSM Layer of Protection Illustration

Both graphics were useful, but did not address the question of "Why should I investigate minor incidents?"

5.2 PSM Events Case for Action

In 2008, with both the BP Texas City incident as well as Pine Bend's own explosion fresh in employees' memories, a poster was developed to show the iceberg theory as it relates to PSM. PSM was compared to personal safety, with the worst consequences 'lurking under the water' if the organization does not learn from the most minor consequences or 'near misses.' This clearly illustrated the similarity between a first aid/fatality comparison vs. a minor PSV overpressure/BLEVE comparison.

The PSM Iceberg Poster was developed using incidents that most employees would be familiar with. The poster was very useful when answering the question of "Why should I investigate a minor PSV overpressure or Class A fire?" Employees could relate to the fact that root causes of minor incidents, such as unexpected vapors present in a process area could be leading indicators of much larger and more hazardous events.

Figure 8: PSM Iceberg Poster

Though Personal Safety and Process Safety have similarities, it was also important for the organization and management to view them as discrete and unique programs with specific metrics and programs all their own.

Figure 9: Safety Binoculars- Seeing Personal Safety and Process Safety as 2 distinct types of Safety

The emphasis on separating personal safety from process safety served as an 'a-ha moment' to the organization, causing them to reflect on the fact that a low recordable incident rate (*personal* safety) has almost nothing to do with your *process* safety performance.

5.3 PSM Overview Training

Each of the above graphics was also included in an overview face-to-face training session with all refinery Supervisors. The strategy was similar to past strategies of educating the Supervisors so they could then teach their teams the basics about PSM.

5.4 PSM Weekly Safety Talks

To compliment the PSM Overview training, the entire PSM Program was broken down into a series of one-page weekly PSM Safety Talks. Two PSM Elements were covered per month, with the first week covering the definition and the second week including an industry incident caused by a weakness in that element.

Figure 10: Sample PSM Safety Talks

5.5 Gaps in PSM Data Access

After some of the first PSM Awareness discussions, it became clear that employees were not aware what PHA was, let alone being aware of how they could access the important information.

PSM information being sought by the organization included:

- Current Process Hazard Analysis Reports for all Units
- Copy of OSHA's PSM Rule (1910.119)
- Criteria for PSM Events & examples of actual events
- Industry event information, such as links to videos by the Chemical Safety Board (CSB)
- Pine Bend PSM Event Investigation Summaries
- Pine Bend PSM Event Performance year to date
- PSM Safety Talks (full series)

This PSM information gap was filled with a custom designed intra-company webpage. Access to the webpage was positioned in a very visible place for employees to see from the central intranet webpage, and the PSM page was promoted as a new source of key safety information.

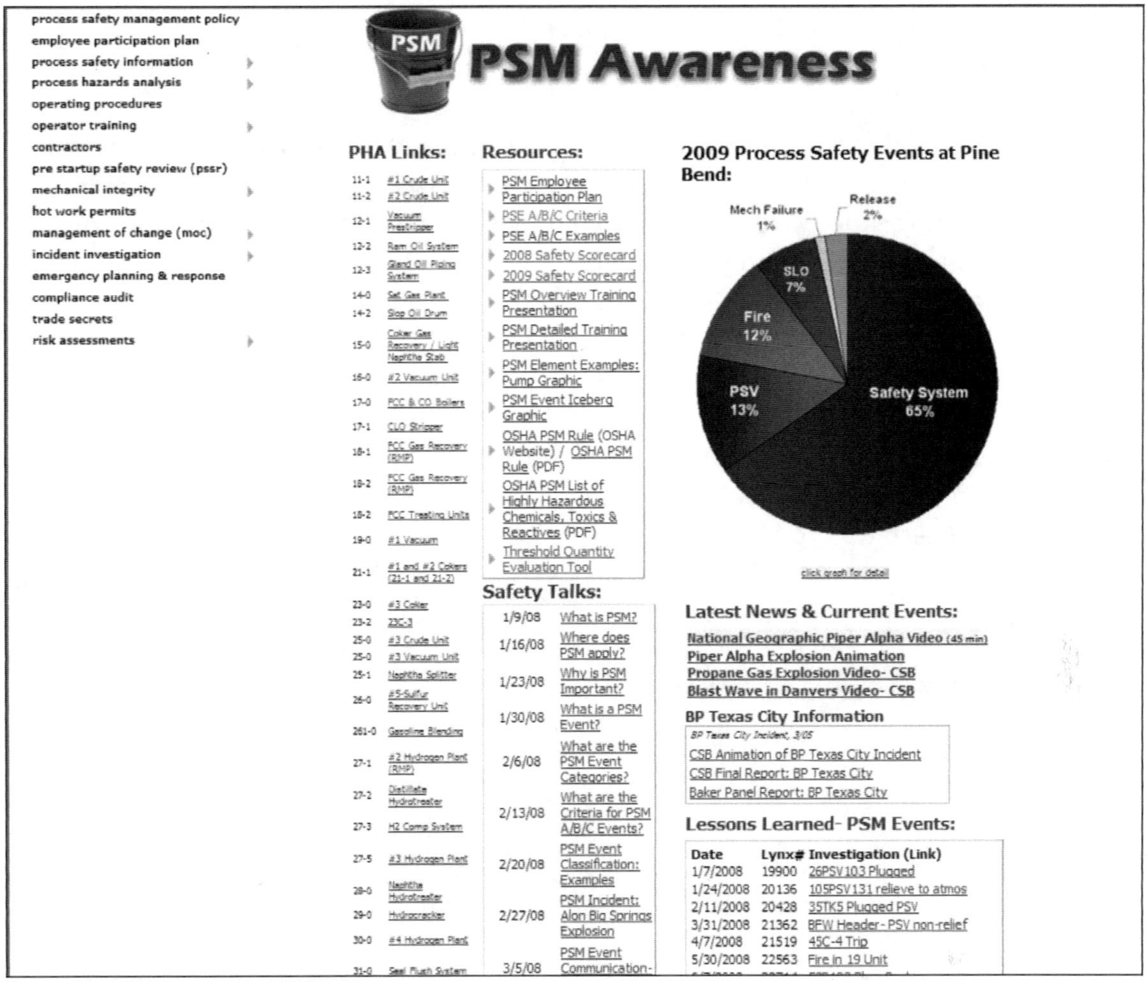

Figure 11: PSM Awareness Webpage

With all of these PSM Awareness tools in place, Pine Bend was then situated well to feed the organization frequent and varied information on PSM that related to their daily actions.

6. The Power of Daily PSM Data

Since 2007, refinery Shift Managers have been using the matrix of process safety event criteria to evaluate the day's incidents, determining if there was any PSM impact, and at what severity (i.e. A/B/C). The daily assessment of events is discussed in Safety Meetings, challenging front line leaders to understand the criteria in order to answer the PSM questions of their teams. This dynamic was and is an extremely powerful tool turning the PSM Awareness effort into a pull to the PSM Group, rather than a push from the PSM Group. The PSM Group members and Safety Leadership were frequently asked to be guest speakers at safety meetings, which served to further educate both the team members and supervisors in a small group setting.

The count of days between moderate and high severity (B and C) events is also included at the bottom of the morning summary. This metric supports the following:

- A-level Events are treated as near misses & no maximum targets are set for A events in order to encourage continued reporting of minor incidents.

- Long-term sustainability of disciplined, event-free operation is a good sign that management systems and the 14 elements of PSM are working as intended.

To ensure that the Process Safety Events were categorized correctly, a full review of all declared PSM Events is held by mid-level management on a monthly basis. This quality control step also served as an impetus for those managers to be knowledgeable on what the criteria were. Those managers were asked to gather additional investigation data in some cases to determine the final classification of the event, which changed the metrics meeting from a presentation to a discussion and team effort.

Mid-level Management was then required to report out to the Refinery Leadership Team on their process safety performance monthly along with their other safety, environmental and operations performance. In order for the Refinery Leadership to hold them accountable, the Refinery Leadership also had to be aware of the criteria and background.

Similar monthly challenges cascaded upwards in the organization, ultimately resulting in reporting of defined high consequence process safety events being reviewed with the Koch Industries Board of Directors during their routine meetings.

7. The Results

With the amount of data collected over 4 years of process safety event tracking, Pine Bend has many examples of solid results that the effort has provided.

7.1 Ownership and Engagement

When Pine Bend started its PSM metrics improvement efforts, PSM was driven by the small PSM Group, with very little engagement by the rest of the facility. After the years of consistent measurement, we see process safety ownership extending far beyond the Safety Department:

- **Process Engineering** has taken on an active role in Process Safety improvement. Examples have included writing a vision statement specifically for their team and how they will contribute to Pine Bend's process safety performance, as well as concrete steps like volunteering to be the investigation leaders on all Process Safety Events, as well as writing weekly PSM Safety Talks for 2 years running.
- **Union Safety Representatives** took on PSM engagement as an important goal for themselves. To promote PSM, the group wrote a PSM Awareness quiz and brought it to every team in the refinery, checking their awareness of PSM.
- **Production Leadership Team** members took on ownership of the PSM Event investigations in their areas, auditing the investigation reports for their units to ensure quality and effective action items were created to address gaps, and also followed the action items to completion.
- **Maintenance Teams** took interest in ways that they could help understand their connection. The PSV Maintenance crew conducted PSV training for all pipefitters, educating them about the importance of careful removal and transport of PSVs so that as-

run conditions could be accurately evaluated and also so that the PSVs could be reinstalled without damage.

- **Operations Teams** drove new awareness of their Process Hazard Analyses (PHA) Reports. Safety Champions in various teams challenged their crews to figure out what the most hazardous pieces of equipment were in their units from a process safety perspective, and communicate that during permitting of Maintenance and Contractor personnel.

7.2 Real Work Leading to Lower Frequency of Incidents

With over 700 Process Safety Event investigations completed, Pine Bend has gained invaluable process safety improvement knowledge.

7.2.1 Data analysis driving to themes for focused improvement

Analysis of events, with both high and low severity, has been conducted at Pine Bend, and has driven projects for mechanical and system improvements. For example, in 2008 Pine Bend had 4 PSM C events. All 4 events were related to pumps. Reviewing the investigations of those failures, Pine Bend launched an effort to improve pump seal mechanical integrity and safer pump operations, including review of double pumping scenarios and vibration limits.

Other examples of distinct efforts which drove Pine Bend to a lower frequency of process safety events include:

- Progress on inspection of Corrosion under Insulation (CUI)
- Pipefitter inspections of PSVs and reports to Reliability Technical Group members for improvements
- Installation of new equipment such as installation of a spill back on an acid loading line in the Alkylation Unit
- Audits of Safe Limits of Operation
- Projects to check sizing of PSVs
- Improvements in Safety Instrumented Systems

7.2.2 New Record Set in Days between B or C Events

A step change in significant (B or C) process safety event frequency occurred in 2010. Though the count of process safety A-level events has grown each successive year, B and C events have decreased. With a specific measure in place regarding days between B and C Events, Pine Bend was able to see how sustainable the event performance could be.

On November 10, 2010, Pine Bend celebrated 200 days since the last B or C Process Safety Event. The organization celebrated with a steak dinner for all employees.

Figure 12: Photos and Graphics from Process Safety Celebration

Ultimately, the facility reached a new record of 240 days between events when the streak ended. Employees were especially proud at the clear improvement in event frequency.

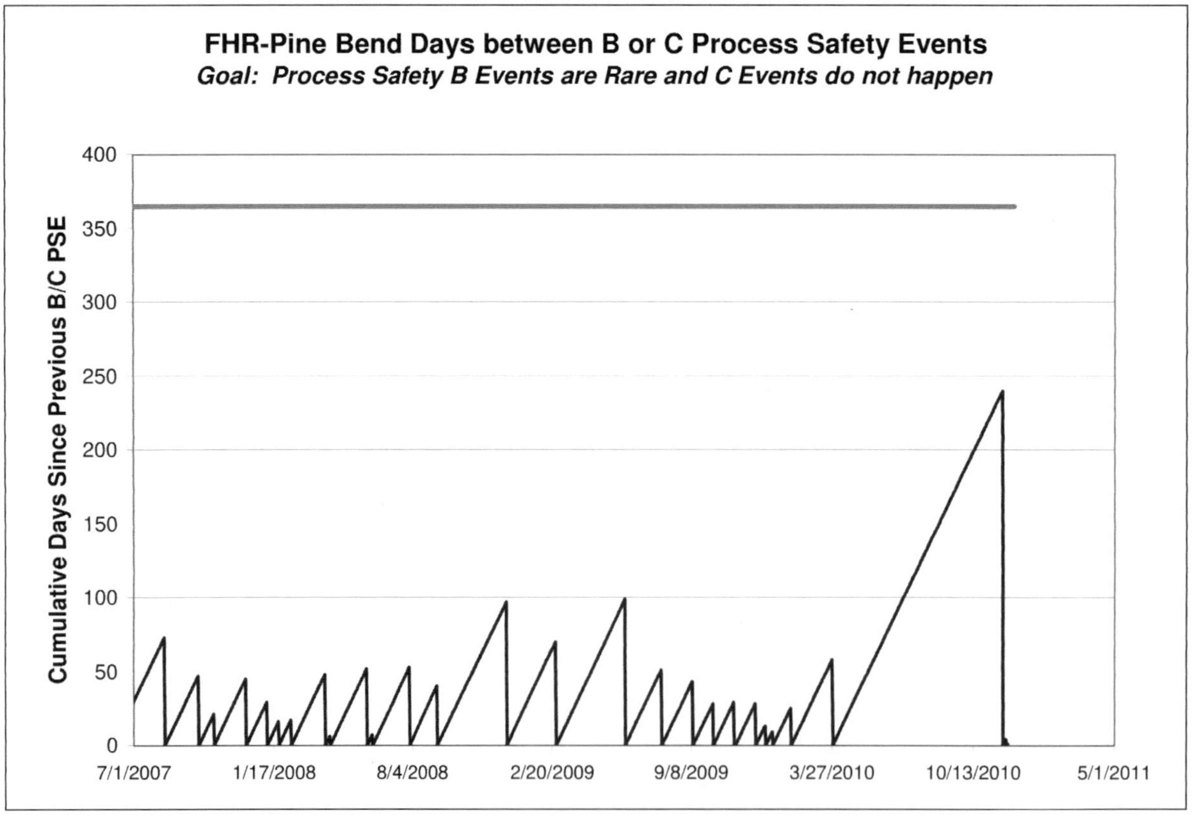

Figure 13: Days between B or C Process Safety Events for the FHR Pine Bend Refinery

7.3 Increased confidence by Stakeholders in PSM Program

7.3.1 Minnesota OSHA

During Pine Bend's Voluntary Protection Program Recertification audit in 2008, the State OSHA Consultation Group was particularly interested in hearing about Pine Bend's efforts to learn from near miss and minor process safety events. OSHA applauded Pine Bend for making the documented effort to record and determine the causes of minor events in an effort to minimize more significant process safety risks. OSHA was also impressed with the fact that the program was implemented in all refinery areas, not only those which were covered by the PSM regulation.

7.3.2 Community Advisory Council

In January, 2010, Safety Management presented a introduction to process safety and Pine Bend's process safety metrics to the Pine Bend Community Advisory Council, which consists of 20 members of the nearby community. With knowledge of the community impact of large incidents such as BP-Texas City in 2005, they were keenly interested in Pine Bend's efforts and performance in process safety. Following the presentation, the group had many positive comments, clearly glad that Pine Bend was looking to 'keep process in the pipe' and learn from any incidents that could have potential to lead to an off-site incident.

7.3.3 Employees and Management

With efforts to increase process knowledge at all levels of the organization, a step change in challenge and questioning occurred regarding process safety status and improvement over the past 3-4 years. Fortunately, the metrics also show that awareness continues to increase with A-level events increasing each year, but B and C level events staying at a steady or decreasing level (see Figure 14). FHR showed their commitment to event tracking by expanding the program to all FHR facilities in 2009.

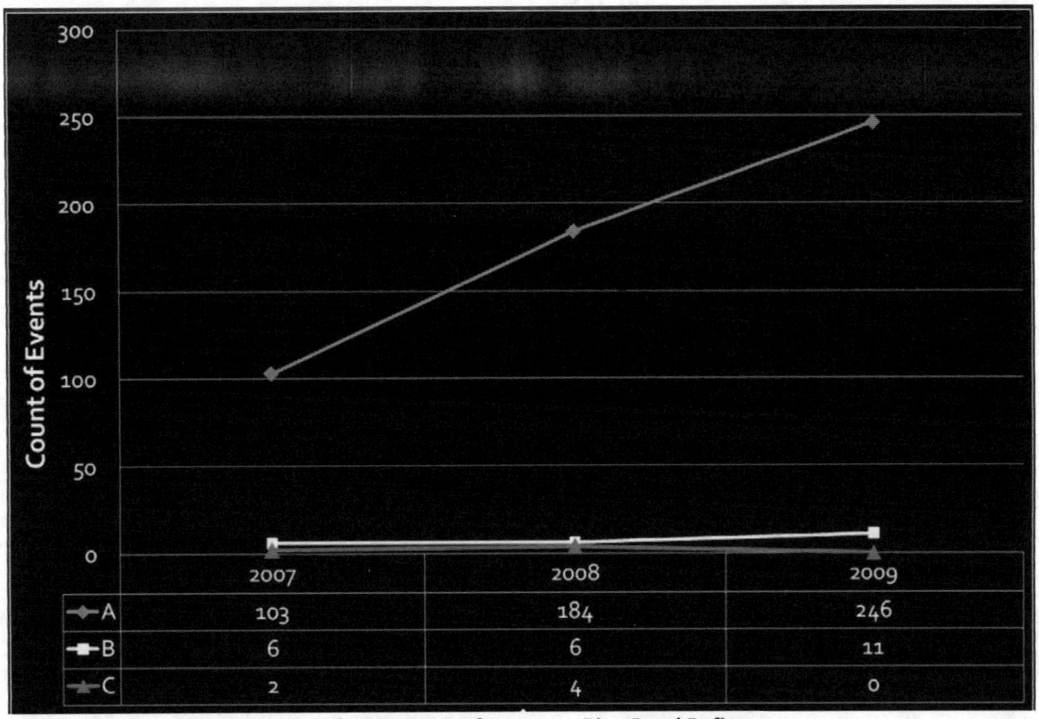

Figure 14: 2007-2009 Process Safety Event Performance- Pine Bend Refinery

Feedback gathered from employees via employee suggestions and informal surveys has encouraged Pine Bend to stay with the program and keep up progress with action item completion to reduce process risk and increase education.

8. Lessons Learned

To summarize the lessons learned by Pine Bend during this process safety improvement effort, ideas for several groups that impact Pine Bend's culture are described below.

Group	What Worked	What Didn't Work
Management	– Process Safety measures inclusion in monthly refinery metrics – Management Review of high severity events – Cross-site roll-out of PSM Metrics for site to site benchmarking	– Metrics roll-out before sufficient orientation on the basis
Process Engineers	– Data behind what is a process safety event (math to be done for justification- release scenario, exceeded design pressure); – 'Show me the math' attitude around PSV sizing/confirmation – Limits emphasis and culture	– Authoring safety talks without sensitivity to broad non-engineer audience
Mid-Level Management (Production Leadership Team)	– Monthly measures by Area – Team by team/ Area by Area/ Unit by Unit breakdown of performance in a consistent format with clear criteria – Smart Cards/Cheat Sheets with the basics of PSM – Ownership of Investigation review and quality of action items	– PSM Compliance measures (i.e. late action items) don't drive culture to drive down PSM- only drives compliance for compliance-sake – Process Safety event data difficult to pull & obtain consistent reports on using existing data collection system
Front Line Workers and Supervisors	– Daily Reporting of Incidents (Morning Summary) – Safety Central & PSM Webpages (one-stop shopping for information) – Posters/Graphics in plain refinery language – Safety Talks with graphics in them – Listing releases as safety events vs. only Environmental releases – PHA 5-year schedule posted near cafeteria for all employees to see – Face to face training for Powerpoint packages intended for sharing	– Training on the PSM Elements rather than PSM impact to day to day work – Safety talks that are all words and longer than a page – Monthly measures if they are not segregated by Team or with no avenue for sharing or discussion. – Computer-based training for PSM background information
PSM Group	– Assistance with measures and orientation by Safety Department Management to minimize additional workload on PSM Group	– Clear segregation of Process Safety Events away from regulatory Process Safety Incidents – Safety Management orientation efforts without constant involvement of PSM Group

9. Conclusion

Pine Bend's efforts to improve process safety and process safety culture have involved a large number of ideas, methods, tools, and people over the past four years. The use of metrics to: (1) identify what we feel is important, (2) measure how we are doing against our baseline, and (3) make adjustments and take action accordingly has improved the status of process safety at this large refinery. Affecting the process safety culture and sustaining the awareness of process safety in daily activities are excellent compliments to the core practices of the 14 PSM Elements.

With continued emphasis and discipline, Pine Bend will continue to strive to achieve our vision of both injury-free and Incident-free operation.

---

[i] Process Safety Leading and Lagging Metrics, CCPS,

Implementing Process Safety In a Petrochemical Complex - 20 Years of History In Camacari, Brazil

Mariana Bardy
Det Norske Veritas
Rua Sete de Setembro, 111/12th floor – Rio de Janeiro, Brazil
mariana.bardy@dnv.com

Aurinézio Calheira
COFIC
Jaime Lima
Det Norske Veritas
Sandro França
Det Norske Veritas

Keywords: process safety, risk analysis, process safety management, risk management

Abstract

The objective of this paper is to share the lessons learned during 20 years of implementing a safety culture in Camaçari Petrochemical Complex, located on the State of Bahia, in Brazil. It started its operation in 1978, with the construction of a basic petrochemical unit and a fertilizers unit. In 1990, in order to renew the operation license of the complex, the environmental agency of the State of Bahia requested a risk assessment to be conducted for the whole complex. It was the first process safety activity developed in Camaçari. The project was called APPOLO 1 Project and involved the existing 25 companies of the Complex. First a qualitative risk analysis was developed, with the technique of PHA – Preliminary Hazard Analysis, being followed by consequence analysis for the main scenarios. The success of this study gave the first incentives to the companies installed there to increase their activities related to process safety, as well as in promoting local process safety culture. Also the administrative agency COFIC - Camaçari Industrial Development Committee, with high influence of the HSE professionals of the several companies, started different initiatives towards increase HSE aspects in the complex, as well on the surroundings communities, including: Mutual Assistance Plan, Mutual Medical Assistance Plan, Contingency Plan of Camaçari, Emergency Plan for Communities, Annual Evacuation Exercise, HSE Award, Project APPOLO 2 (for risk assessment) and Risk Management Program.

1. Introduction

Camaçari Petrochemical Complex started its operation in 1978, with the construction of a basic petrochemical unit and a fertilizers unit. Over the years, it has increased the number of sites installed to 90, being 63 chemical and petrochemical, with a capacity installed of over 11,5 million ton/year, 15 thousand direct employees and 20 thousand employees thought subcontractors. The major features currently presented by the Complex are industrial expansion and diversification.

Today, the largest integrated industrial complex in the southern hemisphere, the majority of the Complex companies is connected through a piping network to a raw material unit. The largest company of the Camaçari Industrial Complex and one of the five major private enterprises of the country, BRASKEM raw material unit receives petroleum-by-products from PETROBRAS, mainly naphtha, and processes them into primary petrochemicals (ethylene, propylene, benzene, toluene, butadiene, xylene, solvents and others).

All of these products plus utilities such as electric power, steam and instrument air are supplied from a central production facility and to the neighboring second generation industries, for manufacturing intermediate and some final petrochemicals, as represented in Figure 1.

Figure 1: Camaçari Complex

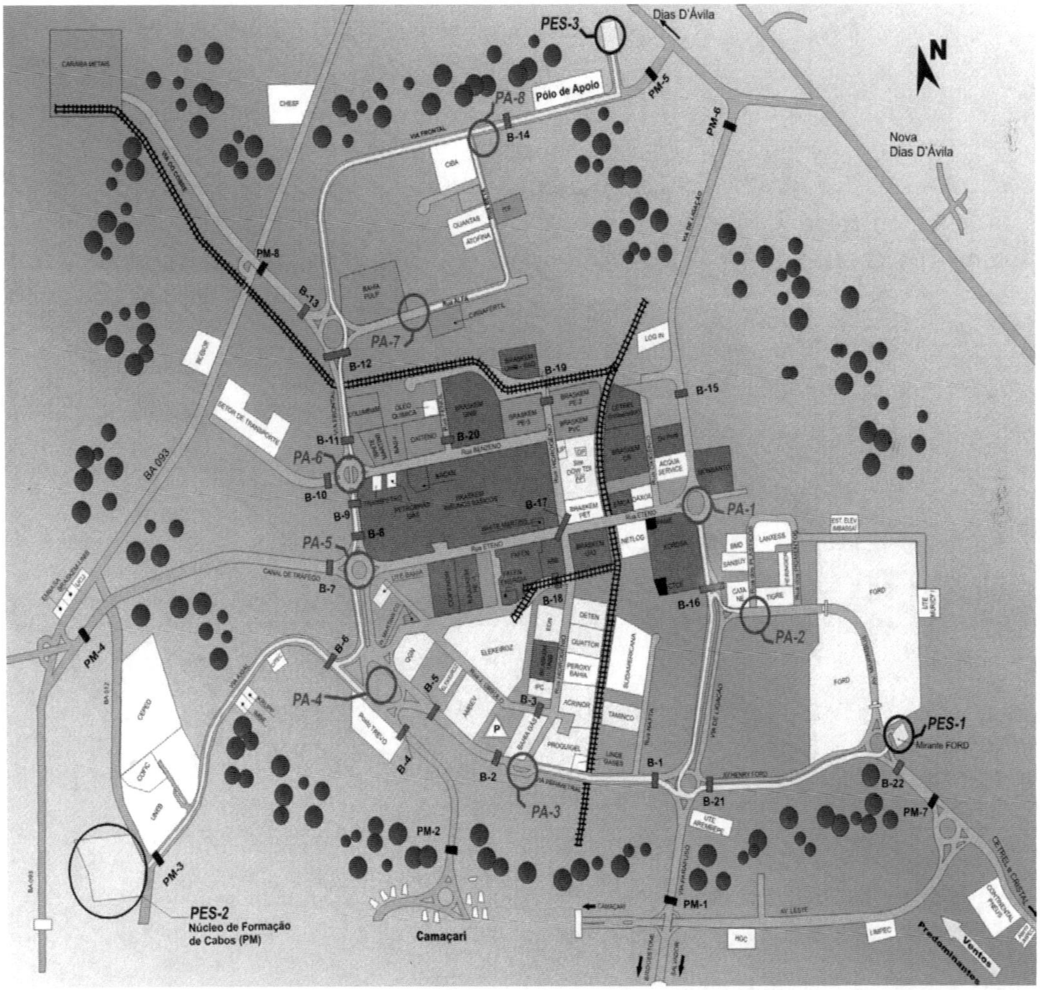

By the end of the 80's an expansion of Camaçari Complex was planned and to accomplish with the environmental law established on the State of Bahia, an Environmental Impact Assessment was performed. This study indicated the need to evaluate the risks of operation of the complex towards the nearby community. With that, the first initiative on Process Safety was established in Camaçari.

In order to coordinate the activities with all companies and decide how to develop the studies, COFIC – Camaçari Industrial Development Committee1, established a Technical Steering Committee with participation of professionals from different companies to define methodology, criteria and how to conduct the study, named APPOLO 1.

Over the years, COFIC has promoted several initiatives towards excellence in Safety in Camaçari Complex (including Process Safety and Non Process Safety) that today has a performance superior of the national average in Brazil and compared to good practices around the word.

This paper presents the evolution of the implantation of Process Safety, starting with fulfillment of mandatory requirements to a sustainable position, over 20 years of work.

2. The Process – Past and Presence

2.1 Over the Years

The process of implementing a Process Safety Culture in Camaçari Complex started in 1990, after the indentified need of an assessment of the risks that the activities of the companies towards the nearby communities. The main milestones of this process are indicated bellow.

1990 → First Requirement of Risk Assessment
1991 → Development of Technical Specification for APPOLO 1
1992–1995 → APPOLO 1 Project
1996 → First NUDEC - Community Development Center
1997 → Development of Award on HSE
1998 → First Award on HSE; First Emergency Drill
1999
2000 → Development of Technical Specification for APPOLO 2
2001–2004 → APPOLO 2 Project
2005 → Initiated Risk Management Program Development and Implementation
2006 → First Emergency Drill with Communities
2007 → Road Transportation included on the RMP
2008
2009 → New Regulation for Risk Management for the State of Bahia
2010

[1] COFIC was created with the mission to promote the development of the complex. It centralizes the administration of common areas in the complex, relationship with govern agencies and communities

The next sections describe the main steps of the process that Camaçari has gone thorough over the years, since 1990, telling the history until the present situation.

2.2 APPOLO 1 Project

As previous indicated, in order to guarantee the operational license for the expansion of Camaçari Complex, it was requested by the Environmental Agency from State of Bahia the development of a risk analysis study for the associated companies.

This was one of the first formal requests of risk assessments for chemical and petrochemical companies in Brazil. A total of 31 companies were involved and the study was conducted by PRINCIPIA (consulting company acquired by Det Norske Veritas – DNV in 2000). The methodology used on the study was the Preliminary Hazard Analysis (PHA) and a risk matrix was used to classified severity, likelihood and risk. During the analysis around 9.000 scenarios were identified and classified, with focus on impact to people. The risk matrix was divided in to regions:

- <u>Acceptable risk:</u> no mitigation needed
- <u>Mandatory mitigation:</u> recommendations mandatory in order to reduce the risk.

Note that ALARP (As Low As Reasonable Possible) approach was not used on this study. From the identified scenarios, around 1% was classified on the region with mandatory mitigation, with 93 recommendations. For the remaining scenarios, around 1,500 suggestions, without mandatory implementation were identified, but also for safety and continuous improvement.

The scenarios with potential to reach the nearby communities were subject to consequence analysis. No quantitative risk assessments were performed during APPOLO 1 Project.

The limited experience of the professionals involved on the study, the lack of similar studies in the country and possible misunderstanding of the results towards the communities and inside companies themselves were some of the difficulties to be transposed during the four years of the project [1]. At that time, no particular effort was given or established in following the implementation of the recommendations and suggestions as well continuing the risk analysis and risk management on a more structured and centralized condition on Camaçari Complex.

APPOLO 1 was the first step towards process safety and after this first initiative several other have followed.

2.3 Award on Health, Safety and Environment

In 1997, the Award on Health, Safety and Environment (HSE) was created to encourage improved performance among the COFIC-associated companies in prevention of accidents, also promoting the exchange of experience in Health, Safety and Environment in the Complex.

It is a volunteer process, where associated companies can candidate themselves for the award. A team composed of 80 auditors assess, in every company, approximately 1135 items addressing the three areas of the award (Health, Safety and Environment) as well as the corresponding evidences of performance. Other associated companies that do not wish to participate on the award are still audit during that year. Audit results are transformed into recommendations for the companies [2].

The protocol for the audit process, first created in 1997 and applied in 1998, has passed through reviews and has now 16 elements, reinforced in 2007 with more complex requirements on Process Safety and Risk Analysis, including Leadership and Administration, Competence and Training, Planned Inspections in HSE, Analysis and Statistics in Incidents and Accidents, Emergency Control, Company Procedures, HSE Audit, Personnel Protection Equipment, Occupational Hygiene, Risk Management and Process Safety, Communication, Workplace Safety Management, Occupational Health, Subcontractors, Environmental Management and Safety on Road Transportation (Products and Waste).

In order to guarantee the classification and the award, a criteria based on rate of accidents with loss of time and level of compliance on the protocol is established. In order to obtain the highest classification the company must [2]:

- Classification of 4★: Achieve compliance between 7.000 and 7.999 points and rate of accidents with loss of time < 3,0;
- Classification of 5★: Achieve compliance between 8.000 and 8.999 points and rate of accidents with loss of time < 2,0;
- Classification of Excellence in HSE: Achieve compliance over 9.000 points and rate of accidents with loss of time < 1,0;

Over the 13 years of the award, an increase of the compliance with the protocol has been achieved by the companies and from the first year with the competing companies with 4 stars to the last years with companies achieving the level of Excellence, demonstrates that an upgrade on performance can be obtained throughout implementation of a management system.

Table 1. Evolution of the Award on HSE [2]

Year	Protocol Elements	Companies Participating	Awarded Companies				Average of Compliance (%)
			4 ★	5★	Excellence	Total	
1998	12	10	10	-		10	76,3
1999	14	10	9	1		10	78,8
2000	14	15	12	3		15	80,5
2001	15	18	9	9		18	86,3
2002	15	22	11	11		22	86,1
2003	15	19	5	14		19	88,5
2004/2005	16	16	-	16		15	83,4
2006	16	18	4	5	3	18	84,4
2007/2008	16	18	3	9	3	18	82,9

2.5 APPOLO 2 Project

In 2000, to obtain a new environmental and operational license for the Complex, a review of the risk analysis performed from 1992 to 1995 (APPOLO 1 Project) was requested by the State of Bahia Environmental Agency.

The Technical Steering Committee from COFIC created a new Technical Specification for the studies to be conducted, including this second time more focus on quantitative studies and new criteria for classification of risks.

The study was conducted by DBNV from 2001 to 2004, including companies in the following main sectors: chemicals, petrochemicals, beverage, cupper and automobilist. The transportation of dangerous goods by pipelines, railway and roads, between the physical limits of the complex, was also included in the analysis.

Representatives of the main companies in the complex, with large experience in process safety, as Braskem, Deten, Petrobras, Du Pont, Monsanto and Dow have participate in the committee and elaborated a technical specification to be followed in the analysis. This specification is based on Brazilian regulations, establish by the States of Rio de Janeiro and São Paulo, and also international regulations from OSHA, EPA, the Council of the European Union (Seveso II) and International Labor Organization (ILO 174).

As established in the technical specification, APPOLO 2 Project was structured in four phases [3]. The revision of the risk assessment performed from 1992 to 1996 was much more structured and a detailed methodology was developed.

- Phase 1: Preliminary Hazard Analysis (PHA)
- Phase 2: Consequence Analysis
- Phase 3: Quantitative Risk Assessment (QRA)
- Phase 4: Process Safety Management Program

PHA was again selected as the primary qualitative approach to be applied to all companies for the identification and classification of risks. A risk matrix, with five likelihood categories and four severity categories was established. The qualitative acceptability criteria for APPOLO 2 project was defined as follow:

- Acceptable Risk: scenarios classified in this category are considered in acceptable condition, not being necessary risk reduction. The measures identified for them are classified as suggestions, without obligation to be implemented.
- Global Analysis: scenarios classified in this category have as characteristics high likelihood and low consequences, the actions normally are related to inspection and

maintenance activities. All measures proposed are considered Recommendations, with obligation of implementation;

- Confirm Risk: in the category, complementary studies (Consequence Analysis and Quantitative Risk Assessment) need to be performed, in order to confirm the risk acceptability;
- Acceptable with Consequence Analysis: for these scenarios, even thought they are considered accepted, a consequence analysis needs to be performed to confirm the severity of the consequences;
- Pre-Analysis: as the category Confirm Risk, for Pre-Analysis, complementary studies are to be performed;
- Not Acceptable Risk: scenarios classified as not acceptable need to have immediate attention and the risk measures have to be implemented in order to eliminate the hazardous condition or to reduce it to an acceptable risk.

During the assessment, over 21.000 accidental scenarios were identified, around 570 recommendations were proposed, to reduce the risk in the industry sites and transportation routes. These recommendations have the implementation status followed by the local environmental agency, according to annual reports presented by COFIC. Besides that, 1,390 suggestions were identified, with the aim to increase the safety on the operations, but without obligation of implementation [4].

From Figure 2, 74% of the scenarios identified were classified as acceptable and 21% as acceptable with need to develop consequence analysis to map out the areas that could be affected by physical effects from the accidents. Around 5% of scenarios were classified in intermediate categories and were subjected to quantitative risk studies.

Figure 2: APPOLO 2 Risk Distribution

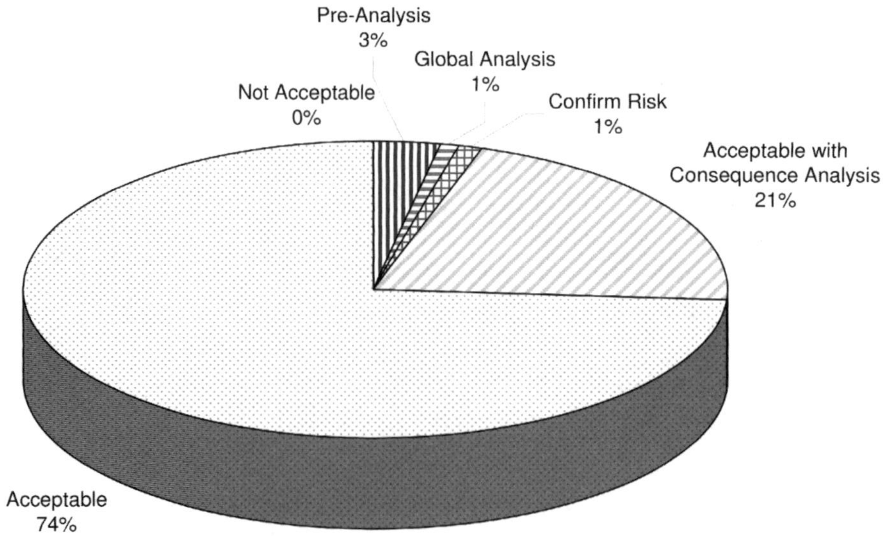

These results represent a significant progress comparing to APPOLO 1, especially regarding the more intense participation of the workers, number of identified scenarios and proposed recommendations and suggestions, showing that the companies saw the study as an opportunity to improve the process safety performance.

According to the results from the PHA, all scenarios with consequences classified as critical or catastrophic were included in the Consequence Analysis. Their effects were quantified, using the software PHAST – Process Hazard Analysis Software Tool, developed by DNV Software [5].

A total of 1,100 simulations were developed, including process units, storage and transportation, The distribution of results shows that only 1% of the accidental scenarios have more than 2,000 m of hazardous distance, with the possibility of reaching the nearby communities, located around Camaçari Industrial Complex. All the remained scenarios, 99%, have their effects limited to the industrial sites in Camaçari.

Finally, for the Phase 3 – Quantitative Risk Analysis (QRA), according to APPOLO 2 risk matrix, all scenarios classified as Pre-Analysis and Confirm Risk, have to be assessed using QRA methodology. It has implied that 30% of the 52 companies initially analyzed by the preliminary hazard analysis (PHA) and also road transportation and pipe ways between the Complex sites have conducted QRAs. The software SAFETI [6] was used on the study.

The results for the QRA showed that, among the 90 process units, storage areas and transportation systems evaluated, 77% of them were classified as acceptable, 16% as ALARP and 7% as not acceptable. Due to these results, for the 23% of the analysis with risk different from acceptable, reduction measures were proposed and the risk was review to consider the implementation of them. Finally, with the mitigation, 86% of the units were considered with acceptable risk (increase of 9% after risk mitigation) and the remaining 14% as ALARP.

No unit has remain with not acceptable risk, after the evaluation of mitigation, that was based on reduction of inventory, installation of new blockage systems and interlocks and improvement of the reliability of safety instrumented systems. All recommendation from the QRA were reported to the local environmental agency and included in the status report, issued annually by the companies.

Additionally, significant efforts were giving on increase the competence in risk assessment and risk management for the Camaçari Complex professionals, with several training classes performed and over 600 trained people.

APPOLO 2 Project has to be considered as one of the most relevant and complete risk assessment studies ever done, due to its centralized approach and coordination that has implied in the use of common methodology for all 52 companies included in the study. A few relevant conclusions can be extracted from the study:

- Proximity of each site in the complex leading to a high probability of dangerous effects from one company reaching neighboring ones, increasing the necessity of constant reviews of the Emergency Plan of each site;

- Considerable distance of 2,000 m, from most of the facilities to the community and the urgent need of actions to preserve the dense vegetation located around the complex, due to its importance in maintaining the gap between industrial and residential areas;
- Centralize management of pipe ways located in the complex, that connect the several companies, with the coordination of raw material unit, and establishment of technical committee to ensure that all safety measures are being take care by;
- New technologies and revision of inspections plans, in order to increase the integrity of process equipments and storage vessels and tanks, considering the 28 years of operation of Camaçari Industrial Complex;
- Increase of the use of risk analysis as a support for the management of process changes; support from all companies, including administration level and technical professionals;
- Increase of safeguards to avoid soil contamination and reduction of existing passive.

APPOLO 2 Project can be considered as a first initiative in Brazil, allowing all companies involved in getting permanent and systematic control of their risks, using the Risk Management Program as a standard.

2.6 Risk Management Program Development and Implementation

Complementing the results from APPOLO 2 Project, during the definition of the Technical Specification of the study, in 2000, the Technical Steering Committee has indicated that the development of the risk assessment studies should become a more continuous activity and a more systematic approach should be used.

It was suggested, and approved by local authorities, that each company should develop and implement a Risk Management Program (RMP). The outline of the program is based on ten critical elements, as following:

1. Process Safety Information
2. Risk Assessment
3. Operational Procedures
4. Training
5. Sub-contractors
6. Integrity and Maintenance
7. Management of Changes
8. Incidents and Accidents Investigation
9. Emergency Planning/Evacuation and Emergency Response Procedures
10. Auditing Process

The process of development and implementation of the RMP started after the conclusion of the risk assessments of risk assessments in 2005. Considering that in Camaçari Complex, there were at that time 52 companies that should implement the RMP and that different levels of culture towards process safety existed (from totally dependent companies, that didn't have any initiative on process safety to international and very well developed sites, with process safety programs already implemented), COFIC has developed with support from the Steering Committee and DNV a guide for implementation of the RMP.

This guide has detail description of each element, with focus given to goals, objectives, documents or procedures related to the implementation of each of the aspects for the protection of safety and environment. Also a PDCA cycle, indicating activities that have to be performed for implementation and an audit protocol, to support the verification and progress review of the implementation are included on the guide.

As part of APPOLO 2 Project and in order to obtain the operational license of the Complex, each company had to present the level of compliance with the RMP. A first gap analysis was performed with most of the companies and action plans were developed, with an aim to have the conclusion of implementation process in a five-year period.

In 2007, the Risk Management Program has been extended to include the pipeline transportation systems in the Complex and main interconnections with export Terminal, located 35km from the Camaçari.

In 2008, the hazardous materials road transportation mode has been included on the RMP, with a new guide developed and gap analysis done for the main transportation companies operating in Camaçari. This program has to be applied to all companies that transport dangerous good on the limits with the complex.

Today, annually, each company has to present to COFIC an update from the status of the implementation of the program, as a requirement of its operational license. The information is then reported to the environmental agency that can follow closely with the companies the actions and verify in loco, if needed.

2.8 Emergency Planning and Contingency

Emergency planning is one of the most important activities conducted by COFIC and centralized on the Complex. Directives with requirements for communication of risk, risk statistics, training for brigades and fire fighting/rescue workers, crisis management and others are in place [7].

A Contingency Plan that contemplates the accidental scenarios with possibility to extend beyond the limits of the complex is developed and implemented by COFIC. Other scenarios that have effects limits to the Complex are under management of each responsible company.

As support to planning for emergencies as well as actively , Camaçari Complex has implemented today:

2.8.1 Emergency Medical Plan (PAME)

This program provides medical emergency assistance for the companies. It is formed of specialized professionals (physicians and nurses), ambulances (IC units), modern equipment for life support and service agreements with hospitals and clinics. It has a centralized emergency center, where initial medical treatment can be given and with adequate transportation provided to local hospitals.

This initiative guarantee an uniform treatment to any injured person, correct treatment in case of contact with chemical toxic products, considered the high level of specialization of the medical team, which, being centralized, minimize local efforts of each company.

2.8.2 Mutual Aid Plan

Through the Mutual Aid Plan (PAM), any company installed within the limits of the Complex can request prompt support from other companies in emergency situations. It involves human and material resources from the companies, such as medical staff, firefighting brigades, emergency equipment, etc. A pioneer in Brazil, such system reflects the high concern with safety issues in the Complex.

2.8.3 Emergency Drill and Evacuation

Starting in 1998, yearly, an emergency drill and evacuation of the whole Camaçari Complex workers is established. The scenarios under evaluation are from the Internal Contingency Plan and each year one of the involved companies in the main scenarios identified is responsible to conduct the drill, with support from COFIC. The evacuation involves more then 18.000 workers and the technical support provided by health and safety professionals from all companies.

2.8.4 Emergency Control Training Center

COFIC is responsible for the Emergency Control Training Center – CTCE, which is the most complete in Brazil, with a firefighting simulation unit and capacity to train about 3.000 individuals per year.

2.9 Initiatives with the Communities

Around Camaçari Complex, are located two main communities, Dias D'Avila and Camaçari, Dias D'Avila are located 4 km from the complex and has 55.000 habitants [8] and Camaçari is 2 km, with 240.000 habitants [9].

COFIC has been promoting the inclusion of these communities in the activities of industrial complex and several initiatives are now in place. The development in 1996 of the first Community Development Center – NUDEC started with the aim of maintain a constant dialogue with the neighboring communities, expand and strengthen the relationship between the Complex and the nearby communities in a very concrete way.

Some of the initiatives are indicated on the next sections.

2.9.1 Community Advisory Council

The Community Advisory Council is a formal channel of dialogue between the companies and the communities surrounding the Camaçari Industrial Complex.

Composed of 24 representatives from different community and society sectors, this council meets on a bimonthly basis to discuss matters of mutual interest related to community, primarily in the areas of safety, health, environment and social responsibility.

2.9.2 Visit Program

The Camaçari Industrial Complex frequently receives visitors from other countries, from other Brazilian states and from the neighboring communities interested in getting acquainted with their activities or interested in establishing business relationship.

The program designed for visits by members of neighbor communities are coordinated by COFIC and involves mainly students, teachers and community leaders from Camaçari and Dias D'Avila.

2.9.3 Emergency Drills with Communities

The first emergency drill with the participation of the communities was held in Camaçari, in December of 2006. It had a participation of govern agencies as hospitals, emergency rescue team, police and local leaderships.

The objective of these drills is to demonstrate for the communities the actions that need to be taken in case of emergency involving chemical products.

3. Conclusions

Looking to the future, over the 20 years of progress, a sustainable and continuous implementation of Risk Management Program has been achieved. It is important to keep with the integrated and common approach for the complex. Maintaining the constant contact and exchange of experiences is very valuable for a loss prevention and much needed philosophy.

Process safety, represented by the RMP developed and implemented by COFIC is now a reality and is being followed by all new companies that want to be installed in Camaçari Complex. As well, it represents a continuous building of sustainable culture for involved companies

The safety index, represented in Figure 3, achieved by Camaçari Complex is lower than the Brazilian average (indicated by Brazilian Association of Chemical Industries – ABIQUIM) and has been reducing, being a clear reflection of the practices implemented on complex.

Figure 3: Rate of Accident with Loss of Time (ABIQUIM x COFIC) [10]

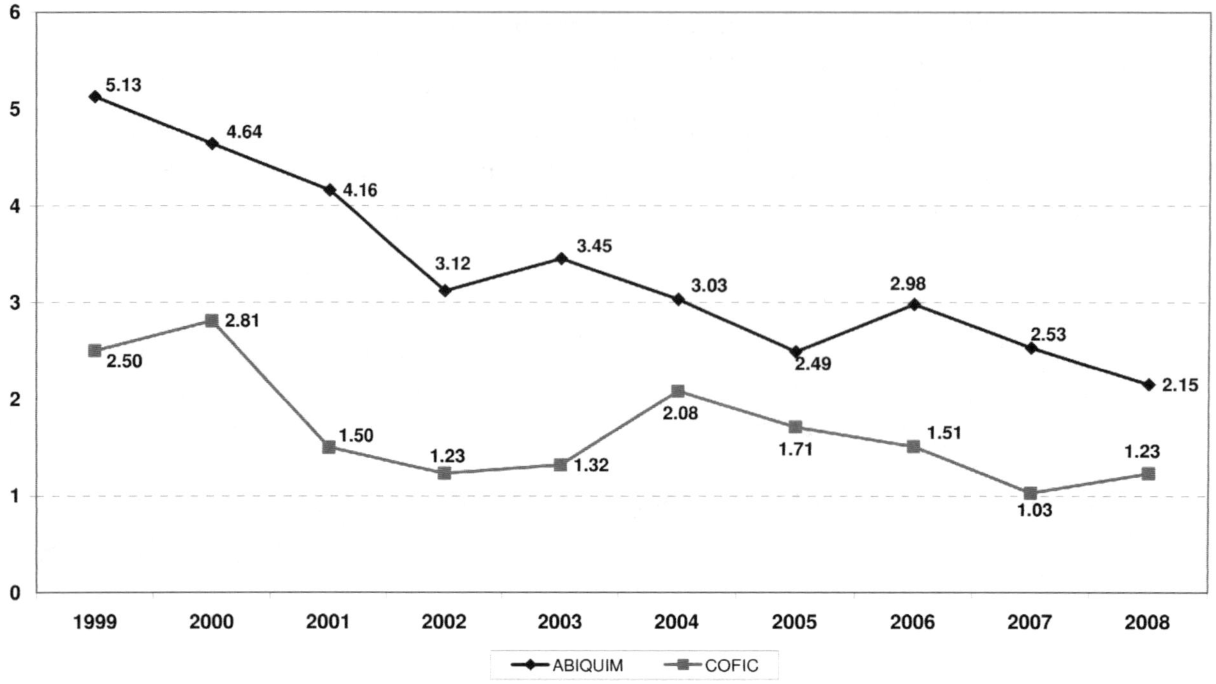

Besides that, looking to the successful case of APPOLO 2 Project, risk assessment and risk management is an established legal requirement for all companies to be installed and under operation in the State of Bahia, showing that this best practice is being followed through the state

Camaçari Complex, with coordination from COFIC, has experienced a process of development of a process safety culture over the last 20 years. The results obtained regarding process safety are a reflection of the high level of commitment from the leadership teams of all associated companies.

As a summary of the main initiatives of the Camaçari Industrial Development Committee, there are:

- Pioneered the implementing of an integrated system for Environmental protection.
- HSE (Health, Safety & Environment) performance exceeds the regulatory standards.
- Most complete risk analysis study carried out in Brazil, involving all companies installed in the Industrial Complex – APPOLO Project.
- HSE audits in all companies (Awards in HSE).
- Risk Management Program is implemented in all companies, including transportation of chemicals.
- Continuous interface with the Environmental agencies.
- Continuous dialogue with neighbor communities through COFIC Community Advisory Council.

Opportunities for upgrades on the processes implemented and under implementation always exist and COFIC is looking to the new technologies, methodologies and best practices that can intensify their leading position in process safety management in Brazil, combining the commitment of both leadership and workers, the successful exchange of experience from all and ability to work as a team towards loss prevention in every aspect.

5. References

1] IBP, Analise e Mapeamento de Riscos do Pólo Petroquímico de Camaçari, 12th Industrial Safety Seminar, Rio, Brazil, 1997.

[2] COFIC, PPSSMA Presentation, 2010.

[3] COFIC, Technical Specification for APPOLO 2 Project, Salvador, Brazil, 2000.

[4] DNV, Summary of APPOLO 2 Project, Salvador, Brazil, 2005.

[5] RMS, PHAST Process Hazard Analysis Software Tool, London, 2001.

[6] RMS, SAFETI Suite for the Assessment of Flammable, Explosive, and Toxic Impact, London, 2001.

[7] COFIC, Annual Technical Report on Environment, Salvador, Brazil, 2010.

[8] "Dias D'Avila", www.ibge.gov.br (accessed in Dec. 01, 2010).

[9] "Camaçari", www.ibge.gov.br (accessed in Dec. 01, 2010).

[10] ABIQUIM, Annual Report,

AUTHOR INDEX